AF581650

Advances in Micro- and Nanomechanics

Advances in Micro- and Nanomechanics

Editor

Victor A. Eremeyev

MDPI • Basel • Beijing • Wuhan • Barcelona • Belgrade • Manchester • Tokyo • Cluj • Tianjin

Editor
Victor A. Eremeyev
Gdansk University of Technology
Poland

Editorial Office
MDPI
St. Alban-Anlage 66
4052 Basel, Switzerland

This is a reprint of articles from the Special Issue published online in the open access journal *Nanomaterials* (ISSN 2079-4991) (available at: https://www.mdpi.com/journal/nanomaterials/special_issues/nano_micro_mechanics).

For citation purposes, cite each article independently as indicated on the article page online and as indicated below:

LastName, A.A.; LastName, B.B.; LastName, C.C. Article Title. *Journal Name* **Year**, *Volume Number*, Page Range.

ISBN 978-3-0365-6291-9 (Hbk)
ISBN 978-3-0365-6292-6 (PDF)

Contents

About the Editor

Victor A. Eremeyev

Victor A. Eremeyev, Ph.D., Dr. hab., is Associate Professor at University of Cagliari. His research focuses on continuum and structural mechanics, including the following topics: nonlinear elasticity, generalized continua (micropolar, micromorphic, stress- and strain- gradient media), metamaterials, surface elasticity, plates and shells, nano- and micromechanics, piezo- and flexoelectricity, flexomagneticity, and surface waves. He graduated from the Department of Mechanics and Applied Mathematics at the Rostov State University (Rostov on Don, Russia) in 1985. In 1990, he received a PhD degree from the same university; he also holds a doctor of science degree from 2004, attaomed at the Institute of Problems of Mechanical Engineering of Russian Academy of Science (St. Petersburg, Russia). He was appointed as assistant, associate and full professor at Rostov State University/South Federal University (1988–2017). Later, he worked as a researcher at Martin Luther University Halle-Wittenberg (2010–2011) and Otto-von-Guericke-University Magdeburg (2012–2015), Germany, as associate professor at Rzeszów University of Technology, Rzeszów (2015–2017), and as a full professor at Gdansk University of Technology, Poland (2017–2021). In 2020, he joined the University of Cagliari, Italy. He is the recipient of several awards, including the International Prize "Tullio Levi-Civita" for the Mathematical and Mechanical Sciences, 2018 (MeMOCS Center, Italy); a Royal Society Wolfson Visiting Fellowship (2021), UK; a professorship awarded by the President of the Republic of Poland, 2021; Abilitazione Scientifica Nazionale, SC: 08/B2, Fascia I. Italy, 2021. He was included in the World's Top 2% Scientists according to citation metrics by Stanford University and Elsevier.(https://elsevier.digitalcommonsdata.com/datasets/btchxktzyw/3) 2020–2022. He has published more than 270 articles in peer-reviewed scientific journals, and he is the co-author and co-editor of 21 research monographs/graduate textbooks.

Preface to "Advances in Micro- and Nanomechanics"

Material science, including the field of nanotechnologies, requires the consistent development of a theoretical and experimental basis for better understanding of small-scale material behaviour. Current advances in the field have resulted in the rapid development of micro- and nano-electromechanical systems (MEMS and NEMS). Among examples of these, it is worth mentioning nanosensors, actuators, filters and energy harvesters.

For the design and modelling of these systems, various enhanced models of continuum and structural mechanics can be applied. Here we mention strain and strain gradient elasticity, other nonlocal models, surface elasticity, micropolar, microstretch, couple-stress and micromorphic continua among others.

Manufacturing and further experimental studies of material at small scales also require advanced techniques, as discussed in this Special Issue of *Nanomaterials*. These techniques include, but are not restricted to, magnetron sputtering, nanoindentation, optical, scanning electron and atomic force microscopy, and X-ray microtomography Raman spectroscopy.

This reprint could be useful for both theoreticians and experimentalists working in the field of micro- and nanomechanics.

Victor A. Eremeyev
Editor

 nanomaterials MDPI

Editorial

Advances in Micro- and Nanomechanics

Victor A. Eremeyev

Department of Civil and Environmental Engineering and Architecture (DICAAR), University of Cagliari, Via Marengo, 2, 09123 Cagliari, Italy; victor.eremeev@unica.it or eremeyev.victor@gmail.com

Citation: Eremeyev, V.A. Advances in Micro- and Nanomechanics. *Nanomaterials* **2021**, *11*, 274. https://doi.org/10.3390/nano12244433

Received: 28 November 2022
Accepted: 8 December 2022
Published: 13 December 2022

Publisher's Note: MDPI stays neutral with regard to jurisdictional claims in published maps and institutional affiliations.

Recent advances in technologies of design, manufacturing and further studies of new materials and structures result in an essential extension of classic models of continuum and structural mechanics. For example, nowadays it is well established that material properties at small scales could be size-dependent. In the literature, various approaches were proposed for studying such phenomena. It is worth mentioning stress and strain gradient elasticity, surface elasticity, media with internal degrees of freedom, and nonlocal continua among others. These models can successfully describe various sized effects.

Another direction in the mechanics of materials, close-related to the aforementioned enhancements, addresses theoretical and experimental studies of material properties for media with complex internal microstructure. In particular, the great interest is the determination of effective properties of new composite materials and further analysis of their dependence on the microstructure.

Let us also note that at micro- and nano-scales one can observe a more rich picture of electromechanical couplings. The latter may play an important role in material response. For example, so-called flexoeffects relate electric polarization or magnetization to gradients of strains. So both flexoelectricity and flexomagneticity bring us another example of strain gradient models. Being sometimes even negligible at the macro-scale, these properties may be dominant at the nanoscale. This gives a possibility to use such materials as elements of MEMS and NEMS, such as energy harvesters, sensors, and actuators.

This special issue "Advances in Micro- and Nanomechanics" collects several papers that have presented theoretical, numerical, and experimental studies of materials and structures at small scales. It is rather natural to expect new phenomena in nanometer-sized thin-walled structures such as nanowires and nanofilms.

The new model of a nanowire embedded into an elastic substrate was proposed in [1]. Here surface energy was taken into account as in the Gurtin–Murdoch surface elasticity as well as a nonlocality according to the strain gradient approach. For the derivation of the governing equations, the virtual force technique was applied.

Experimental studies of thin films were presented in [2–4]. Here, films were produced with magnetron sputtering and further analyzed using various techniques such as atomic force microscopy and nanoindentation. As a result, microstructural, nanomechanical, and tribological properties were discussed in more detail. The residual stress-driven technique was applied to the determination of Young's modulus of nanofilms in [5]. Here authors proposed a new relatively simple approach based on the consideration of deformations of bilayer cantilevers. The analysis of the thermal stability and hardness of nanocrystalline Ni thin films was given in [6]. Here it was shown that the addition of cysteine results in improved hardness of films.

Properties of nanoparticles and related composites were investigated in [7,8]. Fracture strength and local hardness of spherical particles made of B_4C and TiC were estimated in [7]. In [8], nanoparticles of NiO/C applied for the manufacturing of nanocomposites for supercapacitors were investigated using X-ray diffraction and other techniques.

Biomechanical studies of coatings were discussed in [9]. Here authors discussed the mechanical properties, microstructure, and composition of enamel and dentine at the initial stages of caries. Here X-ray microtomography, optical, Raman, atomic force, scanning electron microscopy, and nanoindentation were simultaneously applied.

Magneto-electro-elastic coupling was studied in [10,11]. In [10], linear vibrations and buckling of nanoplates in a hygro-thermal environment were analyzed. To this end, a strain gradient nonlocal approach was used. Finally, using a variational approach nonlinear deformations of a nanobeam considering piezo- and flexomagneticity were studied in [11].

The content of the SI reflects the state of the art in the field of micro- and nanomechanics. It combines new theoretical models with modern experimental studies of materials.

Conflicts of Interest: The author declares no conflict of interest.

References

1. Limkatanyu, S.; Sae-Long, W.; Mohammad-Sedighi, H.; Rungamornrat, J.; Sukontasukkul, P.; Prachasaree, W.; Imjai, T. Strain-gradient bar-elastic substrate model with surface-energy effect: Virtual-force approach. *Nanomaterials* **2022**, *12*, 375. [CrossRef] [PubMed]
2. Melnikova, G.; Kuznetsova, T.; Lapitskaya, V.; Petrovskaya, A.; Chizhik, S.; Zykova, A.; Safonov, V.; Aizikovich, S.; Sadyrin, E.; Sun, W.; et al. Nanomechanical and Nanotribological Properties of Nanostructured Coatings of Tantalum and Its Compounds on Steel Substrates. *Nanomaterials* **2021**, *11*, 2407. [CrossRef] [PubMed]
3. Kuznetsova, T.; Lapitskaya, V.; Solovjov, J.; Chizhik, S.; Pilipenko, V.; Aizikovich, S. Properties of CrSi2 Layers Obtained by Rapid Heat Treatment of Cr Film on Silicon. *Nanomaterials* **2021**, *11*, 1734. [CrossRef] [PubMed]
4. Kuznetsova, T.; Lapitskaya, V.; Khabarava, A.; Chizhik, S.; Warcholinski, B.; Gilewicz, A.; Kuprin, A.; Aizikovich, S.; Mitrin, B. Effect of Metallic or Non-Metallic Element Addition on Surface Topography and Mechanical Properties of CrN Coatings. *Nanomaterials* **2020**, *10*, 2361. [CrossRef] [PubMed]
5. Velosa-Moncada, L.A.; Raskin, J.P.; Aguilera-Cortés, L.A.; López-Huerta, F.; Herrera-May, A.L. Estimation of the Young's Modulus of Nanometer-Thick Films Using Residual Stress-Driven Bilayer Cantilevers. *Nanomaterials* **2022**, *12*, 265. [CrossRef] [PubMed]
6. Kolonits, T.; Czigány, Z.; Péter, L.; Bakonyi, I.; Gubicza, J. Improved hardness and thermal stability of nanocrystalline nickel electrodeposited with the addition of cysteine. *Nanomaterials* **2020**, *10*, 2254. [CrossRef] [PubMed]
7. Nakamura, D.; Koshizaki, N.; Shishido, N.; Kamiya, S.; Ishikawa, Y. Fracture and Embedment Behavior of Brittle Submicrometer Spherical Particles Fabricated by Pulsed Laser Melting in Liquid Using a Scanning Electron Microscope Nanoindenter. *Nanomaterials* **2021**, *11*, 2201. [CrossRef] [PubMed]
8. Chernysheva, D.; Pudova, L.; Popov, Y.; Smirnova, N.; Maslova, O.; Allix, M.; Rakhmatullin, A.; Leontyev, N.; Nikolaev, A.; Leontyev, I. Non-isothermal decomposition as efficient and simple synthesis method of NiO/C nanoparticles for asymmetric supercapacitors. *Nanomaterials* **2021**, *11*, 187. [CrossRef] [PubMed]
9. Sadyrin, E.; Swain, M.; Mitrin, B.; Rzhepakovsky, I.; Nikolaev, A.; Irkha, V.; Yogina, D.; Lyanguzov, N.; Maksyukov, S.; Aizikovich, S. Characterization of enamel and dentine about a white spot lesion: Mechanical properties, mineral density, microstructure and molecular composition. *Nanomaterials* **2020**, *10*, 1889. [CrossRef] [PubMed]
10. Tocci Monaco, G.; Fantuzzi, N.; Fabbrocino, F.; Luciano, R. Critical temperatures for vibrations and buckling of magneto-electro-elastic nonlocal strain gradient plates. *Nanomaterials* **2021**, *11*, 87. [CrossRef] [PubMed]
11. Malikan, M.; Eremeyev, V.A. On nonlinear bending study of a piezo-flexomagnetic nanobeam based on an analytical-numerical solution. *Nanomaterials* **2020**, *10*, 1762. [CrossRef] [PubMed]

MDPI

Article

Strain-Gradient Bar-Elastic Substrate Model with Surface-Energy Effect: Virtual-Force Approach

Suchart Limkatanyu [1], Worathep Sae-Long [2,*], Hamid Mohammad-Sedighi [3,4], Jaroon Rungamornrat [5], Piti Sukontasukkul [6], Woraphot Prachasaree [1] and Thanongsak Imjai [7]

1 Department of Civil and Environmental Engineering, Faculty of Engineering, Prince of Songkla University, Songkhla 90112, Thailand; suchart.l@psu.ac.th (S.L.); woraphot.p@psu.ac.th (W.P.)
2 Civil Engineering Program, School of Engineering, University of Phayao, Phayao 56000, Thailand
3 Mechanical Engineering Department, Faculty of Engineering, Shahid Chamran University of Ahvaz, Ahvaz 6135783151, Iran; hmsedighi@gmail.com
4 Drilling Center of Excellence and Research Center, Shahid Chamran University of Ahvaz, Ahvaz 6135783151, Iran
5 Applied Mechanics and Structures Research Unit, Department of Civil Engineering, Faculty of Engineering, Chulalongkorn University, Bangkok 10330, Thailand; jaroon.r@chula.ac.th
6 Construction and Building Materials Research Center, Department of Civil Engineering, King Mongkut's University of Technology North Bangkok, Bangkok 10800, Thailand; piti.s@eng.kmutnb.ac.th
7 School of Engineering and Technology, Center of Excellence in Sustainable Disaster Management, Walailak University, Nakhon Si Thammarat 80161, Thailand; thanongsak.im@wu.ac.th
* Correspondence: worathep.sa@up.ac.th

Abstract: This paper presents an alternative approach to formulating a rational bar-elastic substrate model with inclusion of small-scale and surface-energy effects. The thermodynamics-based strain gradient model is utilized to account for the small-scale effect (nonlocality) of the bar-bulk material while the Gurtin–Murdoch surface theory is adopted to capture the surface-energy effect. To consider the bar-surrounding substrate interactive mechanism, the Winkler foundation model is called for. The governing differential compatibility equation as well as the consistent end-boundary compatibility conditions are revealed using the virtual force principle and form the core of the model formulation. Within the framework of the virtual force principle, the axial force field serves as the fundamental solution to the governing differential compatibility equation. The problem of a nanowire embedded in an elastic substrate medium is employed as a numerical example to show the accuracy of the proposed bar-elastic substrate model and advantage over its counterpart displacement model. The influences of material nonlocality on both global and local responses are thoroughly discussed in this example.

Keywords: virtual force principle; nanobar; surface-energy effect; thermodynamics-based strain gradient; elastic substrate media

Citation: Limkatanyu, S.; Sae-Long, W.; Mohammad-Sedighi, H.; Rungamornrat, J.; Sukontasukkul, P.; Prachasaree, W.; Imjai, T. Strain-Gradient Bar-Elastic Substrate Model with Surface-Energy Effect: Virtual-Force Approach. *Nanomaterials* **2021**, *11*, 274. https://doi.org/10.3390/nano12030375

Academic Editors: Sergio Brutti and Victor A. Eremeyev

Received: 18 November 2021
Accepted: 20 January 2022
Published: 24 January 2022

1. Introduction

In recent years, enormous research efforts by scientists and engineers worldwide have been dedicated to the understanding and characterization of the unique responses of micro-sized and nano-sized structures. Their superior mechanical properties have attracted a wide spectrum of novel applications in modern science and technology [1]. Examples of novel devices employing small-sized structures are biosensors [2], piezoelectric actuators [3], nanosensors [4], and gyroscopes [5]. Profound understanding and characterizing mechanical properties of small-sized structures are critical to rational design procedure and performance assessment of these devices during their service life. In addition, nano-sized structures are commonly used as reinforcement components in nanocomposites due to their excellent mechanical properties [6,7]. Generally, the response characteristic of structures at microscale and nanoscale is drastically different from the corresponding response at macroscale due to two unique features inherent to micro-sized and nano-sized structures, namely the small-scale effect and the size-dependent effect. The former effect is related to the discrete nature of matter, thus inducing the material nonlocality while the latter is associated with excessive energy stored in the surface due to high surface-to-volume ratio, hence resulting in the size-dependency characteristic.

Experimental studies and atomistic/molecular dynamic modeling have been carried out by researchers to gain a thorough understanding of mechanical responses of structures at microscale and nanoscale. Due to the small-sized nature of specimens, experimental studies generally require high-precision apparatus and special testing procedure [8]. However, atomistic/molecular dynamic modeling is a viable method to characterize mechanical responses of micro-sized and nano-sized structures and can provide comprehensive simulation information [9,10] but high computational expense must be paid [11]. Consequently, only systems with limited amounts of atoms and molecules can be practically investigated, thus an alternative modeling approach with a superior computational efficiency is deemed essential. Collaboration between a structural-mechanics model (bar, beam, plate, and shell) and non-classical elasticity theory has been carried out by researchers to develop an alternative tool to characterize mechanical responses of small-sized structures [12–18]. This integrated modeling approach could account for the small-scale effect as well as the size-dependent effect with good balance between accuracy and computational efficiency.

Long-range inter-atomic forces associated with the discrete nature of materials are more influential when the dimension of a structure is in the range spanning from nanoscale to microscale. In the literature, this phenomenon is often referred to as the "small-scale" effect and induces nonlocality in the material. The assertion of material nonlocality is that dependency of a stress at a generic point is not only on the strain at that particular point, but also on those strains and related quantities at all other points throughout the elastic body. Several amended elasticity models have been proposed in the literatures to account for this discrete nature of materials [19–26]. Most widely used among them is the Eringen differential form of the strain-driven nonlocal elasticity model [20,21]. Consequently, a myriad of structural-mechanics models has been armed with this nonlocal constitutive model to account for the small-scale effect [12,27–32]. Unfortunately, those enhanced structural-mechanics models usually result in debatable and discrepant responses as pointed out by several researchers [27,33,34]. Romano et al. [35] have thoroughly diagnosed the cause of these problematic responses and concluded that an ill-posed mathematical problem is encountered with the adoption of Eringen nonlocal differential model. In addition, the Eringen nonlocal differential model would not accept quadratic energy functional form of elasticity [36] and the work conjugate nature of stress and strain in this nonlocal constitutive model is ambiguous [37].

Other rational nonlocal constitutive models have been proposed and adopted by various researchers to remedy the debatable and discrepant features inherent to the Eringen nonlocal structural-mechanics models [24,38–42]. Among these rational theories, the thermodynamics-based strain gradient model proposed by Barretta and Marotti de Sciarra [24] is of special interest since it could be adopted with reasonable effort. It is worth mentioning here that nanobars and nanobeams based on this thermodynamics-based strain gradient model do not present debatable and discrepant responses [43–45]. Therefore, this study would employ the thermodynamics-based strain gradient model of Barretta and Marotti de Sciarra [24] to represent the material nonlocality.

In opposition to mechanical responses at macroscale, the surface free energy related to surface stress and surface elasticity affects mechanical responses of structures at nanoscale. In literatures, this phenomenon is referred to as the "surface-energy" effect and induces the size dependency of nano-sized structures. To enhance structural-mechanics models with the surface-energy effect, the surface elasticity theory of Gurtin and Murdoch [46,47] has been widely adopted [48–54]. In this surface elasticity model, the surface layer of a solid core is considered a negligibly thin membrane perfectly bonded to the wrapped solid core.

Nanowires have found a wide spectrum of novel applications in nanoscience and nanotechnology covering optoelectronics, biotechnology, biosensors, and micro/nano electro-mechanical systems (M/NEMS) due to their outstanding mechanical, electrical, and thermal performances [55–59]. In these novel applications, nanowires have often been fabricated into larger parts via polymer substrate media. As a result, the interactive mechanism between the nanowire and its surrounding polymer substrate is of practical value, and plays a crucial role in designing and controlling of performance of devices and systems in such novel applications. In the literature, researchers have developed different nanobeam-elastic substrate models to characterize responses of nanowire-elastic substrate systems. For example, Ponbunyanon et al. [60] analytically investigated static flexural responses of silver nanowire-elastic substrate systems; Zhao et al. [61] analytically conducted buckling load analyses of nanowire-elastic substrate systems; Malekzadeh and Shojaee [62] used beam models to study nonlocal and surface-energy effects on vibrating responses of nanowire-elastic substrate systems.

In the literature, analytical models—though limited in number—have been devoted to characterize the tensile response of nanobar-elastic substrate systems [12,63] and the "irrational" Eringen nonlocal differential model has been employed in those models. Recently, Sae-Long et al. [44] has proposed a rational nanobar-substrate model within the framework of the virtual displacement principle. The thermodynamics-based strain-gradient model of Barretta and Marotti de Sciarra [24] was employed to represent the bulk-material nonlocality. The debatable and discrepant characteristics inherent to the Eringen nonlocal differential model were eliminated in this model. As a counterpart of the nanobar-substrate model proposed by Sae-Long et al. [44], the fundamental interest of this research work is to develop the nanobar-substrate model within the framework of the virtual force principle. The general idea of the model formulation stems from the Eringen's nonlocal bar-substrate model proposed by Limkatanyu et al. [63] and Eringen's nonlocal beam-substrate model proposed by Ponbunyanon et al. [60]. To the best knowledge of the authors, this research work presents, for the first time, the formulation of the strain-gradient bar-substrate model within the framework of the virtual force principle and the merit of this formulation framework is discussed. This developing novelty is a more efficient computational platform and is able to remedy several flaws inherent to the standard displacement-based method, as confirmed in the literature [58,61].

Organization of this research work is as follows: first, introductions to thermodynamics-based strain gradient model, surface elasticity model, and Winkler foundation model are briefly presented. The first two models are respectively employed to account for the small-scale and size-dependent effects, while the third is used to represent the interactive mechanism between the bar and its surrounding elastic substrate. Then, the differential equilibrium equation and end-force equilibrium conditions revealed by Sae-Long et al. [44] using the virtual displacement principle are introduced. The system sectional deformation-force (compliance form) relations are subsequently derived. Next, differential compatibility equations, as well as associated classical and non-classical end-displacement compatibility conditions, are consistently derived using the virtual force principle and form the core of the model formulation. The modified Tonti's diagram is employed to illustrate the problem formulation within the framework of the virtual force principle. Finally, a nanowire-substrate system is employed as a numerical example to show the accuracy of the proposed nanobar-substrate model and to present the advantage over its counterpart proposed by Sae-Long et al. [44]. Both global and local responses of the nanowire-substrate system are thoroughly discussed. The computer software Mathematica [64] is used to perform all symbolic calculations.

2. Strain Gradient Bars with Inclusion of Surface-Free Energy

In the present work, the bar section of Figure 1 is considered a composite composing of a bar-bulk material and a mathematically zero-thickness surface. The simplified strain-gradient elasticity theory of Altan and Aifantis [65] is employed to account for the small-scale effect of the bar-bulk material while the surface elasticity theory of Gurtin and Murdoch [46,47] is used to consider the surface-free energy due to the excess energy at the surface of the bar-bulk material.

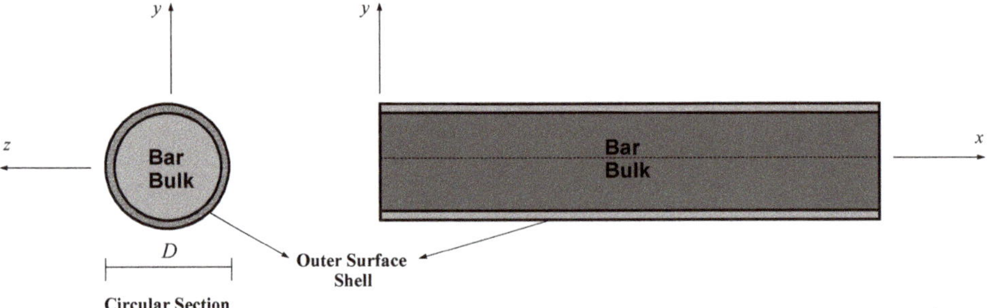

Figure 1. Nanobar section with a warping surface layer.

2.1. Simplified Strain-Gradient Model

As a simple variant of Mindlin's strain-gradient elasticity theory [19], the simplified strain-gradient elasticity model of Altan and Aifantis [65] is adopted herein to represent higher-order deformation mechanism of materials. This simplified variant is of great interest since it contains only one material small-scale parameter, thus rendering the process of the material-parameter determination and calibration simple and expeditious. Therefore, the simplified strain-gradient elasticity model is well suited to the simplest form of structural-mechanics model-like bars in this study.

For a uniaxial response, the degenerated form of the strain energy density functional Ψ is given by Barretta and Marotti de Sciarra [24] as

$$\Psi\left[\varepsilon_{xx},\eta_{xxx}\right]=\underbrace{\frac{1}{2}E_{xx}\left(\varepsilon_{xx}\right)^2}_{\text{Local term}}+\underbrace{\frac{1}{2}E_{xx}l_x^2\left(\eta_{xxx}\right)^2}_{\text{Gradient term}} \tag{1}$$

with E_{xx} being the elastic modulus; $\varepsilon_{xx}(x)$, the axial strain; $\eta_{xxx}(x) = \partial\varepsilon_{xx}(x)/\partial x$, the gradient of axial strain along the x-axis (the axial-strain gradient); and l_x, the material length-scale parameter associated with the axial-strain gradient.

Clearly, the strain energy density functional Ψ of Equation (1) depends not only on the local axial strain $\varepsilon_{xx}(x)$ but also on the axial-strain gradient $\eta_{xxx}(x)$, thus inducing nonlocality of the bar-bulk material. Therefore, nonlocality associated with higher-order deformation mechanism of the strain-gradient materials are accounted for through this amended strain energy density functional Ψ.

Following the strain–displacement compatibility relation [24], the axial strain $\varepsilon_{xx}(x)$ and the axial-strain gradient $\eta_{xxx}(x)$ can be expressed in terms of the bar axial displacement $u_x(x)$ as

$$\varepsilon_{xx}(x) = \frac{\partial u_x(x)}{\partial x}; \text{ and } \eta_{xxx}(x) = \frac{\partial^2 u_x(x)}{\partial x^2} \tag{2}$$

To couple the strain energy density functional Ψ with the principle of thermodynamics, the rate form of Equation (1) is required and can be expressed as

$$\dot{\Psi}[\varepsilon_{xx},\ \eta_{xxx}] = \frac{\partial\Psi}{\partial\varepsilon_{xx}}\dot{\varepsilon}_{xx} + \frac{\partial\Psi}{\partial\eta_{xxx}}\dot{\eta}_{xxx} \tag{3}$$

with $(\cdot)$ denoting the derivative with respect to time t.

Considering the conjugate-work pairs of strain quantities (ε_{xx} and η_{xxx}), the stress quantities can be defined based on Equation (3) as

$$\sigma_{xx}^{L} = \frac{\partial \Psi}{\partial \varepsilon_{xx}} = E_{xx}\varepsilon_{xx} \text{ and } \Sigma_{xx} = \frac{\partial \Psi}{\partial \eta_{xxx}} = l_x{}^2 E_{xx}\eta_{xxx} \tag{4}$$

where σ_{xx}^{L} defines the local axial stress and represents the conjugate-work pair of the axial strain ε_{xx}; and Σ_{xx} defines the higher-order axial stress and represents the conjugate-work pair of the axial-strain gradient η_{xxx}.

Recalling the first law of thermodynamics, the following expression must be satisfied

$$\int_L \left(\int_A \sigma_{xx}\dot{\varepsilon}_{xx} dA \right) dx - \int_L \left(\int_A \dot{\Psi}\, dA \right) dx = 0 \tag{5}$$

where σ_{xx} is the nonlocal axial stress; A, the bar cross-section area; and L, the bar length.

Substituting Equation (3) into Equation (5) leads to the following expression

$$\int_L N(x)\, \dot{\varepsilon}_{xx}\, dx - \int_L N_L(x)\, \dot{\varepsilon}_{xx} dx - \int_L N_H(x)\, \dot{\eta}_{xxx} dx = 0 \tag{6}$$

where the sectional resultant forces ($N(x)$, $N_L(x)$, and $N_H(x)$) are defined as

$$[N(x), N_L(x), N_H(x)] = \int_A \left[\sigma_{xx}, \sigma_{xx}^{L}, \Sigma_{xx}\right] dA \tag{7}$$

As demonstrated by Sae-Long et al. [44], the integral relation of Equation (6) plays an essential role in deriving the governing differential equilibrium equation of the nanobar-elastic substrate system via the virtual displacement principle.

2.2. Surface Elasticity Theory

The size-dependent phenomenon is unique to nano-sized structures. The so-called "surface-free energy" related to excessive energy at the surface atoms is responsible for this unique phenomenon. This study employs the surface elasticity model of Gurtin and Murdoch [46,47] to account for the surface-energy effect on nano-sized bar responses. For the present problem of nanobars, the degenerated form of the Gurtin–Murdoch surface constitutive model can be written as

$$\tau_{xx}^{sur} - \tau_0^{sur} = E^{sur}\varepsilon_{xx}^{sur} \tag{8}$$

where τ_{xx}^{sur} is the axial component of the surface stress tensor; τ_0^{sur}, the residual surface stress under unconstrained conditions; E^{sur}, the surface elastic modulus; u_{xx}^{sur}, the surface axial displacement; and $\varepsilon_{xx}^{sur} = \partial u_x^{sur}/\partial x$, the surface strain.

Following the perfectly bonded interface between the bar bulk and the wrapped surface layer (full composite action), the following relations are obtained

$$u_{xx}^{sur}(x) = u_x(x) \text{ and } \varepsilon_{xx}^{sur}(x) = \varepsilon_{xx}(x) \tag{9}$$

3. Bar-Substrate Medium Interaction

To consider interactive mechanism between the bar and its surrounding substrate medium, the widely used Winkler foundation model [66] is called for. Smeared elastic springs in a series are distributed along the bar length to represent the surrounding substrate medium. The force–deformation relation for these smeared elastic springs is

$$D_S(x) = k_S \Delta_S(x) \tag{10}$$

with $D_S(x)$ being the substrate interactive force; k_S, the elastic substrate stiffness; and $\Delta_S(x)$, the substrate deformation.

Full compatibility between the bar and its surrounding substrate medium results in the following relation

$$\Delta_S(x) = u_x(x) \tag{11}$$

4. Model Formulation

4.1. Differential Equilibrium Equation and End-Force Equilibrium Conditions: The Virtual Displacement Approach

As an alternative to represent the system equilibrium, the virtual displacement principle is called for. The general expression of the virtual displacement principle is

$$\delta W = \delta W_{\text{int}} + \delta W_{\text{ext}} = 0 \tag{12}$$

where δW represents the system total virtual work; δW_{int} represents the system internal virtual work; and δW_{ext} represents the system external virtual work.

For a strain gradient bar-elastic substrate medium system with inclusion of surface-free energy shown in Figure 2, δW_{int} and δW_{ext} are given by Sae-Long et al. [44] as

$$\begin{aligned} \delta W_{\text{int}} = & \underbrace{\int_L \left(\int_A \sigma_{xx}(x)\, dA \right) \delta\varepsilon_{xx}(x)\, dx}_{\text{Bar-Bulk Contribution}} + \underbrace{\int_L D_S(x)\, \delta\Delta_S(x)\, dx}_{\text{Substrate-Medium Contribution}} \\ & + \underbrace{\int_L \left(\oint_\Gamma \left(\left(\tau_{xx}^{sur}(x) - \tau_0^{sur} \right) \delta\varepsilon_{xx}^{sur}(x) \right) d\Gamma \right) dx}_{\text{Surface-Energy Contribution}} \end{aligned} \tag{13}$$

$$\delta W_{\text{ext}} = -\underbrace{\int_L p_x(x)\, \delta u_x(x)\, dx}_{\text{Distributed-Load Contribution}} - \underbrace{\delta \mathbf{U}^{\mathrm{T}} \mathbf{P}}_{\text{End-Load Contribution}} \tag{14}$$

where Γ is the bar perimeter; $p_x(x)$ represents the longitudinal distributed load; the vector $\mathbf{U} = \{ U_1 \quad U_2 \quad U_3 \quad U_4 \}^T$ collects displacements at the bar ends; and the vector $\mathbf{P} = \{ P_1 \quad P_2 \quad P_3 \quad P_4 \}^T$ collects conjugate-work forces at the bar ends.

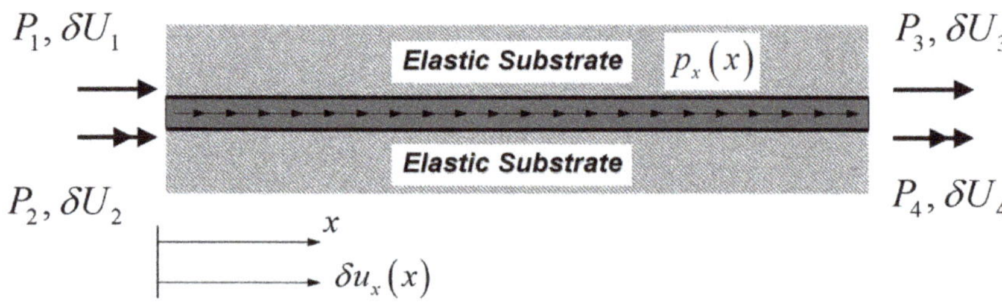

Figure 2. Nanobar-elastic substrate system: the virtual-displacement formulation.

Recalling the definition of the sectional resultant forces of Equation (7), imposing the compatibility conditions of Equations (2), (9) and (11) and subsequently imposing the thermodynamics condition of Equation (6), the system internal virtual work δW_{int} becomes

$$\begin{aligned} \delta W_{\text{int}} = & \int_L N_L(x) \frac{\partial \delta u_x(x)}{\partial x} dx + \int_L N_H(x) \frac{\partial^2 \delta u_x(x)}{\partial x^2} dx + \int_L D_S(x) \delta u_x(x) dx \\ & + \int_L N_{\tau_{xx}-\tau_0}^{sur}(x) \frac{\partial \delta u_x(x)}{\partial x} dx \end{aligned} \tag{15}$$

where the surface axial force $N^{sur}_{\tau_{xx}-\tau_0}(x)$ is defined as

$$N^{sur}_{\tau_{xx}-\tau_0}(x) = \oint\limits_{\Gamma} \left(\tau^{sur}_{xx}(x) - \tau^{sur}_0\right) d\Gamma \tag{16}$$

Consequently, the virtual displacement statement of Equation (12) is rewritten as

$$\begin{aligned} \delta W = \int\limits_L N^{sur}_L(x)\frac{\partial \delta u_x(x)}{\partial x}dx + \int\limits_L N_H(x)\frac{\partial^2 \delta u_x(x)}{\partial x^2}dx + \int\limits_L D_S(x)\delta u_x(x)dx \\ -\int\limits_L p_x(x)\delta u_x(x)dx - \delta \mathbf{U}^{\mathbf{T}}\mathbf{P} = 0 \end{aligned} \tag{17}$$

where $N^{sur}_L(x) = N_L(x) + N^{sur}_{\tau_{xx}-\tau_0}(x)$ is defined as the lower-order composite axial force and is contributed from the full composite action between the bar-bulk material and the wrapped surface layer.

Following the virtual displacement principle employed by Sae-Long et al. [44], the governing differential equilibrium equation (Euler–Lagrange equation) as well as its associated end-boundary force conditions (natural boundary conditions) of the nanobar-elastic substrate system are consistently derived as

$$\frac{\partial^2 N_H(x)}{\partial x^2} - \frac{\partial N^{sur}_L(x)}{\partial x} + D_S(x) - p_x(x) = 0 : \text{for} x \in (0,\, L) \tag{18}$$

$$\begin{aligned} P_1 = -\left(N^{sur}_L(x) - \frac{\partial N_H(x)}{\partial x}\right)_{x=0};\ P_2 = -(N_H(x))_{x=0}; \\ P_3 = \left(N^{sur}_L(x) - \frac{\partial N_H(x)}{\partial x}\right)_{x=L};\ P_4 = (N_H(x))_{x=L} \end{aligned} \tag{19}$$

It is worth remarking that the differential equilibrium equation of Equation (18) is of vital importance when the virtual force principle is employed to reveal the differential compatibility equation of the problem, as will be presented subsequently.

4.2. Sectional Constitutive Relations: Compliance Form

The sectional constitutive relations can be obtained by substituting the stress–strain relations of Equations (4) and (8) into Equations (7) and (16), respectively, and can be written in the compliance form as

$$\varepsilon_{xx}(x) = \frac{N_L(x)}{E_{xx}A};\ \eta_{xxx}(x) = \frac{N_H(x)}{{l_x}^2 E_{xx}A};\ \text{and}\ \varepsilon^{sur}_{xx}(x) = \frac{N^{sur}_{\tau_{xx}-\tau_0}(x)}{E^{sur}\Gamma} \tag{20}$$

Imposing the full composite action between the bar bulk and the wrapped surface layer of Equations (9) and (20) provides the following relations between axial force components

$$\begin{aligned} N_L(x) = \frac{E_{xx}A}{E_{xx}A + E^{sur}\Gamma}N^{sur}_L(x);\ N^{sur}_{\tau_{xx}-\tau_0}(x) = \frac{E^{sur}\Gamma}{E_{xx}A+E^{sur}\Gamma}N^{sur}_L(x);\ \text{and} \\ N_H(x) = \frac{{l_x}^2 E_{xx}A}{E_{xx}A+E^{sur}\Gamma}\frac{\partial N^{sur}_L(x)}{\partial x} \end{aligned} \tag{21}$$

Based on Equation (10), the deformation-force (compliance form) relation for an elastic substrate medium can be expressed as

$$\Delta_S(x) = \frac{D_S(x)}{k_S} \tag{22}$$

4.3. Differential Compatibility Equations and End-Displacement Compatibility Conditions: The Virtual Force Approach

To express the system compatibility conditions in the integral (weak) form, the virtual force principle is applied. The virtual force equation can be written in a general form as

$$\delta W^* = \delta W^*_{\text{int}} + \delta W^*_{\text{ext}} = 0 \tag{23}$$

where δW^* represents the system total complementary virtual work; δW^*_{int} represents the system internal complementary virtual work; and δW^*_{ext} represents the system external complementary virtual work.

For the bar-substrate medium system of Figure 3, δW^*_{int} and δW^*_{ext} are

$$\begin{aligned}\delta W^*_{\text{int}} &= \int_L \delta N_L(x)\varepsilon_{xx}(x)\,dx + \int_L \delta N_H(x)\eta_{xxx}(x)\,dx \\ &+ \int_L \delta N^{sur}_{\tau_{xx}-\tau_0}(x)\varepsilon^{sur}_{xx}(x)dx\ + \int_L \delta D_S(x)\Delta_S(x)\,dx\end{aligned} \tag{24}$$

$$\delta W^*_{\text{ext}} = -\int_L \delta p_x(x)u_x(x)dx - \delta\mathbf{P}^T\mathbf{U} \tag{25}$$

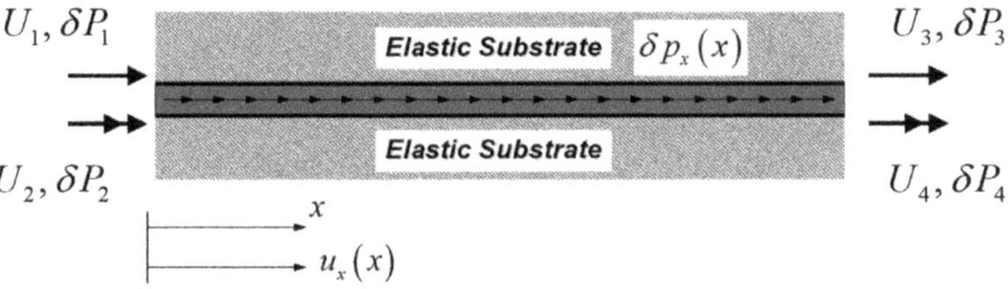

Figure 3. Nanobar-elastic substrate system: the virtual-force formulation.

To eliminate the bar axial displacement $u_x(x)$ from the virtual force statement, the virtual longitudinal distributed load $\delta p_x(x)$ can arbitrarily be chosen to be zero without loss of model generality. Therefore, Equation (23) becomes

$$\begin{aligned}&\int_L \delta N_L(x)\varepsilon_{xx}(x)\,dx + \int_L \delta N_H(x)\eta_{xxx}(x)\,dx + \\ &\int_L \delta D_S(x)\Delta_S(x)\,dx\ + \int_L \delta N^{sur}_{\tau_{xx}-\tau_0}(x)\varepsilon^{sur}_{xx}(x)dx - \delta\mathbf{P}^T\mathbf{U} = 0\end{aligned} \tag{26}$$

Enforcing the compliance-type constitutive relations of Equations (20) and (22), Equation (26) can be rewritten as

$$\begin{aligned}&\int_L \delta N^{sur}_L(x)\frac{N_L(x)}{E_{xx}A}\,dx + \int_L \delta N_H(x)\frac{N_H(x)}{l_x{}^2E_{xx}A}\,dx + \\ &\int_L \delta D_S(x)\frac{D_S(x)}{k_S}\,dx - \delta\mathbf{P}^T\mathbf{U} = 0\end{aligned} \tag{27}$$

Imposing the differential equilibrium relation of Equation (18), the substrate interactive force $D_S(x)$ and its virtual counterpart $\delta D_S(x)$ can be excluded from the virtual force statement. Thus, Equation (24) is rewritten as

$$\begin{aligned}&\int_L \delta N^{sur}_L(x)\frac{N_L(x)}{E_{xx}A}\,dx + \int_L \delta N_H(x)\frac{N_H(x)}{l_x{}^2E_{xx}A}\,dx \\ &+\int_L \left(-\frac{\partial^2\delta N_H(x)}{\partial x^2} + \frac{\partial\delta N^{sur}_L(x)}{\partial x}\right)\left(\frac{1}{k_S}\right)\left(-\frac{\partial^2 N_H(x)}{\partial x^2} + \frac{\partial N^{sur}_L(x)}{\partial x} + p_x(x)\right)dx \\ &-\delta\mathbf{P}^T\mathbf{U} = 0\end{aligned} \tag{28}$$

In order to move differential operators to axial forces $N_L^{sur}(x)$ and $N_H(x)$, integration by parts is called for, thus resulting in the following expression

$$\begin{aligned}
&\int_L \delta N_L^{sur}(x)\left(\frac{N_L(x)}{E_{xx}A}+\left(\frac{1}{k_S}\right)\left(\frac{\partial^3 N_H(x)}{\partial x^3}-\frac{\partial^2 N_L^{sur}(x)}{\partial x^2}-\frac{\partial p_x(x)}{\partial x}\right)\right)dx+\\
&\int_L \delta N_H(x)\left(\frac{N_H(x)}{l_x{}^2E_{xx}A}+\left(\frac{1}{k_S}\right)\left(\frac{\partial^4 N_H(x)}{\partial x^4}-\frac{\partial^3 N_L^{sur}(x)}{\partial x^3}-\frac{\partial^2 p_x(x)}{\partial x^2}\right)\right)dx+\\
&\left[\frac{1}{k_S}\left(-\frac{\partial^2 N_H(x)}{\partial x^2}+\frac{\partial N_L^{sur}(x)}{\partial x}+p_x(x)\right)\left(\delta N_L^{sur}(x)-\frac{\partial \delta N_H(x)}{\partial x}\right)\right]_0^L+\\
&\left[\frac{1}{k_S}\left(-\frac{\partial^3 N_H(x)}{\partial x^3}+\frac{\partial^2 N_L^{sur}(x)}{\partial x^2}+\frac{\partial p_x(x)}{\partial x}\right)\delta N_H(x)\right]_0^L-\delta\mathbf{P}^T\mathbf{U}=0
\end{aligned}\tag{29}$$

The virtual force quantities in the first boundary term of Equation (29) reveal that the total lower-order (local) axial force $N(x)$ is defined in terms of the lower-order composite axial force $N_L^{sur}(x)$ and the higher-order axial force $N_H(x)$ as

$$N(x) \;=\; N_L^{sur}(x)-\frac{\partial N_H(x)}{\partial x}\tag{30}$$

The axial-force relation of Equation (30) was also gained by Sae-Long et al. [44] using the virtual displacement principle as shown in Equation (19).

Following the Cartesian sign convention and recalling the axial-force definition of Equation (30), Equation (29) becomes

$$\begin{aligned}
&\int_L \delta N_L^{sur}(x)\left(\frac{N_L(x)}{E_{xx}A}+\left(\frac{1}{k_S}\right)\left(\frac{\partial^3 N_H(x)}{\partial x^3}-\frac{\partial^2 N_L^{sur}(x)}{\partial x^2}-\frac{\partial p_x(x)}{\partial x}\right)\right)dx+\\
&\int_L \delta N_H(x)\left(\frac{N_H(x)}{l_x{}^2E_{xx}A}+\left(\frac{1}{k_S}\right)\left(\frac{\partial^4 N_H(x)}{\partial x^4}-\frac{\partial^3 N_L^{sur}(x)}{\partial x^3}-\frac{\partial^2 p_x(x)}{\partial x^2}\right)\right)dx+\\
&-\delta P_1\left(U_1+\frac{1}{k_S}\left(-\frac{\partial^2 N_H(x)}{\partial x^2}+\frac{\partial N_L^{sur}(x)}{\partial x}+p_x(x)\right)_{x=0}\right)\\
&-\delta P_2\left(U_2-\frac{1}{k_S}\left(-\frac{\partial^3 N_H(x)}{\partial x^3}+\frac{\partial^2 N_L^{sur}(x)}{\partial x^2}+\frac{\partial p_x(x)}{\partial x}\right)_{x=0}\right)\\
&-\delta P_3\left(U_3+\frac{1}{k_S}\left(-\frac{\partial^2 N_H(x)}{\partial x^2}+\frac{\partial N_L^{sur}(x)}{\partial x}+p_x(x)\right)_{x=L}\right)\\
&-\delta P_4\left(U_4-\frac{1}{k_S}\left(-\frac{\partial^3 N_H(x)}{\partial x^3}+\frac{\partial^2 N_L^{sur}(x)}{\partial x^2}+\frac{\partial p_x(x)}{\partial x}\right)_{x=L}\right)=0
\end{aligned}\tag{31}$$

Accounting for arbitrariness of $\delta N_L^{sur}(x)$ and $\delta N_H(x)$, the governing differential compatibility equations associated with the lower-order and higher-order axial forces are obtained, respectively, as

$$\frac{N_L(x)}{E_{xx}A}+\left(\frac{1}{k_S}\right)\left(\frac{\partial^3 N_H(x)}{\partial x^3}-\frac{\partial^2 N_L^{sur}(x)}{\partial x^2}-\frac{\partial p_x(x)}{\partial x}\right)=0:\text{for}\,x\in(0,\,L)\tag{32}$$

$$\frac{N_H(x)}{l_x{}^2E_{xx}A}+\left(\frac{1}{k_S}\right)\left(\frac{\partial^4 N_H(x)}{\partial x^4}-\frac{\partial^3 N_L^{sur}(x)}{\partial x^3}-\frac{\partial^2 p_x(x)}{\partial x^2}\right)=0:\text{for}\,x\in(0,\,L)\tag{33}$$

It is worth remarking that due to elimination of the substrate interactive force $D_S(x)$ and its virtual counterpart $\delta D_S(x)$ as discussed earlier, the compatibility condition associated with the elastic-substrate medium is not present in the virtual force statement of Equation (31).

Considering the compliance-type constitutive relations of Equations (20) and (22), enforcing the Winkler-foundation assumption of Equation (11), and imposing the equilibrium relation of Equation (18), Equations (20) and (22) simply address the lower-order and higher-order strain–displacement compatibility conditions as

$$\varepsilon_{xx}(x) - \frac{\partial u_x(x)}{\partial x} = 0 \tag{34}$$

$$\eta_{xxx}(x) - \frac{\partial^2 u_x(x)}{\partial x^2} = 0 \tag{35}$$

Considering the first and third axial-force relations of Equation (21), the following relation between the first derivative of the local axial force $N_L(x)$ and higher-order axial force $N_H(x)$ can be established by

$$N_H(x) = {l_x}^2 \frac{\partial N_L(x)}{\partial x} \tag{36}$$

With the axial-force relation of Equation (36), two differential compatibility conditions of Equations (32) and (33) can be combined into a single expression as

$$\frac{N_L(x)}{E_{xx}A} + \left(\frac{1}{k_S}\right)\left({l_x}^2 \frac{\partial^4 N_H(x)}{\partial x^4} - \frac{\partial^2 N_L^{sur}(x)}{\partial x^2} - \frac{\partial p_x(x)}{\partial x}\right) = 0 : \text{for} x \in (0,\, L) \tag{37}$$

It is worth restating that the differential equilibrium equation given by Sae-Long et al. [44] is derived based on the virtual displacement principle while the differential compatibility equation of Equation (37) is derived based on the virtual force principle. Comparison between these two differential equations confirms the dualism of the virtual displacement and virtual force principles.

Accounting for the arbitrariness of $\delta\mathbf{P}$ in Equation (31) yields the following end-boundary displacement conditions (essential boundary conditions).

$$\begin{aligned}
U_1 &= \tfrac{1}{k_S}\left(\tfrac{\partial^2 N_H(x)}{\partial x^2} - \tfrac{\partial N_L^{sur}(x)}{\partial x}\right)_{x=0} - \tfrac{1}{k_S}(p_x(x))_{x=0} \\
U_2 &= \tfrac{1}{k_S}\left(-\tfrac{\partial^3 N_H(x)}{\partial x^3} + \tfrac{\partial^2 N_L^{sur}(x)}{\partial x^2}\right)_{x=0} + \tfrac{1}{k_S}\left(\tfrac{\partial p_x(x)}{\partial x}\right)_{x=0} \\
U_3 &= \tfrac{1}{k_S}\left(\tfrac{\partial^2 N_H(x)}{\partial x^2} - \tfrac{\partial N_L^{sur}(x)}{\partial x}\right)_{x=L} - \tfrac{1}{k_S}(p_x(x))_{x=L} \\
U_4 &= \tfrac{1}{k_S}\left(-\tfrac{\partial^3 N_H(x)}{\partial x^3} + \tfrac{\partial^2 N_L^{sur}(x)}{\partial x^2}\right)_{x=L} + \tfrac{1}{k_S}\left(\tfrac{\partial p_x(x)}{\partial x}\right)_{x=L}
\end{aligned} \tag{38}$$

Compared to the end-boundary force conditions given by Sae-Long et al. [44] using the virtual displacement principle, Equation (38) and those given by Sae-Long et al. [44] are dual. Furthermore, end-displacement components associated with the bar bulk-surface layer composite ($N_L^{sur}(x)$ and $N_H(x)$) and the distributed load $p_x(x)$ are clearly separated in Equation (38).

In summary, a complete set of governing equations of the problem formulated within the framework of virtual force principle are the equilibrium condition of Equation (18), the compliance-type constitutive relations of Equations (20) and (22), and the virtual-force statement (weak form) of the compatibility relations of Equations (32) and (33) together with the end-boundary displacement conditions of Equation (38). The formulation procedure within the framework of the virtual force principle can concisely be presented in the modified Tonti's diagram of Figure 4.

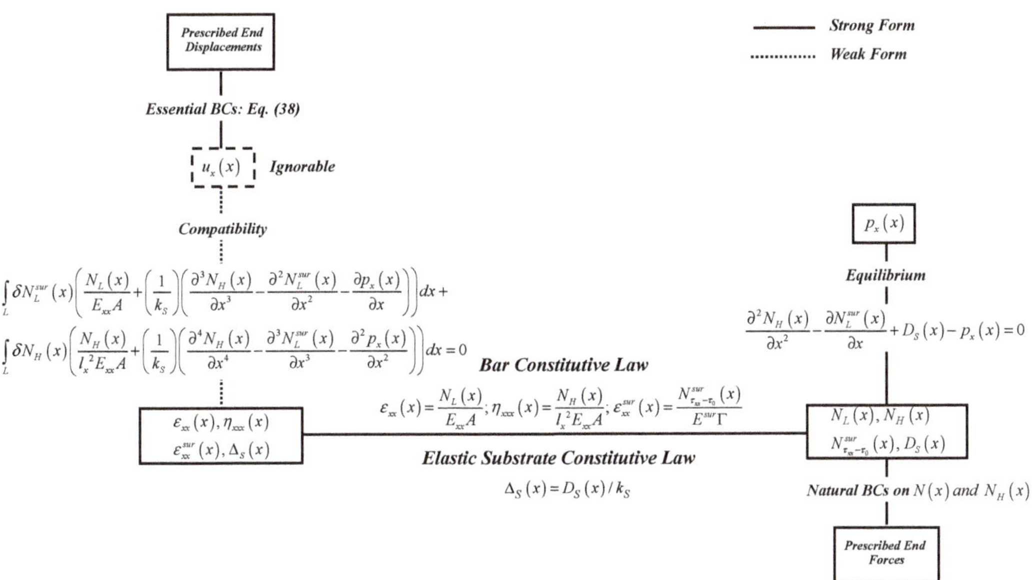

Figure 4. Modified Tonti's diagram for nanobar-elastic substrate system: The virtual-force formulation.

5. Analytical Solution of Differential Compatibility Equation: Axial-Force Solution

Within the framework of the virtual force formulation, the analytical solution to the differential compatibility equation of Equation (37) is expressed in terms of axial force variables. Unfortunately, the present form of Equation (37) cannot be solved readily since it contains multi axial-force variables ($N_L(x)$, $N_L^{sur}(x)$ and $N_H(x)$). Recalling the axial-force relations of Equation (21), the differential compatibility relation of Equation (37) can be expressed in terms of a single axial-force field $N_L^{sur}(x)$ as

$$\left(l_x{}^2 E_{xx}A\right)\frac{\partial^4 N_L^{sur}(x)}{\partial x^4}-(EA)_{xx}^{sur}\frac{\partial^2 N_L^{sur}(x)}{\partial x^2}+k_S N_L^{sur}(x)=(EA)_{xx}^{sur}\frac{\partial p_x(x)}{\partial x}\ \text{for}\ x\in(0,\ L) \tag{39}$$

with the composite bar axial stiffness $(EA)_{xx}^{sur}$ being defined as $E_{xx}A+E^{sur}\Gamma$.

Equation (39) is central to the axial-force determination of the strain-gradient bar-elastic substrate system with inclusion of surface-energy effect. The strain-gradient nature of the bar-bulk material induces the higher-order derivative (fourth order) while the surface-free energy induces the lower-order derivative (second order). Furthermore, it is observed from Equation (39) that the surface-energy effect influences both homogeneous and particular solutions while the strain-gradient effect only affects the homogeneous solution. It is worth pointing out that with the presence of a uniformly distributed load $p_x(x)=p_{x0}$, only the homogeneous solution is required since the term on the right-hand side of Equation (39) vanishes. In other words, the axial-force response $N_L^{sur}(x)$ is not influenced with the presence of a uniformly distributed load $p_x(x)=p_{x0}$. This unique feature makes the proposed bar-elastic substrate model desirable since there is no need for the particular solution with this specific loading case. However, the presence of the uniformly distributed load $p_x(x)=p_{x0}$ affects the system response through the system equilibrium condition of Equation (18).

As suggested by Gülkan and Alemdar [67], the homogeneous solution to Equation (39) can be written in the general form as

$$N_L^{sur}(x)=\phi_1(x)c_1+\phi_2(x)c_2+\phi_3(x)c_3+\phi_4(x)c_4 \tag{40}$$

where c_1, c_2, c_3, and c_4 are constants of integration and can be determined from the imposed boundary conditions; and ϕ_1, ϕ_2, ϕ_3, and ϕ_4 are the basic functions which forms are governed by the system parameters (see Appendix A).

To analytically determine the axial-force solution, essential and natural boundary conditions are both required. Investigating the first boundary term of Equation (29) reveals the following classical essential and natural boundary conditions

$$\left.\begin{array}{l} \text{specify } \overline{u}_x = \frac{1}{k_S}\left(-\frac{\partial^2 N_H(x)}{\partial x^2} + \frac{\partial N_L^{sur}(x)}{\partial x} + p_x(x)\right) \\ \text{specify } \overline{N} = N_L^{sur}(x) - \frac{\partial N_H(x)}{\partial x} \end{array}\right\} \text{at } x = 0, L \tag{41}$$

Similarly, considering the second boundary term of Equation (29) reveals the following non-classical essential and natural boundary conditions

$$\left.\begin{array}{l} \text{specify } \overline{u}'_x = \frac{\partial u_x(x)}{\partial x} = \frac{1}{k_S}\left(-\frac{\partial^3 N_H(x)}{\partial x^3} + \frac{\partial^2 N_L^{sur}(x)}{\partial x^2} + \frac{\partial p_x(x)}{\partial x}\right) \\ \text{specify } \overline{N}_H = N_H(x) \end{array}\right\} \text{at } x = 0, L \tag{42}$$

Unfortunately, boundary conditions of Equations (41) and (42) cannot be readily employed since the lower-order composite axial force $N_L^{sur}(x)$ is only the single variable present in the governing differential compatibility equation of Equation (39). However, boundary conditions of Equations (41) and (42) can be expressed in terms of the lower-order composite axial force $N_L^{sur}(x)$ by employing the axial-force relations of Equation (21). Consequently, Equations (41) and (42) become

$$\left.\begin{array}{l} \text{specify } \overline{u}_x = \frac{1}{k_S}\left(-\frac{l_x{}^2 E_{xx} A}{(EA)_{xx}^{sur}}\frac{\partial^3 N_L^{sur}(x)}{\partial x^3} + \frac{\partial N_L^{sur}(x)}{\partial x} + p_x(x)\right) \\ \text{specify } \overline{N} = N_L^{sur}(x) - \frac{l_x{}^2 E_{xx} A}{(EA)_{xx}^{sur}}\frac{\partial^2 N_L^{sur}(x)}{\partial x^2} \end{array}\right\} \text{at } x = 0, L \tag{43}$$

$$\left.\begin{array}{l} \text{specify } \overline{u}'_x = \frac{\partial u_x(x)}{\partial x} = \frac{1}{k_S}\left(\begin{array}{c} -\frac{l_x{}^2 E_{xx} A}{(EA)_{xx}^{sur}}\frac{\partial^4 N_L^{sur}(x)}{\partial x^4} + \\ \frac{\partial^2 N_L^{sur}(x)}{\partial x^2} + \frac{\partial p_x(x)}{\partial x} \end{array}\right) \\ \text{specify } \overline{N}_H = \frac{l_x{}^2 E_{xx} A}{(EA)_{xx}^{sur}}\frac{\partial N_L^{sur}(x)}{\partial x} \end{array}\right\} \text{at } x = 0, L \tag{44}$$

6. Numerical Example

In the present study, a nanowire-elastic substrate system of Figure 5 is employed as a numerical example to investigate the characteristics and to assess the accuracy of the proposed nanobar-substrate model. An end force P of 2400 nN and a uniformly distributed load p_{x0} of 2.4 nN/nm are exerted on this bar-substrate system. It is noted that this numerical example had been employed to present the characteristics of Eringen's nonlocal bar-substrate model proposed by Limkatanyu et al. [63]. The nanowire is made of silver material with the bulk modulus E_{xx} = 76 GPa and has the following geometric properties: diameter D = 50 nm and length L = 1000 nm. These mechanical and geometric properties follow values given by Juntarasaid et al. [68] and He and Lilley [48], respectively. The material length-scale parameter l_x = 200 nm is assumed as provided by Yang and Lim [69]. As suggested by He and Lilley [48], the surface elastic modulus E^{sur} of 1.22 nN/nm is employed. A stiffness coefficient K_S of 95×10^{-3} nN/nm^3 is employed for the surrounding substrate medium, thus resulting in an elastic substrate stiffness k_S of 14.92 nN/nm^2. This particular value for the surrounding substrate is provided by Liew et al. [70] to represent the surrounding substrate medium as polymer.

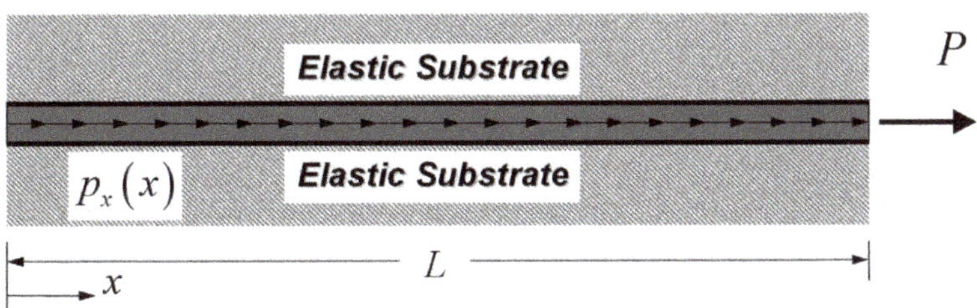

$$P = 2400 \text{ nN};\ p_x(x) = 2.4 \text{ nN/nm};\ L = 1000 \text{ nm};\ D = 50 \text{ nm};$$

$$E_{xx} = 76 \text{ GPa};\ E^{sur} = 1.22 \text{ nN/nm};\ l_x = 200 \text{ nm};\ k_s = 14.92 \text{ nN/nm}^2$$

Figure 5. Nanowire-elastic substrate system under axial force loadings: Numerical example.

Two different bar-substrate systems are employed to simulate the responses of the silver nanowire-substrate system of Figure 5. The first is based on the classical (local) bar-substrate model [71], thus excluding both nonlocal and surface-energy effects while the second is based on the proposed strain-gradient bar-substrate model, thus including both nonlocal and surface-energy effects. Furthermore, the responses of the silver nanowire-substrate system are also simulated by the strain-gradient bar-substrate model of Sae-Long et al. [44] to confirm the accuracy and efficiency of the proposed bar-substrate model. It is worth mentioning that the strain-gradient bar-substrate model of Sae-Long et al. [44] provides the axial displacement $u_x(x)$ as the basic solution while the proposed strain-gradient bar-substrate model yields the lower-order composite axial force $N_L^{sur}(x)$ as the basic solution. Unlike the strain-gradient bar-substrate model of Sae-Long et al. [44], the proposed strain-gradient bar-substrate model does not require the particular solution for the present case of the uniformly distributed load p_{x0}, thus showing the merit of the present model formulation.

Figure 6 plots and compares axial-displacement profiles obtained from classical and two bar-substrate models. Clearly, the axial-displacement profile obtained from the proposed bar-substrate model is identical to that obtained from the bar-substrate model of Sae-Long et al. [44], thus confirming validity of the proposed model. Compared with the classical model, a stiffer nanowire-substrate system response is obtained with the proposed model. Considering the coefficient of the lower-order derivative (second order) term in Equation (39), the system stiffness enhancement related to the surface-free energy can clearly be noticed. However, this stiffening effect of the surface-free energy is minimal for specific values of system parameters herein since the composite bar axial stiffness $(EA)_{xx}^{sur}$ increases merely 0.128% with inclusion of the surface-energy effect. Therefore, the stiffening effect associated with the bar-bulk nonlocality is much more pronounced than that associated with the surface-free energy. With the classical model, the axial displacement remains approximately constant at 0.16 nm along half of the nanowire (see the inset in Figure 6) and then drastically increases to reach its maximum value of 1.8 nm at the loading end. With the proposed model, the left half of the nanowire experiences a gradual decrease in the axial displacement between its free end (0.16 nm) and its middle region (0.13 nm) while the right half of the nanowire also encounters a drastic increase in axial displacement between its middle region and its loading end (1.36 nm). This particular displacement characteristic is associated with the higher-order derivative (fourth order) term in Equation (39) and the statically indeterminate nature of the bar-substrate system.

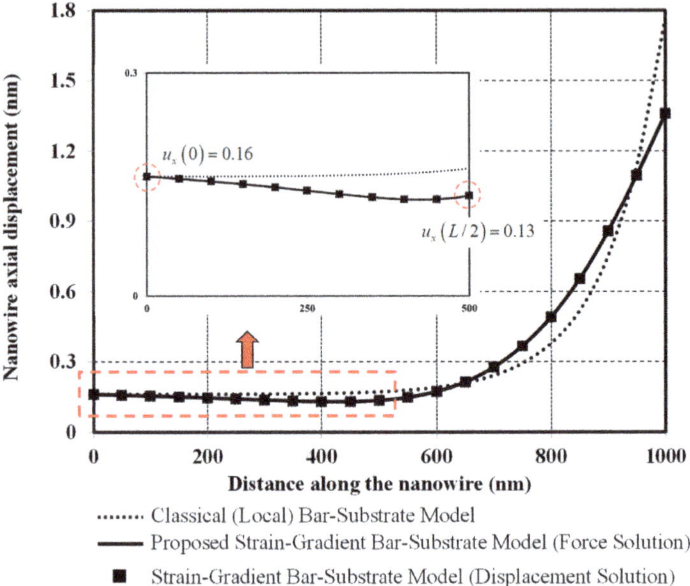

Figure 6. Axial displacement versus distance along the nanowire.

Figure 7a compares axial-strain distributions obtained from classical and two bar-substrate models while Figure 7b plots the axial-strain gradient distribution obtained from the bar-substrate models. It is clear from Figure 7a that the strain-gradient nature of the bar-bulk material drastically alters the distribution characteristics of axial-strain responses. With the classical model, the axial strain remains approximately zero along half of the nanowire (see the inset in Figure 7a) and then rapidly increases toward the loading end. In other words, the axial-strain distribution of the classical model appears to be localized in the neighborhood of the loading end. With the proposed model, the axial strain along approximately half of the nanowire is in compression (negative value) and then smoothly increases to reach its maximum positive value at the loading end. The maximum axial strain obtained with the proposed model is about three times less than that obtained with the classical model. This peculiar but unique axial-strain response complies with the axial-displacement response presented in Figure 6. The axial-strain gradient distribution is shown in Figure 7b. Vanishing of the axial-strain gradient at either nanowire end ($\eta_{xxx}(0) = \eta_{xxx}(L) = 0$) is associated with the imposed higher-order force boundary conditions at both the nanowire ends ($N_H(0) = N_H(L) = 0$) through the constitutive relation of Equation (20).

Figure 8 compares the axial-force distributions obtained from classical and two bar-substrate models. With the classical model, the axial force remains approximately zero along the left half of the nanowire (see the inset in Figure 8) and then drastically increases toward the loading end, thus implying that only the right half of the nanowire takes part in the axial-force resistance. With the proposed model, a whole portion of the nanowire participates in the axial-force resistance. The axial force is in compression (negative value) along around three quarters of the whole length (see the inset in Figure 8) and drastically increases to reach its maximum in tension at the loading end. This unique but rather peculiar axial-force response is induced by the higher-order axial force solution to the governing differential equation of Equation (39) and the statical indeterminacy inherent to the nanowire-elastic substrate system.

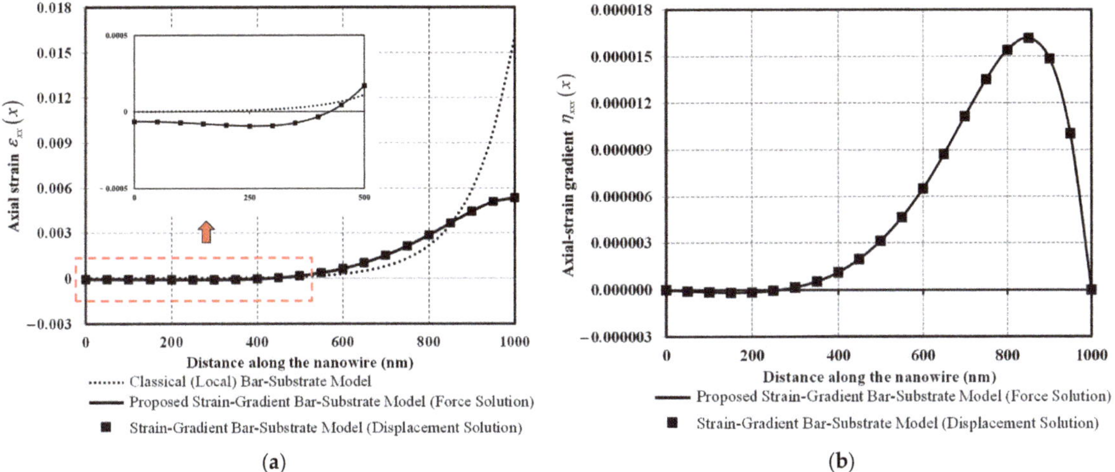

Figure 7. Axial strain and axial-strain gradient versus distance along the nanowire: (**a**) Axial strain; (**b**) Axial-strain gradient.

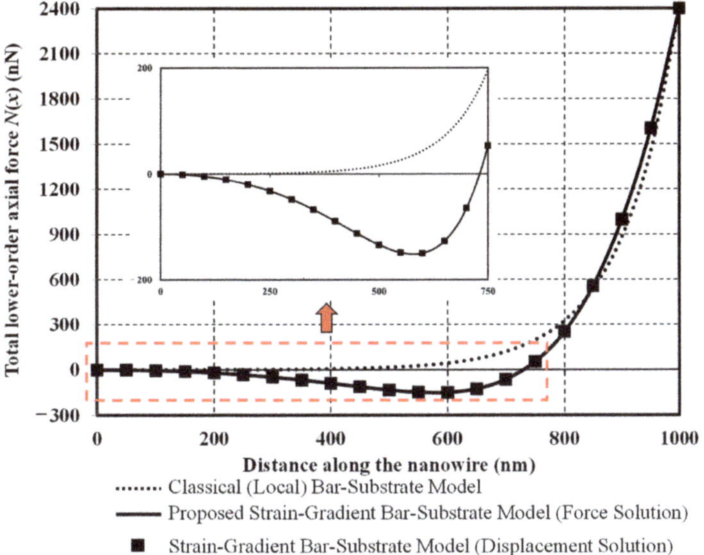

Figure 8. Total lower-order axial force versus distance along the nanowire.

To scrutinize the axial-force distribution nature of the proposed bar-substrate model, distribution diagrams of lower-order composite axial force $N_L^{sur}(x)$, higher-order axial force $N_H(x)$, and higher-order axial-force gradient $\partial N_H(x)/\partial x$ are respectively plotted in Figure 9a–c. All axial-force diagrams obtained from the bar-substrate model of Sae-Long et al. [44] are also superimposed to confirm validity of the proposed bar-substrate model.

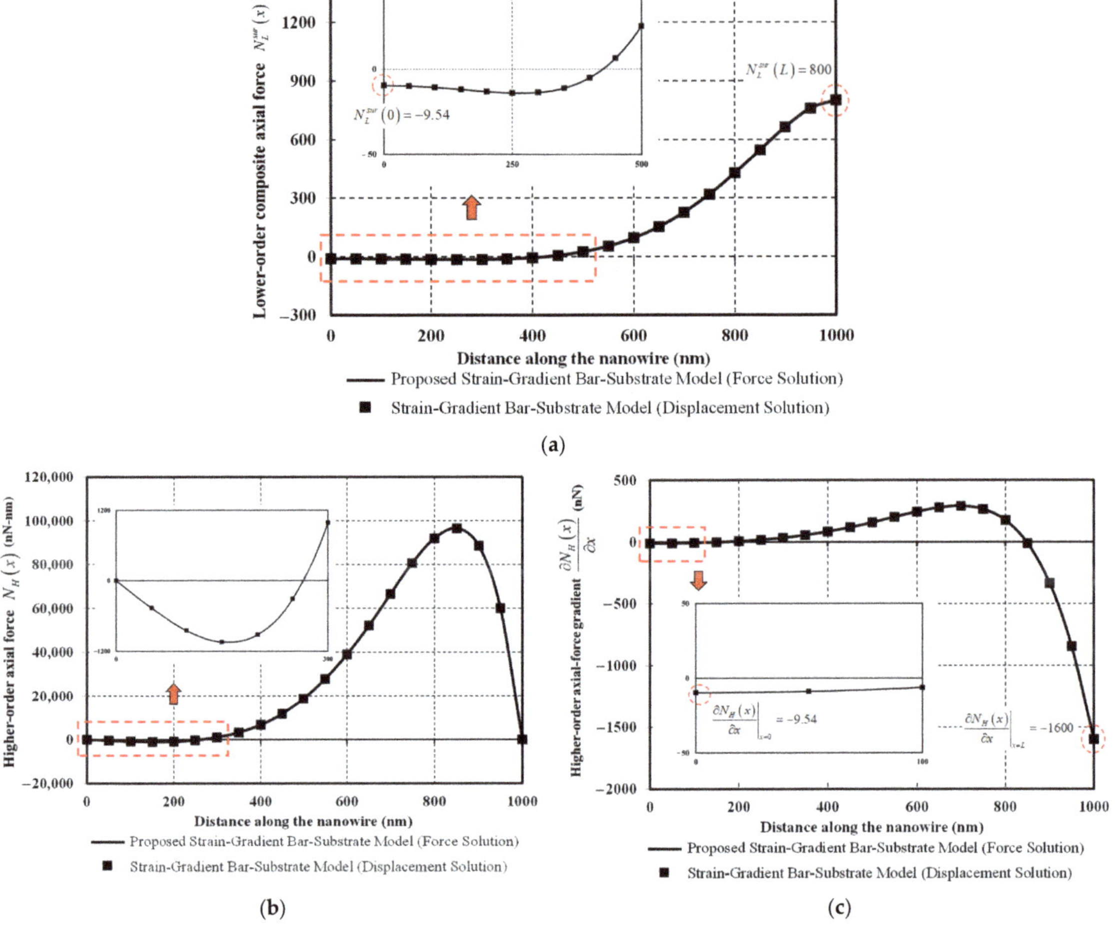

Figure 9. Axial force responses versus distance along the nanowire: (**a**) Lower-order composite axial force $N_L^{sur}(x)$; (**b**) Higher-order axial force $N_H(x)$; (**c**) Higher-order axial-force gradient $\partial N_H(x)/\partial x$.

Figure 9a clearly shows that the lower-order composite axial force $N_L^{sur}(x)$ does not satisfy the end-force boundary conditions, and its distribution nature is much smoother than that of its local counterpart present in the classical model presented in Figure 8. It is worth remarking that the lower-order composite axial force $N_L^{sur}(x)$ is the fundamental solution to the fourth order differential equation of Equation (39) while its local counterpart is the fundamental solution to the second order differential equation. The higher-order axial force $N_H(x)$ is related to the derivative of the lower-order composite axial force $N_L^{sur}(x)$ through Equation (36).

Figure 9b shows the higher-order axial-force distribution and confirms satisfaction of the imposed higher-order end-force boundary conditions ($N_H(0) = N_H(L) = 0$). It is observed that the distribution nature of the higher-order axial force is much more rapid than that of the lower-order composite axial force due to the differentiation relation between these two axial-force quantities in Equation (36).

Figure 9c presents the distribution diagram of the higher-order axial-force gradient. As indicated in Equation (30), the lower-order composite axial force $N_L^{sur}(x)$ and the higher-order axial-force gradient $\partial N_H(x)/\partial x$ both contribute to the total lower-order axial force $N(x)$. Therefore, the end-value combinations of axial-force diagrams in Figure 9a,c satisfy the end-force boundary conditions imposed on the total lower-order axial force $N(x)$ as shown in Figure 8 ($N(0)$ = 0 and $N(L)$ = 2400 nN). Furthermore, it is observed from Figure 9a,c that the higher-order axial-force gradient $\partial N_H(x)/\partial x$ participates more in contributing to the total lower-order axial force $N(x)$ for the present nanowire-elastic substrate system.

The substrate interactive-force diagrams obtained from classical and two bar-substrate models are shown in Figure 10. Complying with the Winkler-foundation hypothesis, the shapes of the substrate interactive-force diagrams resemble those of the axial-displacement diagrams shown in Figure 6. With the classical model, the substrate interactive force remains approximately constant at the value of the uniformly distributed load p_{x0} of 2.4 nN/nm along half of the nanowire (see the inset in Figure 10). This observation has the physical interpretation as the bar component has no contribution to the system resistance to externally applied loads along the left half of the nanowire, thus complying with the axial-force distribution presented in Figure 8. With the proposed model, the substrate interactive force continuously varies along the length of the nanowire, thus implying the whole part of the bar component participates in the system resistance to the externally applied loads.

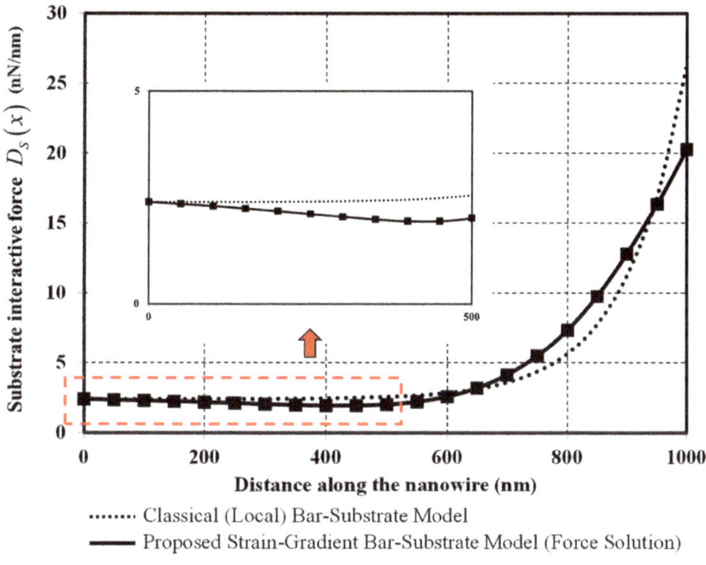

Figure 10. Substrate interactive force versus distance along the nanowire.

7. Summary and Conclusions

In the present work, a rational bar-elastic substrate model with inclusion of small-scale and surface-energy effects is alternatively formulated within the framework of virtual force principle. The small-scale (nonlocal) effect of the bar bulk is introduced through the thermodynamics-based strain-gradient model while the surface-energy-dependent size effect is included using the Gurtin–Murdoch surface model. To account for the bar-elastic substrate interaction, Winkler foundation model is called for. The higher-order differential compatibility equation of the problem and its associated classical and non-classical end-displacement compatibility conditions are consistently derived from the virtual force principle and form the core of the proposed bar-elastic substrate model. The axial force field serves as a basic solution to the higher-order differential compatibility equation.

To show the accuracy and merit of the proposed bar-elastic substrate model, a nanowire-elastic substrate system under axial loadings is employed as a numerical example. Under a uniformly distributed loading, the proposed model requires no particular solution. This is in opposition to its counterpart proposed by Sae-Long et al. [44]. Considering the small-scale and surface-energy effects consistently leads to a stiffer bar-elastic substrate system in similar way as enhancement of the bar axial rigidity when compared to the classical bar-elastic substrate model. This system stiffness enhancement has been confirmed by both theoretical studies and experimental evidence available in the literature [72]. Peculiar but specific response distributions along the nanowire length are observed at both global and local levels and they are associated with the higher-order governing differential equation and the statistical indeterminacy of the nanowire-elastic substrate system. It is anticipated that the bar-elastic substrate model proposed herein will be especially useful to scientists and engineers working in the area of nanoscience and nanoengineering.

Author Contributions: Conceptualization, S.L. and W.S.-L.; Methodology, S.L. and W.S.-L.; Software, W.S.-L.; Validation, S.L., W.S.-L., J.R. and P.S.; Formal analysis, S.L. and W.S.-L.; Investigation, S.L., H.M.-S., W.P. and T.I.; Data curation, S.L. and W.S.-L.; Writing—original draft preparation, S.L. and W.S.-L.; Writing—review and editing, S.L., W.S.-L. and H.M.-S.; Visualization, S.L. and W.S.-L.; Supervision, S.L., H.M.-S. and J.R.; Project administration, S.L. and W.S.-L.; Funding acquisition, S.L. and H.M.-S. All authors have read and agreed to the published version of the manuscript.

Funding: This research was funded by TRF Senior Research Scholar, Grant RTA 6280012 and by Research Council of Shahid Chamran University of Ahvaz, Grant No. SCU.EM1400.98.

Institutional Review Board Statement: Not applicable.

Informed Consent Statement: Not applicable.

Data Availability Statement: The data presented in this study are available on request from the corresponding authors.

Acknowledgments: This study was supported by TRF Senior Research Scholar under Grant RTA 6280012. H.M. Sedighi is grateful to the Research Council of Shahid Chamran University of Ahvaz for its financial support (Grant No. SCU.EM1400.98). Any opinions expressed in this paper are those of the authors and do not reflect the views of the sponsoring agencies. Special thanks go to a senior lecturer Wiwat Sutiwipakorn for reviewing and correcting the English of this paper.

Conflicts of Interest: The authors declare no conflict of interest.

Appendix A

The general form of the homogeneous solution to Equation (39) can be written as

$$N_L^{sur}(x) = \phi_1(x)c_1 + \phi_2(x)c_2 + \phi_3(x)c_3 + \phi_4(x)c_4 \tag{A1}$$

The homogeneous solution $N_L^{sur}(x)$ in Equation (A1) is derived stem from the analytical solution of the beam on a two-parameter foundation as introduced by Gülkan and Alemder [67]. Thus, the basic functions in each solution cases can be expressed as:

Case I: $\lambda_2 < 2\sqrt{\lambda_1}$

$$\begin{aligned} \phi_1 &= \cosh[\alpha x]\cos[\beta x];\ \phi_2 = \sinh[\alpha x]\cos[\beta x] \\ \phi_3 &= \cosh[\alpha x]\sin[\beta x];\ \phi_4 = \sinh[\alpha x]\sin[\beta x] \end{aligned} \tag{A2}$$

Case II: $\lambda_2 > 2\sqrt{\lambda_1}$

$$\begin{aligned} \phi_1 &= \cosh[\alpha x]\cosh[\beta x];\ \phi_2 = \sinh[\alpha x]\cosh[\beta x] \\ \phi_3 &= \cosh[\alpha x]\sinh[\beta x];\ \phi_4 = \sinh[\alpha x]\sinh[\beta x] \end{aligned} \tag{A3}$$

Case III: $\lambda_2 = 2\sqrt{\lambda_1}$

$$\phi_1 = e^{\sqrt[4]{\lambda_1}x}; \phi_2 = xe^{\sqrt[4]{\lambda_1}x}; \phi_3 = e^{-\sqrt[4]{\lambda_1}x}; \phi_4 = xe^{-\sqrt[4]{\lambda_1}x} \quad \text{(A4)}$$

with the auxiliary variables as

$$\begin{aligned} &\lambda_1 = \frac{k_S}{(I_x{}^2 E_{xx} A)}; \lambda_2 = \frac{(EA)_{xx}^{sur}}{(I_x{}^2 E_{xx} A)}; \alpha = \sqrt{\frac{\sqrt{\lambda_1}}{2} + \frac{\lambda_2}{4}}; \\ &\beta = \sqrt{\frac{\sqrt{\lambda_1}}{2} - \frac{\lambda_2}{4}} \text{ for Case I; and } \beta = \sqrt{\frac{\lambda_2}{4} - \frac{\sqrt{\lambda_1}}{2}} \text{ for Case II} \end{aligned} \quad \text{(A5)}$$

References

1. Awrejcewicz, J.; Krysko, A.V.; Zhigalov, M.V.; Krysko, V.A. Size-dependent theories of beams, plates and shells. In: Mathematical modelling and numerical analysis of size-dependent structural members in temperature fields. *Adv. Struct. Mater.* **2021**, *142*, 25–78.
2. Soltan Rezaee, M.; Bodaghi, M. Simulation of an electrically actuated cantilever as a novel biosensor. *Sci. Rep.* **2020**, *10*, 3385. [CrossRef] [PubMed]
3. Yan, Z.; Jiang, L. Modified continuum mechanics modeling on size-dependent properties of piezoelectric nanomaterials: A review. *Nanomaterials* **2017**, *7*, 27. [CrossRef] [PubMed]
4. Lei, X.-W.; Bando, K.; Shi, J.-X. Vibration control of diamond nanothreads by lattice defect introduction for application in nanomechanical sensors. *Nanomaterials* **2021**, *11*, 2241. [CrossRef]
5. Larkin, K.; Ghommem, M.; Serrano, M.; Abdelkefi, A. A review on vibrating beam-based micro/nano-gyroscopes. *Microsyst. Technol.* **2021**, *27*, 4157–4181. [CrossRef]
6. Tharu, S.A.; Panchal, M.B. Effect of interphase on elastic and shear moduli of metal matrix nanocomposites. *Eur. Phys. J. Plus* **2020**, *135*, 121. [CrossRef]
7. Malikan, M.; Eremeyev, V.A. Effect of surface on the flexomagnetic response of ferroic composite nanostructures; nonlinear bending analysis. *Compos. Struct.* **2021**, *271*, 114179. [CrossRef]
8. Abazari, A.M.; Fotouhi, M.; Tavakkoli, H.; Rezazadeh, G. An experimental study for characterization of size-dependence in microstructures via electrostatic pull-in instability technique. *Appl. Phys. Lett.* **2020**, *116*, 244102. [CrossRef]
9. Liu, Z.; Zhang, Y.; Wang, B.; Cheng, H.; Cheng, X.; Huang, Z. DFT study on Al-doped defective graphene towards adsorption of elemental mercury. *Appl. Surf. Sci.* **2018**, *427*, 547–553. [CrossRef]
10. Madani, S.H.; Sabour, M.H.; Fadaee, M. Molecular dynamics simulation of vibrational behavior of annular graphene sheet: Identification of nonlocal parameter. *J. Mol. Graph. Modell.* **2018**, *79*, 264–272. [CrossRef]
11. Shahabodini, A.; Gholami, Y.; Ansari, R.; Rouhi, H. Vibration analysis of graphene sheets resting on Winkler/Pasternak foundation: A multiscale approach. *Eur. Phys. J. Plus* **2019**, *134*, 510. [CrossRef]
12. Limkatanyu, S.; Damrongwiriyanupap, N.; Prachasaree, W.; Sae-Long, W. Modeling of axially loaded nanowires embedded in elastic substrate media with inclusion of nonlocal and surface effects. *J. Nanomater.* **2013**, *2013*, 635428. [CrossRef]
13. Gao, X.-L.; Zhang, G.Y. A non-classical Mindlin plate model incorporating microstructure, surface energy and foundation effects. *Proc. Math. Phys. Eng. Sci.* **2016**, *472*, 20160275. [CrossRef]
14. Malikan, M.; Jabbarzadeh, M.; Dastjerdi, S. Non-linear static stability of bi-layer carbon nanosheets resting on an elastic matrix under various types of in-plane shearing loads in thermo-elasticity using nonlocal continuum. *Microsyst. Technol.* **2017**, *23*, 2973–2991. [CrossRef]
15. Demir, C.; Mercan, K.; Numanoglu, H.M.; Civalek, O. Bending response of nanobeams resting on elastic foundation. *J. Appl. Comput. Mech.* **2018**, *4*, 105–114.
16. Panyatong, M.; Chinnaboon, B.; Chucheepsakul, S. Nonlinear bending analysis of nonlocal nanoplates with general shapes and boundary conditions by the boundary-only method. *Eng. Anal. Bound. Elem.* **2018**, *87*, 90–110. [CrossRef]
17. Sarafraz, A.; Sahmani, S.; Aghdam, M.M. Nonlinear primary resonance analysis of nanoshells including vibrational mode interactions based on the surface elasticity theory. *Appl. Math. Mech.* **2020**, *41*, 233–260. [CrossRef]
18. Russillo, A.F.; Failla, G.; Alotta, G.; Marotti de Sciarra, F.; Barretta, R. On the dynamics of nano-frames. *Int. J. Eng. Sci.* **2021**, *160*, 103433. [CrossRef]
19. Mindlin, R.D. On the equations of elastic materials with micro-structure. *Int. J. Solids Struct.* **1965**, *1*, 73–78. [CrossRef]
20. Eringen, A.C. Nonlocal polar elastic continua. *Int. J. Eng. Sci.* **1972**, *10*, 1–16. [CrossRef]
21. Eringen, A.C. On differential equations of nonlocal elasticity and solutions of screw dislocation and surface waves. *J. Appl. Phys.* **1983**, *54*, 4703–4710. [CrossRef]
22. Yang, F.; Chong, A.C.M.; Lam, D.C.C.; Tong, P. Couple stress based strain gradient theory for elasticity. *Int. J. Solids Struct.* **2002**, *39*, 2731–2743. [CrossRef]
23. Papargyri-Beskou, S.; Tsepoura, K.G.; Polyzos, D.; Beskos, D.E. Bending and stability analysis of gradient elastic beams. *Int. J. Solids Struct.* **2003**, *40*, 385–400. [CrossRef]

24. Barretta, R.; Marotti de Sciarra, F. A nonlocal model for carbon nanotubes under axial loads. *Adv. Mater. Sci. Eng.* **2013**, *2013*, 360935. [CrossRef]
25. Lazopoulos, A.K.; Lazopoulos, K.A.; Palassopoulos, G. Nonlinear bending and buckling for strain gradient elastic beams. *Appl. Math. Modell.* **2014**, *38*, 253–262. [CrossRef]
26. Zhang, G.Y.; Gao, X.-L. A new Bernoulli–Euler beam model based on a reformulated strain gradient elasticity theory. *Math. Mech. Solids* **2020**, *25*, 630–643. [CrossRef]
27. Peddieson, J.; Buchanan, G.R.; McNitt, R.P. Application of nonlocal continuum models to nanotechnology. *Int. J. Eng. Sci.* **2003**, *41*, 305–312. [CrossRef]
28. Reddy, J.N. Nonlocal theories for bending, buckling, and vibration of beams. *Int. J. Eng. Sci.* **2007**, *45*, 288–307. [CrossRef]
29. Pradhan, S.C.; Reddy, G.K. Buckling analysis of single walled carbon nanotube on Winkler foundation using nonlocal elasticity theory and DTM. *Comput. Mater. Sci.* **2011**, *50*, 1052–1056. [CrossRef]
30. Limkatanyu, S.; Sae-Long, W.; Horpibulsuk, S.; Prachasaree, W.; Damrongwiriyanupap, N. Flexural responses of nanobeams with coupled effects of nonlocality and surface energy. *ZAMM* **2018**, *98*, 1771–1793. [CrossRef]
31. Shishesaz, M.; Shariati, M.; Yaghootian, A. Nonlocal elasticity effect on linear vibration of nano-circular plate using adomian decomposition method. *J. Appl. Comput. Mech.* **2020**, *6*, 63–76.
32. Nasr, M.E.; Abouelregal, A.E.; Soleiman, A.; Khalil, K.M. Thermoelastic vibrations of nonlocal nanobeams resting on a Pasternak foundation via DPL model. *J. Appl. Comput. Mech.* **2021**, *7*, 34–44.
33. Challamel, N.; Wang, C.M. The small length scale effect for a non-local cantilever beam: A paradox solved. *Nanotechnology* **2008**, *19*, 345703. [CrossRef]
34. Lim, C.W. On the truth of nanoscale for nanobeams based on nonlocal elastic stress field theory: Equilibrium, governing equation and static deflection. *Appl. Math. Mech.* **2010**, *31*, 37–54. [CrossRef]
35. Romano, G.; Barretta, R.; Diaco, M.; Marotti de Sciarra, F. Constitutive boundary conditions and paradoxes in nonlocal elastic nanobeams. *Int. J. Mech. Sci.* **2017**, *121*, 151–156. [CrossRef]
36. Koutsoumaris, C.C.; Eptaimeros, K.G.; Zisis, T.; Tsamasphyros, G.J. A straightforward approach to Eringen's nonlocal elasticity stress model and applications for nanobeams. *AIP Conf. Proc.* **2016**, *1790*, 150018.
37. Ma, H.M.; Gao, X.-L.; Reddy, J.N. A microstructure-dependent Timoshenko beam model based on a modified couple stress theory. *J. Mech. Phys. Solids* **2008**, *56*, 3379–3391. [CrossRef]
38. Gao, X.-L.; Zhou, S.-S. Strain gradient solutions of half-space and half-plane contact problems. *Z. Angew. Math. Phys.* **2013**, *64*, 1363–1386. [CrossRef]
39. Challamel, N.; Reddy, J.N.; Wang, C.M. Eringen's stress gradient model for bending of nonlocal beams. *J. Eng. Mech.* **2016**, *142*, 04016095. [CrossRef]
40. Apuzzo, A.; Barretta, R.; Fabbrocino, F.; Faghidian, S.A.; Luciano, R.; Marotti de Sciarra, F. Axial and torsional free vibrations of elastic nano-beams by stress-driven two-phase elasticity. *J. Appl. Comput. Mech.* **2019**, *5*, 402–413.
41. Vaccaro, M.S.; Marotti de Sciarra, F.; Barretta, R. On the regularity of curvature fields in stress-driven nonlocal elastic beams. *Acta Mech.* **2021**, *232*, 2595–2603. [CrossRef]
42. Pinnola, F.P.; Vaccaro, M.S.; Barretta, R.; Marotti de Sciarra, F. Finite element method for stress-driven nonlocal beams. *Eng. Anal. Bound. Elem.* **2022**, *134*, 22–34. [CrossRef]
43. Marotti de Sciarra, F.; Barretta, R. A new nonlocal bending model for Euler-Bernoulli nanobeams. *Mech. Res. Commun.* **2014**, *62*, 25–30. [CrossRef]
44. Sae-Long, W.; Limkatanyu, S.; Prachasaree, W.; Rungamornrat, J.; Sukontasukkul, P. A thermodynamics-based nonlocal bar-elastic substrate model with inclusion of surface-energy effect. *J. Nanomater.* **2020**, *2020*, 8276745. [CrossRef]
45. Sae-Long, W.; Limkatanyu, S.; Rungamornrat, J.; Prachasaree, W.; Sukontasukkul, P.; Sedighi, H.M. A rational beam-elastic substrate model with incorporation of beam-bulk nonlocality and surface-free energy. *Eur. Phys. J. Plus* **2021**, *136*, 80. [CrossRef]
46. Gurtin, M.E.; Murdoch, A.I. A continuum theory of elastic material surface. *Arch. Ration. Mech. Anal.* **1975**, *57*, 291–323. [CrossRef]
47. Gurtin, M.E.; Murdoch, A.I. Surface stress in solids. *Int. J. Solids Struct.* **1978**, *14*, 431–440. [CrossRef]
48. He, J.; Lilley, C.M. Surface effect on the elastic behavior of static bending nanowires. *Nano Lett.* **2008**, *8*, 1798–1802. [CrossRef]
49. Gao, X.-L.; Mahmoud, F.F. A new Bernoulli-Euler beam model incorporating microstructure and surface energy effects. *Z. Angew. Math. Phys.* **2014**, *65*, 393–404. [CrossRef]
50. Gao, X.-L. A new Timoshenko beam model incorporating microstructure and surface energy effects. *Acta Mech.* **2015**, *226*, 457–474. [CrossRef]
51. Sourki, R.; Hosseini, S.A. Coupling effects of nonlocal and modified couple stress theories incorporating surface energy on analytical transverse vibration of a weakened nanobeam. *Eur. Phys. J. Plus* **2017**, *132*, 184. [CrossRef]
52. Sapsathiarn, Y.; Rajapakse, R.K.N.D. Mechanistic models for nanobeams with surface stress effects. *J. Eng. Mech.* **2018**, *144*, 04018098. [CrossRef]
53. Yekrangi, A.; Yaghobi, M.; Riazian, M.; Koochi, A. Scale-dependent dynamic behavior of nanowire-based sensor in accelerating field. *J. Appl. Comput. Mech.* **2019**, *5*, 486–497.
54. Sae-Long, W.; Limkatanyu, S.; Sukontasukkul, P.; Damrongwiriyanupap, N.; Rungamornrat, J.; Prachasaree, W. A fourth-order strain gradient bar-substrate model with nonlocal and surface effects for the analysis of nanowires embedded in substrate media. *Facta Univ. Ser. Mech. Eng.* **2021**, *19*, 657–680.

55. Wang, Z.L.; Song, J. Piezoelectric nanogenerators based on zinc oxide nanowire arrays. *Science* **2006**, *312*, 242–246. [CrossRef]
56. Feng, X.L.; He, R.; Yang, P.; Roukes, M.L. Very high frequency silicon nanowire electromechanical resonators. *Nano Lett.* **2007**, *7*, 1953–1959. [CrossRef]
57. Fang, M.; Han, N.; Wang, F.; Yang, Z.; Yip, S.; Dong, G.; Hou, J.J.; Chueh, Y.; Ho, J.C. III–V nanowires: Synthesis, property manipulations, and device applications. *J. Nanomater.* **2014**, *2014*, 702859. [CrossRef]
58. Ambhorkar, P.; Wang, Z.; Ko, H.; Lee, S.; Koo, K.; Kim, K.; Cho, D.D. Nanowire-based biosensors: From growth to applications. *Micromachines* **2018**, *9*, 679. [CrossRef]
59. Sohn, H.; Park, C.; Oh, J.-M.; Kang, S.W.; Kim, M.-J. Silver nanowire networks: Mechano-electric properties and applications. *Materials* **2019**, *12*, 2526. [CrossRef]
60. Ponbunyanon, P.; Limkatanyu, S.; Kaewjuea, W.; Prachasaree, W.; Chub-Uppakarn, T. A novel beam-elastic substrate model with inclusion of nonlocal elasticity and surface energy effects. *Arab. J. Sci. Eng.* **2016**, *41*, 4099–4113. [CrossRef]
61. Zhao, T.; Luo, J.; Xiao, Z. Buckling analysis of a nanowire lying on Winkler–Pasternak elastic foundation. *Mech. Adv. Mater. Struct.* **2015**, *22*, 394–401. [CrossRef]
62. Malekzadeh, P.; Shojaee, M. Surface and nonlocal effects on the nonlinear free vibration of non-uniform nanobeams. *Compos. Part B Eng.* **2013**, *52*, 84–92. [CrossRef]
63. Limkatanyu, S.; Prachasaree, W.; Damrongwiriyanupap, N.; Kwon, M. Exact stiffness matrix for nonlocal bars embedded in elastic foundation media: The virtual-force approach. *J. Eng. Math.* **2014**, *89*, 163–176. [CrossRef]
64. Wolfram, S. *Mathematica Reference Guide*; Addison-Wesley Publishing Company: Redwood City, CA, USA, 1992.
65. Altan, B.S.; Aifantis, E.C. On some aspects in the special theory of gradient elasticity. *J. Mech. Behav. Mater.* **1997**, *8*, 231–282. [CrossRef]
66. Argatov, I. From Winkler′s foundation to Popov′s foundation. *Facta Univ. Ser. Mech. Eng.* **2019**, *17*, 181–190.
67. Gülkan, P.; Alemdar, B.N. An exact finite element for a beam on a two-parameter elastic foundation: A revisit. *Struct. Eng. Mech.* **1999**, *7*, 259–276. [CrossRef]
68. Juntarasaid, C.; Pulngern, T.; Chucheepsakul, S. Bending and buckling of nanowires including the effects of surface stress and nonlocal elasticity. *Phys. E Low-Dimens. Syst. Nanostruct.* **2012**, *46*, 68–76. [CrossRef]
69. Yang, Y.; Lim, C.W. Non-classical stiffness strengthening size effects for free vibration of a nonlocal nanostructure. *Int. J. Mech. Sci.* **2012**, *54*, 57–68. [CrossRef]
70. Liew, K.M.; He, X.Q.; Kitipornchai, S. Predicting nanovibration of multi-layer graphene sheets embedded in an elastic matrix. *Acta Mater.* **2006**, *54*, 4229–4236. [CrossRef]
71. Buachart, C.; Hansapinyo, C.; Kanok-Nukulchai, W. Analysis of axial loaded pile in multilayered soil using nodal exact finite element model. *Int. J. GEOMATE* **2018**, *14*, 1–7. [CrossRef]
72. Li, X.-F.; Wang, B.-L.; Lee, K.Y. Size effects of the bending stiffness of nanowires. *J. Appl. Phys.* **2009**, *105*, 074306. [CrossRef]

MDPI

Article

Estimation of the Young's Modulus of Nanometer-Thick Films Using Residual Stress-Driven Bilayer Cantilevers

Luis A. Velosa-Moncada [1], Jean-Pierre Raskin [2], Luz Antonio Aguilera-Cortés [3], Francisco López-Huerta [4] and Agustín L. Herrera-May [1,5,*]

1 Micro and Nanotechnology Research Center, Universidad Veracruzana, Boca del Rio 94294, Mexico; luchoa_23@outlook.com
2 Institute of Information and Communication Technologies, Electronics and Applied Mathematics (ICTEAM), Université Catholique de Louvain (UCL), 1348 Louvain-la-Neuve, Belgium; jean-pierre.raskin@uclouvain.be
3 Departamento de Ingeniería Mecánica, DICIS, Universidad de Guanajuato, Salamanca 36885, Mexico; aguilera@ugto.mx
4 Facultad de Ingeniería Eléctrica y Electrónica, Universidad Veracruzana, Boca del Rio 94294, Mexico; frlopez@uv.mx
5 Maestría en Ingeniería Aplicada, Facultad de Ingeniería de la Construcción y el Hábitat, Universidad Veracruzana, Boca del Rio 94294, Mexico
* Correspondence: leherrera@uv.mx

Abstract: Precise prediction of mechanical behavior of thin films at the nanoscale requires techniques that consider size effects and fabrication-related issues. Here, we propose a test methodology to estimate the Young's modulus of nanometer-thick films using micromachined bilayer cantilevers. The bilayer cantilevers which comprise a well-known reference layer and a tested film deflect due to the relief of the residual stresses generated during the fabrication process. The mechanical relationship between the measured residual stresses and the corresponding deflections was used to characterize the tested film. Residual stresses and deflections were related using analytical and finite element models that consider intrinsic stress gradients and the use of adherence layers. The proposed methodology was applied to low pressure chemical vapor deposited silicon nitride tested films with thicknesses ranging from 46 nm to 288 nm. The estimated Young's modulus values varying between 213.9 GPa and 288.3 GPa were consistent with nanoindentation and alternative residual stress-driven techniques. In addition, the dependence of the results on the thickness and the intrinsic stress gradient of the materials was confirmed. The proposed methodology is simple and can be used to characterize diverse materials deposited under different fabrication conditions.

Keywords: bilayer cantilever; deflections; thin films; residual stresses; young's modulus

Citation: Velosa-Moncada, L.A.; Raskin, J.-P.; Aguilera-Cortés, L.A.; López-Huerta, F.; Herrera-May, A.L. Estimation of the Young's Modulus of Nanometer-Thick Films Using Residual Stress-Driven Bilayer Cantilevers. *Nanomaterials* **2021**, *11*, 274. https://doi.org/10.3390/nano12020265

Academic Editor: Victor A. Eremeyev

Received: 30 November 2021
Accepted: 11 January 2022
Published: 14 January 2022

1. Introduction

Accurate values of the Young's modulus are essential to correctly quantify the stiffness of the structures under different loading conditions. A comprehensive understanding of this mechanical property in very thin films is critical to the proper design of small-scale micromachined devices. The Young's modulus of materials with nanometric dimensions (especially below 100 nm) can vary due to the effect of small size [1] and the fabrication process [2]. Therefore, its correct estimation is an area of great interest in many fields such as microelectronics, protective coatings, and nanoelectromechanical systems (NEMS). At this level, the determination of the Young's modulus requires very different procedures to the traditional uniaxial tensile tests of bulk samples. Several techniques, including nanoindentation [3–5], bulge test [6–8], electrostatic pull-in experiments [9,10], or resonant-based methods [11,12] have been developed for this purpose. However, they often require complex experimental setups and extraction procedures that complicate the replicability of the experimental results. The main difficulty of these tests lies in the need to apply external loads that can eventually disturb the samples and produce noise in the measurements.

The residual stresses of materials deposited by surface micromachining processes can be used as a means of actuation to extract the Young's modulus of thin films without external manipulation of the samples. An example is the on-chip nanomechanical testing laboratory developed to apply a uniaxial load to the tested films using the internal stress present in a well-characterized reference material [13]. The test methodology is simple, which allows the study of the stress-strain response of a wide range of materials. Thus, different properties such as Young's modulus, yield stress or fracture stress, fracture strain, and strain hardening can be estimated. In addition, this procedure can also be implemented in a semiconductor device production line [14]. Although ultrathin films can be evaluated with this technique, long actuator beams are needed to provide accurate measurements and large strains. This is an important challenge because of the difficulty of fabricating structures with high aspect ratios (length/thickness). Another proposal consists of combining the measurement of the internal stress after film deposition with the measurement of the corresponding internal elastic strain of freestanding beams [15]. This method is simple and easy to implement but is limited to thin films that have high internal stresses. Furthermore, the accuracy of the strain measurement requires long beams that can suffer from stiction in the release steps. Recently, the deflections of residual stress-driven bilayer cantilevers integrated by a tested film and a well-characterized reference layer have been used to study the elastic properties of ultrathin films [16,17]. In these cases, the parameters that are evaluated are both Young's modulus and mismatch strain of the tested film. Due to these reasons, several cantilevers with different thickness ratios between the tested film and the reference layer must be fabricated. Thus, the intersection of the corresponding stress-deflection curves gives the solution under the assumption that the mismatch strain is the same in all specimens. Accurate measurements can be achieved by applying this method on tested films with thicknesses less than 100 nm due to the large curvatures of the deflected cantilevers. However, the control of thickness in the micromachining process is a critical issue that could limit the reproducibility of the results and hinder the evaluation of materials with high etching rates. Moreover, the presence of intrinsic stress gradients and the use of an adherence layer seriously influence the precision of the results.

In this work, we propose the use of residual stress-driven bilayer cantilevers to estimate the Young's modulus of thin films with nanometer thicknesses. Our proposal is motivated by the large curvature variations of fully released cantilevers, resulting in increased sensitivity to tested film characteristics. However, we consider wafer-level measurement of residual stresses in both the reference layer and the tested film in comparison to the investigations reported in previous reports [16,17]. The in-situ measurement of these residual stresses allows the characterization of the elastic mismatch strain. Therefore, it is not required to vary the thickness ratio among the fabricated cantilevers, since the Young's modulus of the tested film is the only parameter to evaluate. Fabricating cantilevers with common thicknesses eliminates the impact that changes in material sizes have on the results and simplifies thickness control in the fabrication process. The tested film is characterized by relating the measured residual stresses to the corresponding deflections. For this, we propose analytical and finite element models from a static analysis of the cantilever. The models consider the effects of intrinsic stress gradients through the thickness of the materials and the possible use of an adhered layer to strengthen the bond between the reference layer and the tested film. The consideration of these effects is relevant for the correct estimation of the elastic properties of the thin films. The results of our methodology are in good agreement with experimental data reported in the literature.

2. Estimation Methodology

Bilayer cantilevers deflect when released from their base substrate due to the difference in the residual stresses in the materials used to fabricate them. The magnitude of the deflection depends on the elastic properties of the materials and the intensity of the residual stresses stored during the fabrication process. Therefore, the Young's modulus of the tested film can be deduced if the mechanical properties of the reference layer are known and both deflections and residual stresses are measured. Occasionally, an additional layer is deposited to promote adhesion between the reference layer and the tested film. In such cases, it is also necessary to know the residual stresses and the mechanical properties of this adherence layer to correctly characterize the tested film. The mechanical relationship between deflections and residual stresses was found through analytical and finite element models. The analytical model was developed from a large deflection analysis of the flexible structure, while the finite element model (FEM) was carried out in the ANSYS® Workbench software using nonlinear static structural analysis.

2.1. Analytical Modeling

The relaxation of the residual stresses stored in the materials during the fabrication process causes an internal bending moment that deflects the cantilever to an equilibrium position. The radius of curvature R of the deflected cantilever can be related to the internal bending moment M from the Euler–Bernoulli beam equation [18,19]:

$$\frac{1}{R} = \frac{M}{(EI)_e}, \tag{1}$$

where $(EI)_e$ is the equivalent bending rigidity of the cantilever.

The horizontal (a) and vertical (b) deflections of the cantilever are determined as shown in Figure 1a:

$$a = R\sin\left(\frac{L}{R}\right), \tag{2}$$

$$b = R\left[1 - \cos\left(\frac{L}{R}\right)\right], \tag{3}$$

where L is the length of the cantilever.

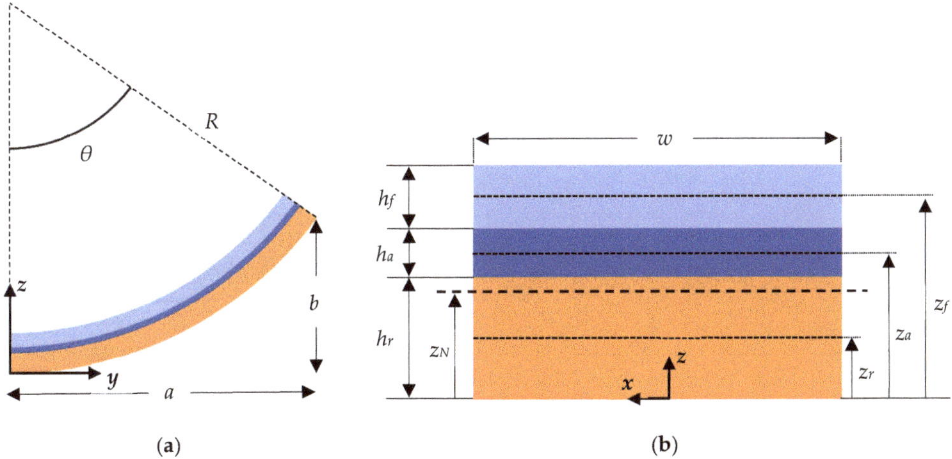

Figure 1. Schematic representation of the residual stress-driven cantilever. (**a**) Deflection profile after released. θ is the angular deflection of the cantilever. (**b**) Geometrical parameters of the cross-section. The reference layer is supposed to be the material located at the bottom of the structure.

2.1.1. Bending Rigidity of the Cantilever

The equivalent bending rigidity of the cantilever $(EI)_e$ is given as reported in a previous report [20]:

$$(EI)_e = \overline{E}_r[I_r + z_r A_r(z_r - z_N)] + \overline{E}_f\left[I_f + z_f A_f\left(z_f - z_N\right)\right] + \overline{E}_a[I_a + z_a A_a(z_a - z_N)], \tag{4}$$

where $z_{r,f,a}$ is the distance between the neutral axis of each material and the bottom of the cantilever. The subscripts r, f, a denote the reference layer, the thin film, and the adherence layer, respectively. In case of homogeneous cross sections, z_r, z_f, and z_a are calculated as shown in Figure 1b:

$$z_r = \frac{h_r}{2}, \tag{5}$$

$$z_a = h_r + \frac{h_a}{2}, \tag{6}$$

$$z_f = h_r + h_a + \frac{h_f}{2}, \tag{7}$$

where $h_{r,f,a}$ is the thickness of each individual material.

Considering rectangular cross-sections, the cross-sectional area ($A_{r,f,a}$) and moment of inertia ($I_{r,f,a}$) of the three materials are determined as:

$$A_{r,f,a} = wh_{r,f,a}, \tag{8}$$

$$I_{r,f,a} = \frac{1}{12}wh^3_{r,f,a}, \tag{9}$$

where w is the width of the cantilever. The moment of inertia of each individual material is calculated with respect to its own neutral axis, that is, passing through its center of symmetry.

The position of the neutral axis of the entire structure (z_N) is obtained as mentioned in previous reports [20]:

$$z_N = \frac{z_r\overline{E}_rA_r + z_f\overline{E}_fA_f + z_a\overline{E}_aA_a}{\overline{E}_rA_r + \overline{E}_fA_f + \overline{E}_aA_a} \tag{10}$$

The biaxial Young's modulus of each material ($\overline{E}_{r,f,a}$) is:

$$\overline{E}_{r,f,a} = \frac{E_{r,f,a}}{1 - v_{r,f,a}}, \tag{11}$$

where $E_{r,f,a}$ and $\nu_{r,f,a}$ are the Young's modulus and the Poisson ratio, respectively.

2.1.2. Internal Bending Moment

The residual stresses in each material (Figure 2a) can be divided into a uniform component and an intrinsic stress gradient (Figure 2b). Uniform residual stresses are positive if they cause compression in the materials once the cantilever is released from its base substrate. On the other hand, intrinsic stress gradients are positive if they produce out-of-plane deflection towards the positive z-axis. The uniform stress and the intrinsic stress gradient can be represented as an axial force and a moment load, respectively, both acting uniformly over the material cross-section (Figure 2c). By the moment equilibrium around z_N,

$$M = F_a(z_a - z_N) + F_f\left(z_f - z_N\right) + F_r(z_r - z_N) + M_a + M_r + M_f, \tag{12}$$

where M is the internal bending moment that deflects the fully released cantilever. The axial force of each material $F_{r,f,a}$ expressed in terms of the uniform stress $\sigma_{r,f,a}$ is given by:

$$F_{r,f,a} = \sigma_{r,f,a}A_{r,f,a} \tag{13}$$

Generally, $\sigma_{r,f,a}$ is determined using the Stoney formula by the measurement of the radius of curvature of the base substrate before and after deposition of each material [21]. Changes in substrate curvatures can be accurately measured using mechanical, capacitive, or optical methods [21]. This technique is quick and practical for the estimation of the uniform residual stresses of wafer-level thin films. However, it has the limitation of being accurate only if the deposited material produces substantial changes in the initial curvature. Curvature changes are practically imperceptible when the thickness of the deposited material is extremely small compared to that of the substrate. In such cases, alternative experimental techniques, such as X-ray diffraction, ultrasound, or Raman spectroscopy should be considered [21,22].

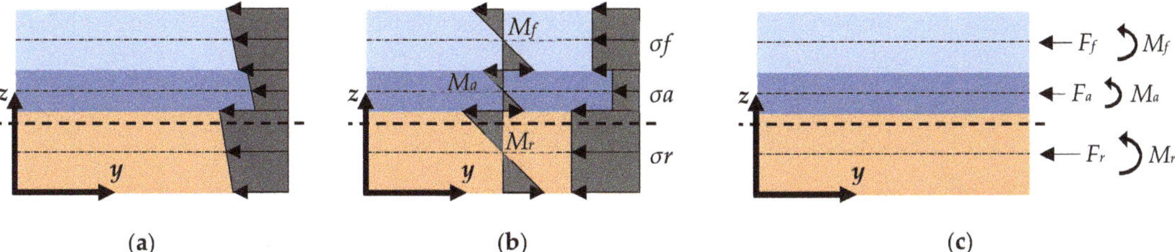

Figure 2. (**a**) Residual stresses stored in the materials during the fabrication process of the cantilever. (**b**) Uniform stresses and intrinsic stress gradients that form the total residual stresses. (**c**) Equivalent system of axial forces and moments.

The moment load of the individual materials $M_{r,f,a}$ can be experimentally estimated by applying Equation (1) in corresponding freestanding monolayer beams. Monolayer beams can be fabricated alongside bilayer cantilevers over the same base substrate following a single fabrication process. Then,

$$M_{r,f,a} = \frac{\overline{E}_{r,f,a} I_{r,f,a}}{R_{r,f,a}}, \tag{14}$$

where $R_{r,f,a}$ is the radius of curvature of each monolayer beam. The radii of curvature of the cantilevers (R and $R_{r,f,a}$) are estimated from the deflection profile using the Taubin method [23]. Nevertheless, Equations (2) and (3) can be used to estimate each radius of curvature in the cases in which only the horizontal or vertical deflections of the cantilever tip are measured.

2.1.3. Solution and Simplification Procedures

The Young's modulus of the tested film E_f is obtained by numerically solving the system of linear equations composed of Equations (1) and (4)–(14). However, the analytical model can be reduced to a single expression:

$$k = \frac{-6C_1 + h_r\overline{E}_r C_2}{h_r\overline{E}_r\left[m_a^4\lambda_a^2 + 4m_a^3\lambda_a\left(m_f\lambda_f + 1\right) + 6m_a^2\left(m_f^2\lambda_a\lambda_f + 2m_f\lambda_f + \lambda_a\right) + 4m_a\left(m_f^3\lambda_a\lambda_f + 3m_f^2\lambda_f + 3m_f\lambda_f + \lambda_a\right)\right] + h_r\overline{E}_r C_3}, \tag{15}$$

where,

$$C_1 = \left[m_a m_f\left(m_a + m_f\right)\left(\lambda_f\sigma_a - \lambda_a\sigma_f\right)\right] + \left[m_a(m_a+1)(\lambda_a\sigma_r - \sigma_a)\right] + \left[m_f\left(m_f + 2m_a + 1\right)\left(\lambda_f\sigma_r - \sigma_f\right)\right], \tag{16}$$

$$C_2 = \left(k_a m_a^3\lambda_a + k_f m_f^3\lambda_f + k_r\right)\left(m_a\lambda_a + m_f\lambda_f + 1\right), \tag{17}$$

$$C_3 = 1 + m_f\lambda_f\left(4 + 6m_f + 4m_f^2\right) + m_f^4\lambda_f^2, \tag{18}$$

where $\lambda_f = \overline{E}_f/\overline{E}_r$ is the biaxial modulus ratio of the tested film and the reference layer, and $\lambda_a = \overline{E}_a/\overline{E}_r$ is the biaxial modulus ratio of the adherence layer and the reference layer. $m_f = h_f/h_r$ is the thickness ratio of the tested film and the reference layer, and $m_a = h_a/h_r$ is the thickness ratio of the adherence layer and the reference layer. $k = 1/R$, $k_r = 1/R_r$,

$k_f = 1/R_f$ and $k_a = 1/R_a$ are defined as the curvature of the bilayer cantilever, reference layer, tested film, and adherence layer, respectively.

The terms related to the adherence layer are overridden in the analytical model if the deposition of this material is not required during the fabrication process. In those cases, Equation (15) can be simplified as:

$$k = \frac{\left[-6m_f\left(m_f+1\right)\left(\lambda_f\sigma_r-\sigma_f\right)\right]+\left[h_r\overline{E}_r\left(m_f\lambda_f+1\right)\left(k_f m_f^3\lambda_f+k_r\right)\right]}{h_r\overline{E}_r\left[1+m_f\lambda_f\left(4+6m_f+4m_f^2\right)+m_f^4\lambda_f^2\right]} \tag{19}$$

Equation (19) can be further simplified if the reference layer and the tested film do not develop intrinsic stress gradients ($M_{r,f} = 0$):

$$k = \left[\frac{-6m_f\left(\lambda_f\sigma_r-\sigma_f\right)}{h_r\overline{E}_r}\right]\frac{\left(m_f+1\right)}{1+m_f\lambda_f\left(4+6m_f+4m_f^2\right)+m_f^4\lambda_f^2} \tag{20}$$

Equation (20) can be also expressed in terms of the uniform residual strain of the reference layer (e_r) and the tested film (e_f) using the Hooke's law ($\sigma_{r,f} = e_{r,f}\,\overline{E}_{r,f}$):

$$k = \left[\frac{-6\lambda_f m_f\left(e_r-e_f\right)}{h_r}\right]\frac{\left(m_f+1\right)}{1+m_f\lambda_f\left(4+6m_f+4m_f^2\right)+m_f^4\lambda_f^2} \tag{21}$$

This expression is the generalization of the Stoney formula for uniform mismatch strain in the tested film [24]. It was used in the aforementioned bilayer cantilever-based methods to extract the Young modulus of ultrathin films [16,17]. However, Equation (21) can be only used in the cases where the contribution of the intrinsic stress gradients to the internal bending moment is very small compared to that from the uniform stress components.

Finally, if the stiffness of the reference layer is much larger than the tested film ($h_r \gg h_f$) such that it does not develop uniform residual stresses (i.e., $\sigma_r = 0$), the second part of Equation (20) is simplified, obtaining the classical Stoney formula:

$$k = \frac{6m_f\sigma_f}{h_r\overline{E}_r} = \frac{6h_f\sigma_f}{h_r^2\overline{E}_r} \tag{22}$$

2.2. Finite Element Modeling

The cantilever is modeled by grouping the materials into a multibody solid that is then fixed at one end (Figure 3a). Each material is split into two symmetrical sections over its thickness. The symmetry of the structure in the *yz*-plane is exploited to simplify the model to half the width. All parts are connected using shared topologies rather than contact regions to have a continuous mesh across the model. The model is meshed with SOLID186 higher-order 3D 20-node hexagonal elements that exhibit quadratic displacement behavior.

Deflections are induced in the model by applying a uniform temperature gradient ΔT to the multilayer material of the solid body. The application of the uniform temperature gradient produces thermal strains in the sections according to the following expression:

$$e_{1(r,f,a)} = \alpha_{1(r,f,a)}\Delta T, \tag{23}$$

$$e_{2(r,f,a)} = \alpha_{2(r,f,a)}\Delta T, \tag{24}$$

where $e_{1(r,f,a)}$ and $e_{2(r,f,a)}$ are the thermal strain of the bottom and top sections of each material, respectively. $\alpha_{1(r,f,a)}$ and $\alpha_{2(r,f,a)}$ are the specific thermal expansion coefficients of the bottom and top sections of each material, respectively.

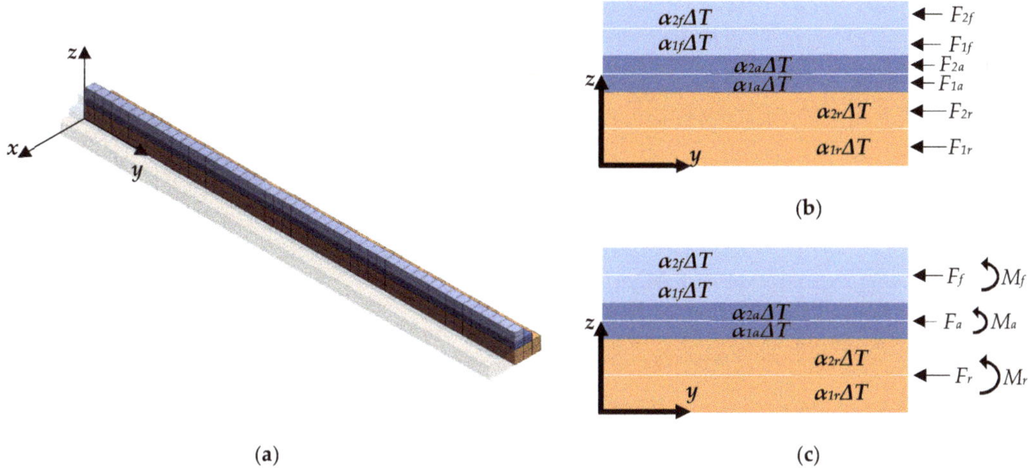

Figure 3. Finite element model of the bilayer cantilever. (**a**) 3D multibody solid meshed with hexagonal elements. (**b**) The thermal strains applied to the two sections of each material produce axial forces due to the interaction between the parts. (**c**) The combined action of the two axial forces in each material produces the uniform stresses and intrinsic stress gradients.

Thermal strains produce axial forces in the sections due to the interaction between the parts (Figure 3b). The values of the thermal strains expressed in terms of their respective axial forces are:

$$e_{1(r,f,a)} = \frac{F_{1(r,f,a)}}{\overline{E}_{r,f,a} A_{r,f,a}}, \tag{25}$$

$$e_{2(r,f,a)} = \frac{F_{2(r,f,a)}}{\overline{E}_{r,f,a} A_{r,f,a}}, \tag{26}$$

where $F_{1(r,f,a)}$ and $F_{2(r,f,a)}$ are the axial forces in the bottom and top sections of each material, respectively.

The required axial forces are determined by applying the following two boundary conditions to each material. First, the sum of the axial forces of the material must be equal to the equivalent force produced by the uniform residual stress (Figure 3c):

$$F_{1(r,f,a)} + F_{2(r,f,a)} = -F_{r,f,a} \tag{27}$$

Second, the sum of the moments around the neutral axis of the material must be equal to the moment load produced by the intrinsic stress gradient (Figure 3c):

$$F_{2(r,f,a)} - F_{1(r,f,a)} = \frac{-4M_{r,f,a}}{h_{r,f,a}} \tag{28}$$

The minus sign appears in Equations (27) and (28) due to the relaxation of positive and negative uniform stresses produce compression and tension in the materials, respectively.

Once the values of the thermal strains are estimated using Equations (25)–(28), the thermal expansion coefficients are specified through Equations (23) and (24) by setting the ΔT value. Negative thermal expansion coefficients or with a value of zero can be required to perform the simulations correctly. If a material does not develop intrinsic stress gradients, it is not mandatory to split it into two sections, since its respective thermal expansion coefficients will have the same value. The combined action of all axial forces causes the deflection of the cantilever until an equilibrium position is reached.

The Young's modulus of the tested film (E_f) is found by varying its magnitude in the model until the simulated deflections coincide with those experimentally measured. The simulated deflection profile is extracted from the plane of symmetry of the structure and can be quantified in terms of a, b, or R (Figure 1a).

2.3. Accuracy of the Proposed Models

The results of the analytical model may differ from those obtained using FEM if the deflection of the cantilever becomes extremely large. To illustrate this, the deflections of three different residual stress-driven cantilevers (S1, S2, and S3) are considered as examples. The geometric dimensions of the cantilevers, the elastic properties of the materials, the curvatures of the monolayer beams, and the uniform residual stresses are indicated in Tables 1 and 2. The parameters indicated in Table 1 are common in the examples, whereas the parameters indicated in Table 2 vary according to the cantilever. The uniform residual stress of the tested film is varied in each example to cover a broad range of deflections in the analysis. The cantilevers exhibit similar deflections with the indicated parameters. The examples include the use of adherence layers and the development of intrinsic stress gradients in the materials.

Table 1. Common parameters in the analyzed examples.

Parameter	Value
Cantilever width, w	12 µm
Cantilever length, L	150 µm
Thickness of the adherence layer, h_a	30 nm
Thickness of the tested film, h_f	60 nm
Biaxial Young's modulus of the reference layer, $\overline{E}_r$	250 GPa
Biaxial Young's modulus of the adherence layer, $\overline{E}_a$	125 GPa
Young's modulus of the tested film, E_f	160 GPa
Poisson ratio of the tested film, v_f	0.2
Curvature of the reference layer beam, k_r	-890 m^{-1}
Curvature of the adherence layer beam, k_a	0
Curvature of the tested film beam, k_f	1790 m^{-1}
Uniform residual stress of the reference layer, σ_r	-50 MPa
Uniform residual stress of the adherence layer, σ_a	0

Table 2. Parameters for cantilevers S1, S2 and S3.

Parameter	S1	S2	S3
Thickness of the reference layer, h_r [nm]	120	240	480
Uniform residual stress of the tested film, σ_f [MPa]	40–200	100–500	300–1500

First, the given parameters are used to calculate the deflection of the cantilevers through the proposed models. In these calculations, the Young's modulus of the tested film has been considered as an input value. The evolution of the normalized curvature (k^*) with respect the normalized internal bending moment (M^*) in each cantilever is shown in Figure 4. The expressions for k^* and M^* are derived following the normalization procedure proposed in previous report [16]:

$$M^* = \frac{M}{1 \times 10^{-8}[N \cdot m]} \frac{w^2}{\left(h_r + h_a + h_f\right)^2}, \tag{29}$$

$$k^* = k\frac{w^2}{h_r + h_a + h_f}, \tag{30}$$

where M and k are the internal bending moment and the curvature of each bilayer cantilever, respectively. The analytical model solutions indicate that curvatures are directly proportional to the internal bending moments. However, the results estimated by FEM show that the relationship between the two parameters evolves nonlinearly as the deflections increase beyond their initial values. Nonlinearity can be attributed to stress hardening (which usually appears in structures with very low bending stiffness) and shear deformations. These effects are not considered in the analytical model as it is derived from the Euler-Bernoulli beam theory (Equation (1)). Nevertheless, it is observed that the nonlinear effects drop considerably as the total thickness of the structures increases (results for cantilever S3).

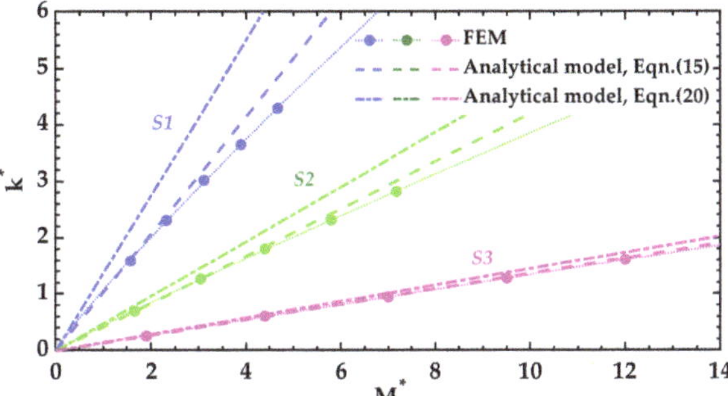

Figure 4. Normalized curvature k^* vs. normalized internal bending moment M^* for the cantilevers S1, S2, and S3. The dotted curve with circular markers corresponds to the results estimated by FEM. The dashed line represents the results obtained by the analytical model (Equation (15)). The dash-dotted line shows the results obtained by the analytical model if the parameters related to the adherence layer and intrinsic stress gradients are not considered (Equation (20)).

Nonlinear deflection of cantilevers can have a significant impact on the accuracy of the analytical model. The Young's modulus of the tested film (E_f) is calculated from the curvatures obtained by FEM using Equation (15). Figure 5a shows the calculated E_f values with respect to the curvatures exhibited in each cantilever (k). As expected, the analytical model solutions deviate from the real value of E_f as the deflections increase. However, it is confirmed that the errors produced by nonlinear effects are greater on the cantilevers with smaller thickness (cantilever S1). In other words, for cantilevers with very thin tested films and adherence layers, the accuracy of the analytical model improves with higher h_r values.

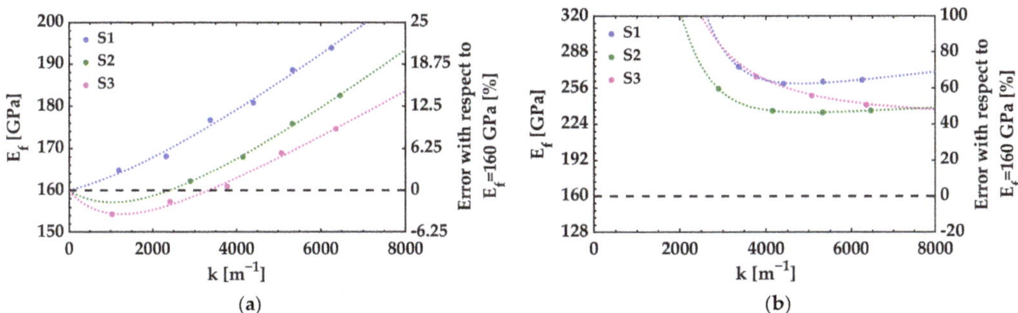

Figure 5. Effect of nonlinear deflections on the estimation of the Young's modulus of the tested films (E_f). The results were calculated from the curvatures obtained from FEM solutions using: (**a**) Equation (15) and (**b**) Equation (20). The vertical axis on the left side of both graphs represents the results of E_f. The vertical axis on the right side of the graphs represents the error of the results with respect to the real value of E_f (indicated in Table 1).

The accuracy of the analytical model also decreases when the effects of adherence layers and intrinsic stress gradients are neglected. If the parameters related to these characteristics are not considered in the calculations, Equation (20) can be used to obtain the results. Based on this mathematical expression, the slope of the curves M^* versus k^* are greater than those obtained using Equation (15) (Figure 4). As a result, the values calculated for the Young's modulus of the tested films from the curvatures obtained by FEM are unrealistic, especially at small deflections (Figure 5b). A comparative analysis of Equations (15) and (20) shows that the effects of the adherence layer can be omitted in the analyzed examples if h_a is equal to or less than 3 nm, 6 nm, and 3 nm of the cantilevers S1, S2, and S3, respectively. Likewise, it can be determined that the effects of intrinsic stress gradients can be neglected when the relative difference between the internal bending moment (M) and the moment produced by the uniform stresses and is less than 2%. However, this last condition is satisfied only if σ_f is greater than 360 MPa, 1300 MPa, and 5100 MPa of the cantilevers S1, S2, and S3, respectively. These values of uniform residual stresses of the tested film are very far from those indicated in Table 2.

It is difficult to establish the validity range of the analytical model proposed in this work due to the large number of variables involved in the equations. However, good results were observed for cantilevers with normalized curvatures below 1.7 ($k^* \leq 1.7$). This condition is generally fulfilled in cantilevers with angular deflections (Figure 1a) less than ninety degrees ($\theta < 90°$) and with a reference layer of thickness greater than or equal to four hundred nanometers ($h_r \geq 400$ nm). Finite element models are generally more accurate because they consider the nonlinear response of the cantilevers when they experience very large deflections. However, some effects that can limit the accuracy of the results are excessive etching of materials and non-homogeneous deposition of the thin films. Furthermore, it is important to note that the models consider initially straight cantilevers with linearly elastic and inextensible materials.

3. Results and Discussions

We use the methodology proposed in this work to estimate the Young's modulus of previously reported silicon nitride (Si_3N_4) tested films in two different cases. In the first case, the bilayer cantilever comprises a reference layer of silicon oxide (SiO_2) [25]. In the second case, the reference layer is made of silicon (Si) [16]. Originally, the research reported by Laconte et al. in [25] aimed to estimate the residual stresses generated in materials during the fabrication process. On the other hand, in the report reported by Favache et al. [16], the bilayer cantilevers were also used to determine the Young's modulus of Si_3N_4 by applying a different procedure. In both reports, the samples were fabricated in the WINFAB (Wallonia Infrastructure Nano Fabrication) cleanroom facilities at Université catholique de Louvain, Louvain-la-Neuve, Belgium (https://sites.uclouvain.be/winfab/NEW_website/. Retrieved 30 November 2021) under similar conditions using different substrates. The results are validated with those previously obtained in the same laboratory using alternative techniques.

3.1. Case 1: SiO_2/Si_3N_4 Bilayer Cantilever

The fabrication process of the SiO2/Si_3N_4 bilayer cantilever started with the growth of a thermal SiO_2 layer on a silicon substrate at 1000 °C under a mixed O_2/H_2 atmosphere. Afterward, Si_3N_4 was deposited over the thermal SiO_2 at 800 °C by low pressure chemical vapor deposition (LPCVD) with a stoichiometric mixture of dichlorosilane with an ammonia (SiH_2Cl_2/NH_3) ratio of 1:3. Individual layers of SiO_2 and Si_3N_4 were separately deposited on two different silicon wafers to obtain the monolayer beams. After growing the thin layers, the structures were defined by photolithography and patterned using a plasma etching for the silicon nitride and hydrofluoric acid (HF) for the thermal silicon oxide. The thick silicon wafers were etched in 20% tetramethyl ammonium hydroxide (TMAH) solution at 90 °C for one hour to release the cantilevers and then rinsed in de-ionized water and dried in methanol to avoid structural damage. The deposition of an adherence layer was not considered in the fabrication process because the materials showed high adhesion. A schematic diagram of the fabrication steps is shown in Figure 6.

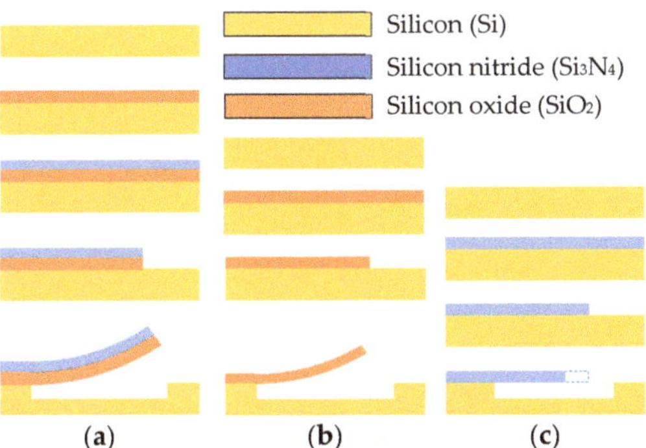

Figure 6. Schematic representation of the main fabrication steps for the case 1. (**a**) SiO_2/Si_3N_4 bilayer cantilever: thermal growth of the SiO_2 layer, deposition of the Si_3N_4 layer, patterning of the deposited layers and release from the Si substrate. (**b**) SiO_2 beam: growth and patterning of the SiO_2 layer and release from the Si substrate. (**c**) Si_3N_4 beam: deposition and patterning of the Si_3N_4 film and release from the Si substrate.

Corresponding uniform components of the residual stresses were measured using the Stoney formula by wafer curvature measurements [21]. The thicknesses of the deposited materials were verified by ellipsometry while the in-plane dimensions of the cantilevers were measured by SEM. Figure 7 shows the SEM views of the fabricated cantilevers after being released from the Si substrate. The vertical deflection for 100 µm length and 10 µm wide cantilevers was measured using an optical microscope comparing the focus on both ends.

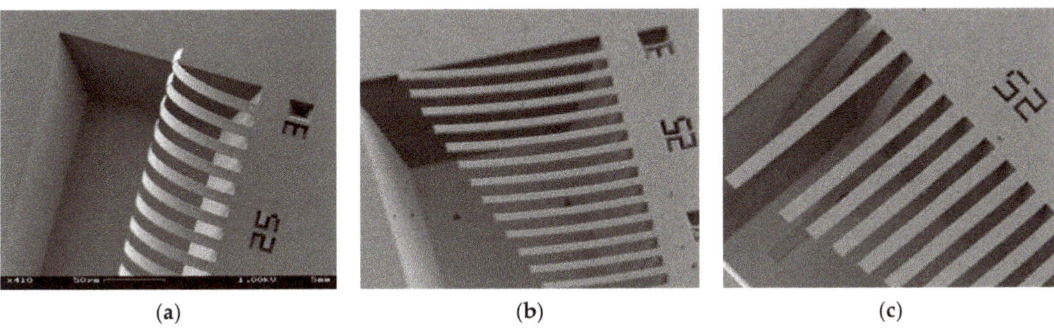

Figure 7. SEM views of the released cantilevers in case 1 (longest showed cantilevers are 150 µm long). (**a**) SiO_2/Si_3N_4 Bilayer Cantilever. (**b**) SiO_2 beam. Deflection reveals the presence of strain gradients over film thickness. (**c**) Si_3N_4 beam. Some stiction appeared after rinsing with water and drying in methanol. Reprinted with permission from [25]. Copyright©2006, Springer Nature.

Monolayer SiO_2 beams exhibited a notable vertical deflection that revealed the presence of intrinsic stress gradients in this material (Figure 7b). In contrast, the tested film appears to be free of the intrinsic stress gradients as their respective freestanding beams remained straight after being released (Figure 7c). The Young's modulus and the Poisson ratio of the SiO_2 reference layer and the Poisson ratio of the tested Si_3N_4 material were reported in the literature [25]. The dimensions of the structures, the uniform residual stresses, the vertical deflections of the cantilevers, and the elastic properties of the materials are indicated in Table 3.

Table 3. Parameters for case 1 (SiO_2/Si_3N_4 bilayer cantilever).

Parameter	Value	Measurement Technique
Cantilever width, w	10 µm	SEM
Cantilever length, L	100 µm	SEM
Thickness SiO_2 layer, h_r	433 nm	Ellipsometry
Thickness Si_3N_4 film, h_f	288 nm	Ellipsometry
Vertical deflection bilayer cantilever, b	50.51 µm	Optical microscopy
Vertical deflection SiO_2 beam, b_r	12 µm	Optical microscopy
Vertical deflection Si_3N_4 beam, b_f	≈0	Optical microscopy
Young's modulus SiO_2, E_r	70 GPa	Reported in [25]
Poisson ratio SiO_2, v_r	0.2	Reported in [25]
Poisson ratio Si_3N_4, v_f	0.27	Reported in [25]
Uniform residual stress SiO_2 layer, σ_r	−281 MPa	Wafer curvature measurement and Stoney formula
Uniform residual stress Si_3N_4 film, σ_f	914 MPa	Wafer curvature measurement and Stoney formula

The analytical solution was estimated using Equation (19) since the SiO_2 layer developed intrinsic stress gradients during the fabrication process. The radii of curvature of the bilayer cantilever and the monolayer beams were calculated from their respective vertical deflections using Equation (3). For the FEM solution, the 3D model was meshed with 250 elements over the length, 15 elements over the half width, 1 element over the thickness of the two sections of the SiO_2 layer, and 1 element over the thickness of the Si_3N_4 film. The silicon nitride tested film was not split into two symmetrical sections since it was free of intrinsic stress gradients. Several simulations were conducted varying the E_f value from 284 to 292 GPa to obtain different Young's Modulus versus deflection responses. Subsequently, a linear regression was performed on the recorded data to approximate the relationship between the two variables (Figure 8). Then, the correct value of E_f was obtained by evaluating the experimentally measured bilayer cantilever deflection on the fitted linear function. Both silicon oxide and silicon nitride were considered isotropic materials [25].

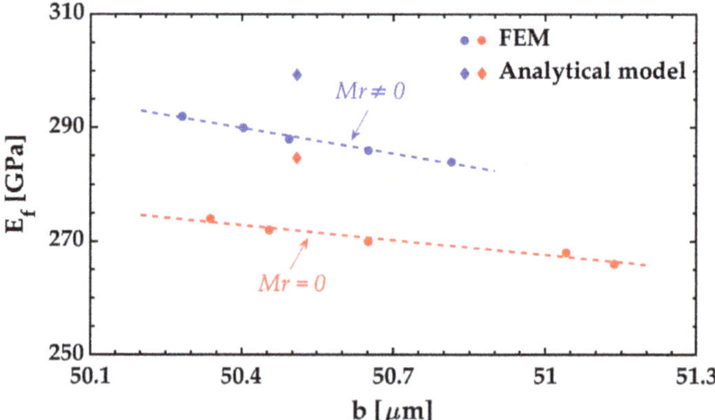

Figure 8. Estimated Young's modulus of the Si_3N_4 tested films for case 1 (SiO_2/Si_3N_4 bilayer cantilever). The colored dots indicate the FEM results for different values of E_f from which linear fits (dashed lines) are made. The correct value of E_f is found by evaluating the measured vertical deflection on the fitted linear function. The incidence of the intrinsic gradient of the SiO_2 layer on the results is evidenced in the lower values of the Young's modulus of the Si_3N_4 when it is assumed that $M_r = 0$.

The Young's modulus values of the 288 nm thick Si_3N_4 tested films found through the models proposed in this work are listed in Table 4. The results of the analytical model agree well with those obtained by FEM. The incidence of the intrinsic gradient of the SiO_2 layer has on the results was studied by repeating the calculations with $M_r = 0$. The elimination of M_r leads to an underestimation of 4.9 and 5.7% in the E_f values estimated by the analytical and FEM models, respectively.

Table 4. Estimated values of the Young's modulus of the Si_3N_4 tested films for case 1 (SiO_2/Si_3N_4 bilayer cantilever).

Result	Analytical Model	FEM	Relative Difference
E_f [GPa]	299.3 [1]	288.3	3.8%
E_f ($M_r = 0$) [GPa]	284.7 [2]	271.9	4.7%

[1] Equation (19). [2] Equation (20).

The results of the Young's modulus of the Si_3N_4 films for case 1 are slightly above the upper range of the reported values for the same lab (summarized in Table 5). The lack of resolution in the optical microscope and the ineffectiveness of the strategy used to estimate vertical deflection may be the reason. Furthermore, edge effects at the free end of the cantilevers can cause miscalculation of the respective radii of curvature. Nevertheless, it is worth mentioning that Young's modulus of the silicon nitride can vary from 193 GPa to 338.5 GPa according to the review of existing data reported in the previous report [16].

Table 5. Young's modulus of silicon nitride thin films deposited in the WINFAB laboratory.

Method	Thickness (nm)	Value (GPa)	Reference
Nanoindentation	250	235 ± 10	[13]
Stoney and freestanding beams	301	233	[15]
Nanoindentation	301	241	[15]
Bilayer cantilever	55	270 ± 20	[16]
This work (case 1)	288	288.3	–
This work (case 2)	46	242.9	–
	63	236.2	
	102	222.4	–
	133	213.9	–

3.2. *Case 2: Si/Si_3N_4 Bilayer Cantilever*

In this case, the structures were fabricated on a silicon-on-insulator (SOI) wafer following the process shown in Figure 9. The first step was the patterning of the upper Si layer of the SOI wafer by reactive ion etching (RIE) with a sulfur hexafluoride (SF_6)-based plasma. The Si_3N_4 film was then deposited at 790 °C through LPCVD and then patterned by RIE using a mixture of sulfur hexafluoride and silicon tetrachloride ($SF_6/SiCl_4$)-based plasma. At this point, the wafer was cut into four samples (G1, G2, G3, and G4) before releasing the structures by the etching of the SiO_2 sacrificial layer using HF (73 vol.%). Since HF also etches Si_3N_4 at a slower rate than SiO_2, the release time was varied in each sample to expect obtain bilayer cantilevers with different tested film thicknesses h_f (Table 6). The fabrication process allows the production of Si/Si_3N_4 bilayer cantilevers and freestanding monolayer Si and Si_3N_4 beams of several lengths.

Table 6. Thicknesses and Curvatures of the Si/Si_3N_4 bilayer cantilevers.

Dimension	G1	G2	G3	G4
Thickness Si layer [nm], h_r	400	400	400	400
Thickness Si_3N_4 film [nm], h_f	46	63	102	133
Curvature bilayer cantilever [m^{-1}], k	5479	6250	7143	7407

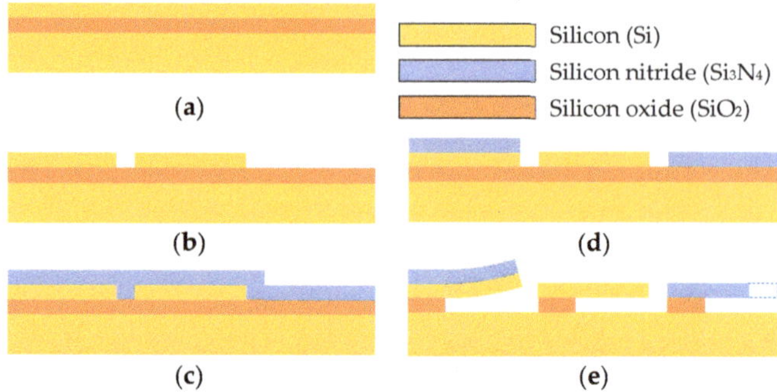

Figure 9. Schematic representation of the main fabrication steps for the case 2 (Si/Si_3N_4 bilayer cantilever). (**a**) SOI wafer; (**b**) patterning of the top Si; (**c**) deposition of Si_3N_4; (**d**) patterning of Si_3N_4; (**e**) etching of the SiO_2 sacrificial layer.

The thickness of the Si reference layer was obtained from the SOI wafer specifications, whereas the thickness of the Si_3N_4 films was measured by ellipsometry using the Cauchi model [26]. On the other hand, the deflection profile shown by the bilayer cantilevers after being released (Figure 10) was measured using SEM and interferometry. Subsequently, the respective radii of curvature were estimated by interpolating a circle on the deformed shape applying the Taubin method [23].

Figure 10. SEM view of the fully released Si/Si_3N_4 (sample G1) bilayer cantilevers. Reprinted with permission from [16]. Copyright©2016, AIP Publishing.

Freestanding monolayer Si and Si_3N_4 beams did not exhibit perceptible deflections, indicating the absence of intrinsic stress gradients. Therefore, the in-plane deformations were used to estimate the uniform residual strains. The residual strains of the Si_3N_4 and Si beams were $e_f = 0.0032$ and $e_r \approx 0$ (indicating the absence of uniform residual stresses at the top of the Si layer of the SOI wafer), respectively. The measured thicknesses and curvatures of the Si/Si_3N_4 bilayer cantilevers are given in Table 6, while the geometrical and elastic properties required in the models are indicated in Table 7.

Table 7. Dimensions and elastic properties for case 2 (Si/Si_3N_4 bilayer cantilever).

Parameter	Value	Measurement Technique
Cantilever width, w	10 μm	Optical microscopy
Cantilever length, L	200 μm	Optical microscopy
Poisson ratio Si_3N_4 film, v_f	0.27	reported in [16]
Uniform residual strain Si layer, e_r	≈ 0	SEM
Uniform residual strain Si_3N_4 film, e_f	0.0032	SEM

The analytical solution was estimated from the uniform residual strains using Equation (21) since the materials were free of intrinsic stress gradients. The FEM solution was found following the same extraction methodology of case 1 (Figure 11). The model was meshed with 160 elements over the length, 15 elements over the half width, and 1 element over the thickness of the Si layer and the Si_3N_4 film. The materials were not split into two symmetrical sections as they did not develop intrinsic stress gradients during the fabrication process. Experimentally, it was observed that the radius of curvature does not have significant changes in the bilayer cantilevers with lengths ranging between 100 μm and 1.9 mm. Therefore, a length of 200 μm is appropriate to perform the simulations with better results. The thermal expansion coefficients of the materials were calculated from the uniform residual strains using Equations (23) or (24). The simulated deflection profile was extracted in the range of 10 μm to 190 μm across the long axis to avoid edge effects.

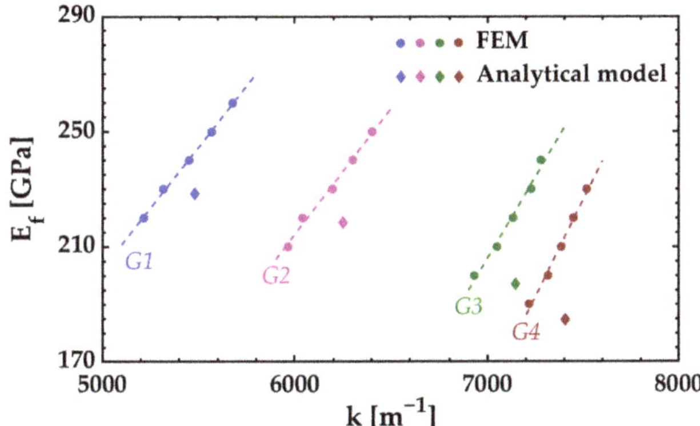

Figure 11. Estimated Young's modulus of the Si_3N_4 tested films for case 2 (Si/Si_3N_4 bilayer cantilever).

In the solutions, Si_3N_4 was considered isotropic while Si was considered orthotropic [16]. According to the global coordinate system of Figure 3, the elastic constants of the orthotropic silicon are [27]:

$$\begin{aligned} [E_x, E_y, E_z] &= [169, 169, 130] \quad [\text{GPa}] \\ [v_{xy}, v_{yz}, v_{zx}] &= [0.064, 0.36, 0.28] \\ [G_{xy}, G_{yz}, G_{zx}] &= [50.9, 79.6, 79.6] \quad [\text{GPa}] \end{aligned} \tag{31}$$

where E, v, and G refer to Young's modulus, Poisson's ratio, and shear modulus, respectively. For the analytical solution, the biaxial Young's modulus of silicon was taken from the elastic constants in the xy–plane:

$$\overline{E}_r = \frac{E_x}{1-v_{xy}} = \frac{E_y}{1-v_{xy}} = 180.55 \quad [\text{MPa}] \tag{32}$$

Table 8 presents the Young's modulus values of the Si_3N_4 tested films for each of the four samples. The results are in good agreement with those values obtained in silicon nitride films fabricated under similar experimental conditions (Table 5).

Table 8. Estimated values of the Young's modulus of the Si_3N_4 tested films for case 2 (Si/Si_3N_4 bilayer cantilever).

Sample	Analytical Model [GPa] [1]	FEM [GPa]	Relative Difference
G1	228.5	242.9	5.9%
G2	218.3	236.2	7.6%
G3	197.1	222.4	11.4%
G4	184.7	213.9	13.7%

[1] Equation (21).

The values obtained using the analytical models agree well with those obtained by FEM, but the relative difference between them increases with higher tested film thickness h_f. The increase of h_f has more incidence on the internal bending moment M than on the bending rigidity $(EI)_e$ due to the high magnitude of σ_f. Nonproportional increase in the values of M and $(EI)_e$ results in larger deflections of the bilayer cantilever. Under large deflections, the curvature k does not vary linearly with the internal bending moment M and the accuracy of the analytical model decreases (as explained in Section 2.3).

The dependence of the results on the size of the tested film is seen in the E_f versus h_f plot shown in Figure 12. Specifically, it is noticed that the Young's modulus of the Si_3N_4 tested film increases as its thickness decreases. The estimated Young's modulus of Si_3N_4 tested film with a thickness of 40 nm (sample G1) is about 14% higher than that of Si_3N_4 tested film with a thickness of 133 nm (sample G4). Nevertheless, it should be considered that the estimation of the Young's modulus of ultrathin layered materials can be influenced by defects in the test material or by internal factors such as roughness. The study of the influence of these internal parameters on the results is not part of the objectives of this work.

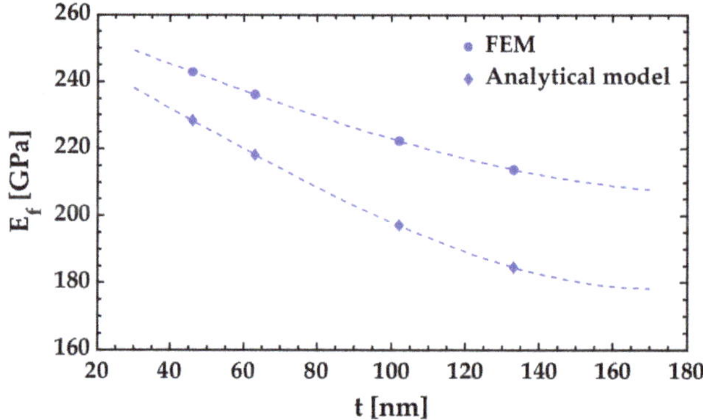

Figure 12. Young's modulus E_f vs. thickness h_f for case 2 (Si/Si_3N_4 bilayer cantilever).

Finally, it was found that the Si_3N_4 films tested in case 1 (SiO_2/Si_3N_4 bilayer cantilever) have a higher Young's modulus than Si_3N_4 films tested in case 2 (Si/Si_3N_4 bilayer cantilever). This may be because the structural properties of the Si_3N_4 deposited on SiO_2 can be different from the structural properties of the Si_3N_4 deposited on Si [25]. In addition, some specific parameters of the fabrication process, such as deposition time, etching time, or etch solutions (which are different in the two investigated cases) can alter the properties of the tested films.

4. Conclusions

A methodology to predict the Young's modulus of nanometer-thick films using residual stress-driven bilayer cantilevers was reported. The bilayer cantilever consists of a well-known reference layer and a tested film that store residual stresses during the fabrication process. The fully released cantilever deflects due to the difference in the residual stresses of the two materials. The measured deflections and residual stresses are related using analytical or finite element models to calculate the Young's modulus of the tested film. The proposed models include the intrinsic stress gradients and the use of adherence layers. Our methodology was applied to previously reported silicon nitride (Si_3N_4)-tested films deposited on silicon oxide (SiO_2) and silicon (Si) reference layers. The estimated Young's modulus for the 288 nm thick Si_3N_4 tested films deposited on SiO_2 was 288.3 GPa. On the other hand, the estimated Young's modulus for the Si_3N_4 tested films deposited on Si with thicknesses ranging between 43 and 133 nm varied from 242.9 GPa to 213.9 GPa. The results obtained in this work were in good agreement with reported literature data of Si_3N_4 films fabricated under similar conditions. This methodology can be easily used for thin films of different materials. However, it is limited by the resolution of the techniques employed to estimate the residual stresses, deflections, and in-plane dimensions. Future research will focus on estimating the Young's modulus of thin films using residual stress-driven structures but reducing the dependence of the methodology on the Poisson's ratio of the tested film and the elastic properties of the reference layer.

Author Contributions: Conceptualization, L.A.V.-M. and J.-P.R.; methodology, L.A.V.-M. and A.L.H.-M.; software, L.A.V.-M. and L.A.A.-C.; validation, L.A.V.-M. and F.L.-H.; formal analysis, L.A.A.-C.; investigation, J.-P.R.; resources, A.L.H.-M.; data curation, F.L.-H.; writing—original draft preparation, L.A.V.-M.; writing—review and editing, L.A.V.-M., J.-P.R., L.A.A.-C., F.L.-H. and A.L.H.-M.; visualization, L.A.A.-C. and F.L.-H.; supervision, J.-P.R.; project administration, A.L.H.-M. All authors have read and agreed to the published version of the manuscript.

Funding: This research received no external funding.

Data Availability Statement: Data is contained within the article.

Acknowledgments: Luis A. Velosa-Moncada thanks the Graduate Program in Materials and Nanoscience from the Universidad Veracruzana for the training received during his doctoral studies. Luis A. Velosa-Moncada also thanks for the scholarship received by CONACYT.

Conflicts of Interest: The authors declare no conflict of interest.

References

1. Babaei Gavan, K.; Westra, H.J.R.; van der Drift, E.W.J.M.; Venstra, W.J.; van der Zant, H.S.J. Size-dependent effective Young's modulus of silicon nitride cantilevers. *Appl. Phys. Lett.* **2009**, *94*, 233108. [CrossRef]
2. Huang, H.; Winchester, K.J.; Suvorova, A.; Lawn, B.R.; Liu, Y.; Hu, X.Z.; Dell, J.M.; Faraone, L. Effect of deposition conditions on mechanical properties of low-temperature PECVD silicon nitride films. *Mater. Sci. Eng. A* **2006**, *435*, 453–459. [CrossRef]
3. Twardowska, A.; Kopia, A.; Malczewski, P. The Microstructure, Mechanical and Friction-Wear Properties of (TiBx/TiSiyCz)x3 Multilayer Deposited by PLD on Steel. *Coatings* **2020**, *10*, 621. [CrossRef]
4. Ke, Y.-E.; Chen, Y.-I. Effects of Nitrogen Flow Ratio on Structures, Bonding Characteristics, and Mechanical Properties of ZrNx Films. *Coatings* **2020**, *10*, 476. [CrossRef]
5. Nagy, P.; Rohbeck, N.; Hegedűs, Z.; Michler, J.; Pethö, L.; Lábár, J.L.; Gubicza, J. Microstructure, Hardness, and Elastic Modulus of a Multibeam-Sputtered Nanocrystalline Co-Cr-Fe-Ni Compositional Complex Alloy Film. *Materials.* **2021**, *14*, 3357. [CrossRef]
6. Vlassak, J.J.; Nix, W.D. A new bulge test technique for the determination of Young's modulus and Poisson's ratio of thin films. *J. Mater. Res.* **1992**, *7*, 3242–3249. [CrossRef]
7. Österlund, E.; Kinnunen, J.; Rontu, V.; Torkkeli, A.; Paulasto-Kröckel, M. Mechanical properties and reliability of aluminum nitride thin films. *J. Alloys Compd.* **2019**, *772*, 306–313. [CrossRef]
8. Hensel, A.; Schröter, C.J.; Schlicke, H.; Schulz, N.; Riekeberg, S.; Trieu, H.K.; Stierle, A.; Noei, H.; Weller, H.; Vossmeyer, T. Elasticity of Cross-Linked Titania Nanocrystal Assemblies Probed by AFM-Bulge Tests. *Nanomaterials* **2019**, *9*, 1230. [CrossRef]
9. Poelma, R.H.; Sadeghian, H.; Noijen, S.P.M.; Zaal, J.J.M.; Zhang, G.Q. A numerical experimental approach for characterizing the elastic properties of thin films: Application of nanocantilevers. *J. Micromechanics Microengineering* **2011**, *21*, 65003. [CrossRef]
10. Chuang, W.-C.; Hu, Y.-C.; Chang, P.-Z. CMOS-MEMS Test-Key for Extracting Wafer-Level Mechanical Properties. *Sensors* **2012**, *12*, 17094–17111. [CrossRef]
11. Guo, X.-G.; Zhou, Z.-F.; Sun, C.; Li, W.-H.; Huang, Q.-A. A Simple Extraction Method of Young's Modulus for Multilayer Films in MEMS Applications. *Micromachines* **2017**, *8*, 201. [CrossRef]
12. Behera, A.R.; Shaik, H.; Rao, G.M.; Pratap, R. A Technique for Estimation of Residual Stress and Young's Modulus of Compressively Stressed Thin Films Using Microfabricated Beams. *J. Microelectromechanical Syst.* **2019**, *28*, 1039–1054. [CrossRef]
13. Gravier, S.; Coulombier, M.; Safi, A.; Andre, N.; BoÉ, A.; Raskin, J.-P.; Pardoen, T. New On-Chip Nanomechanical Testing Laboratory—Applications to Aluminum and Polysilicon Thin Films. *J. Microelectromechanical Syst.* **2009**, *18*, 555–569. [CrossRef]
14. Cuddalorepatta, G.K.; Li, H.; Pantuso, D.; Vlassak, J.J. Measurement of the stress-strain behavior of freestanding ultra-thin films. *Materialia* **2020**, *9*, 100502. [CrossRef]
15. Boé, A.; Safi, A.; Coulombier, M.; Pardoen, T.; Raskin, J.-P. Internal stress relaxation based method for elastic stiffness characterization of very thin films. *Thin Solid Films* **2009**, *518*, 260–264. [CrossRef]
16. Favache, A.; Ryelandt, S.; Melchior, M.; Zeb, G.; Carbonnelle, P.; Raskin, J.-P.; Pardoen, T. A generic "micro-Stoney" method for the measurement of internal stress and elastic modulus of ultrathin films. *Rev. Sci. Instrum.* **2016**, *87*, 15002. [CrossRef]
17. Cuddalorepatta, G.K.; Sim, G.-D.; Li, H.; Pantuso, D.; Vlassak, J.J. Residual stress–driven test technique for freestanding ultrathin films: Elastic behavior and residual strain. *J. Mater. Res.* **2019**, *34*, 3474–3482. [CrossRef]
18. Timoshenko, S.P. Strength of materials Part 1. In *Elementary Theory and Problems*; D. Van Nostrand Company, Inc.: New York, NY, USA, 1955; pp. 88–97.

19. Howell, L.L. *Compliant Mechanisms*; John Wiley & Sons, Inc.: Hoboken, NJ, USA, 2001; pp. 43–45.
20. Lobontiu, N.; Garcia, E. *Mechanics of Microelectromechanical Systems*; Springer: Boston, MA, USA, 2004; pp. 43–46.
21. Freund, L.B.; Suresh, S. *Thin Film Materials: Stress, Defect Formation and Surface Evolution*; Cambridge University Press: Cambridge, UK, 2004.
22. Huang, X.; Liu, Z.; Xie, H. Recent progress in residual stress measurement techniques. *Acta Mech. Solida Sin.* **2013**, *26*, 570–583. [CrossRef]
23. Taubin, G. Estimation of planar curves, surfaces, and nonplanar space curves defined by implicit equations with applications to edge and range image segmentation. *IEEE Trans. Pattern Anal. Mach. Intell.* **1991**, *13*, 1115–1138. [CrossRef]
24. Freund, L.B.; Floro, J.A.; Chason, E. Extensions of the Stoney formula for substrate curvature to configurations with thin substrates or large deformations. *Appl. Phys. Lett.* **1999**, *74*, 1987–1989. [CrossRef]
25. Thin dielectric films stress extraction. In *Micromachined Thin-Film Sensors for SOI-CMOS Co-Integration*; Springer: Boston, MA, USA, 2006.
26. Fujiwara, H. *Spectroscopic Ellipsometry: Principles and Applications*; John Wiley & Sons: Hoboken, NJ, USA, 2007; p. 170.
27. Hopcroft, M.A.; Nix, W.D.; Kenny, T.W. What is the Young's Modulus of Silicon? *J. Microelectromechanical Syst.* **2010**, *19*, 229–238. [CrossRef]

MDPI

Article

Nanomechanical and Nanotribological Properties of Nanostructured Coatings of Tantalum and Its Compounds on Steel Substrates

Galina Melnikova [1], Tatyana Kuznetsova [1], Vasilina Lapitskaya [1], Agata Petrovskaya [1], Sergei Chizhik [1], Anna Zykova [2,3], Vladimir Safonov [2,3], Sergei Aizikovich [4,*], Evgeniy Sadyrin [4,*], Weifu Sun [5,6] and Stanislav Yakovin [3]

1 Laboratory of Nanoprocesses and Technologies, A.V. Luikov Heat and Mass Transfer Institute of the National Academy of Sciences of Belarus, 15 P. Brovki Str., 220072 Minsk, Belarus; galachkax@gmail.com (G.M.); kuzn06@mail.ru (T.K.); vasilinka.92@mail.ru (V.L.); agata.petrovskaya@gmail.com (A.P.); chizhik_sa@tut.by (S.C.)
2 National Science Center "Kharkov Institute of Physics and Technology", 1 Akademicheskaya Str., 61108 Kharkov, Ukraine; zykova.anya@gmail.com (A.Z.); safonov600@gmail.com (V.S.)
3 V.N. Karazin Kharkiv National University, 4 Svobody Sq., 61022 Kharkov, Ukraine; stanislav.yakovin@gmail.com
4 Research and Education Center "Materials", Don State Technical University, 1 Gagarin Sq., 344003 Rostov-on-Don, Russia
5 State Key Laboratory of Explosion Science and Technology, Beijing Institute of Technology, Beijing 100081, China; weifu.sun@bit.edu.cn
6 Beijing Institute of Technology Chongqing Innovation Center, Chongqing 401120, China
* Correspondence: saizikovich@gmail.com (S.A.); e.sadyrin@sci.donstu.ru (E.S.); Tel.: +7-863-238-15-58 (S.A. & E.S.)

Citation: Melnikova, G.; Kuznetsova, T.; Lapitskaya, V.; Petrovskaya, A.; Chizhik, S.; Zykova, A.; Safonov, V.; Aizikovich, S.; Sadyrin, E.; Sun, W.; et al. Nanomechanical and Nanotribological Properties of Nanostructured Coatings of Tantalum and Its Compounds on Steel Substrates. *Nanomaterials* **2021**, *11*, 274. https://doi.org/10.3390/nano11092407

Academic Editor: Victor A. Eremeyev

Received: 29 July 2021
Accepted: 13 September 2021
Published: 15 September 2021

Abstract: The present paper addresses the problem of identification of microstructural, nanomechanical, and tribological properties of thin films of tantalum (Ta) and its compounds deposited on stainless steel substrates by direct current magnetron sputtering. The compositions of the obtained nanostructured films were determined by energy dispersive spectroscopy. Surface morphology was investigated using atomic force microscopy (AFM). The coatings were found to be homogeneous and have low roughness values (<10 nm). The values of microhardness and elastic modulus were obtained by means of nanoindentation. Elastic modulus values for all the coatings remained unchanged with different atomic percentage of tantalum in the films. The values of microhardness of the tantalum films were increased after incorporation of the oxygen and nitrogen atoms into the crystal lattice of the coatings. The coefficient of friction, CoF, was determined by the AFM method in the "sliding" and "plowing" modes. Deposition of the coatings on the substrates led to a decrease of CoF for the coating-substrate system compared to the substrates; thus, the final product utilizing such a coating will presumably have a longer service life. The tantalum nitride films were characterized by the smallest values of CoF and specific volumetric wear.

Keywords: atomic force microscopy; friction coefficient; magnetron sputtering; nanoindentation; nanostructured coatings; tantalum

1. Introduction

The stainless steel (for example, 316L SS), platinum iridium alloys, tantalum, nitinol, cobalt–chrome alloys, titanium and its alloys, and pure iron and magnesium alloys are the basic materials for the production of stents. Stainless steel is the most common material for the production of stents with and without coatings. The stents made of stainless steel demonstrate suitable mechanical properties and excellent corrosion resistance. However, the clinical application of steel is limited by the ferromagnetic nature of the alloy and its low density. Due to these properties, the steel is poorly visible in X-ray and magnetic resonance imaging [1]. The implants of the stainless steel may cause an allergy to nickel, chromium, and molybdenum, leading to local immune responses and inflammation. Different materials are used as coatings on stainless steel stents that lead to improvements of X-ray visibility and biocompatibility. Titanium and its alloys have excellent biocompatibility, high corrosion resistance, and are intensively used in orthopedics and dentistry. Pure titanium is not suitable for the stent production due to rather low values of the mechanical properties; however, it can be used as a coating for stainless steel stents to improve biocompatibility. Such stents show excellent results in clinical trials [1,2].

Usage of stents that represent a system of the stainless steel substrate with titanium or tantalum coating allows combining the suitable mechanical properties and bioinertness [3–5]. Tantalum is characterized by a good plasticity, high strength, wear resistance, weldability, corrosion resistance, infusibility, biocompatibility, and it is clearly visible in X-rays and magnetic resonance imaging [6,7]. In addition, due to its properties, tantalum is widely used not only in electronics [8,9], protective coatings [10,11], anti-corrosion coatings [12,13], optical coatings [14–17], chemical industry [18], but also in biomedicine [19–24] (orthopedics and dentistry [25–29], endovascular stents, and neurosurgical implants [30]). However, the use of tantalum is difficult due to its high density, manufacturing complexity, and the relatively high cost. For these reasons, a various approaches are currently being proposed to modify the surface properties of metal substrates for the improvement of biological responses by applying coatings based on Ta. The modern processing methods allow obtaining the tantalum coatings with a fine-grained structure and optimal properties (tensile strength up to 600 MPa, and elongation of about 30%) for the stents production [31]. In combination with increased strength, tantalum has a high protective ability that prevents active corrosion processes and the electrochemical destruction of metal surface structures in various environments. For example, β-Ta nanocrystalline coatings on Ti-6Al-4V substrates showed high hardness in combination with good resistance to contact damage [32]. Thus, the corrosion resistance of 316 L stainless steel was significantly improved due to the TaC_xN_{1-x} coatings [33]. These coatings also demonstrated good adhesion characteristics. Thus, depending on the formation conditions, the properties of the coatings change. However, the best conditions for tantalum coatings formation have not yet been determined.

The development of technologies for the formation of functional tantalum nanostructured coatings is currently of high interest for both medical and material science community [34]. The perspective ion–plasma spraying method for creating nanostructured coatings allows to form nanocoatings to change surface properties and obtain biocompatible materials with desired properties. Thus, in [35], it was found that the TaNx coating applied by high-frequency magnetron sputtering at a relatively high bias voltage of 200 V demonstrates good tribological characteristics, hardness, and adhesive strength. The authors of [36] synthesized tantalum nitride films onto silicon using magnetron sputtering (the N_2 content in the gas mixture was changed). As a result, it was revealed which content of N_2 allows obtaining films with the highest hardness, low friction coefficient, and low wear rate. In the present research, a new efficient technology has been developed for the deposition of nanostructured coatings based on tantalum by magnetron sputtering. The optimal modes of metal substrates modification with nanostructured coatings based on Ta were developed in order to create new improved devices for medical applications (cardiovascular surgery, endoscopy, and orthopedics). Remarkably, nanostructured materials have larger surface energy than typical materials enhancing the adhesion of bone cells and producing higher osseointegration [32]. The surface nature of a biomaterial (relief, hydrophobic–hydrophilic properties, chemical composition, etc.) plays an important role in regulating the cellular response of a biological organism to the biomaterial. Since the approach for the deposition of the Ta coatings involves changing the structure and properties of materials at both the micro- and nanoscales, it is advisable to assess changes in the structure and properties at these levels. Such instrumental research methods as atomic force microscopy (AFM) and nanoindentation (NI) allow studying the surface properties of the films of tantalum and its compounds at the nanoscale, as well as to evaluate the nature of their changes (roughness, friction coefficient) in the biological medium and estimate the possibility of using these nanocoatings as biocompatible materials [37–40].

The aim of the present work was to study the physical, mechanical, and tribological properties of nanostructured tantalum oxynitride coatings on the stainless steel substrates. This complex of characteristics helps to assess the performance characteristics of coatings more accurately than the traditionally used microhardness. Estimation of such a complex is vital for the production of devices for various medical applications.

2. Materials and Methods

The coatings of Ta, Ta_2O_5, TaN, and TaON were deposited on the polished stainless steel (type is 316 L SS) substrates using reactive direct current planar magnetron sputtering. Deposition process was performed on the experimental set-up (KhNU and NSC KIPT NASU, Kharkov, Ukraine) [41,42]. The physical and chemical processes in plasma were investigated during the reactive magnetron deposition of tantalum oxynitride [43]. Prior to the deposition, substrates were cleaned in an ultrasonic bath, then the ion cleaning was performed (Hall type ion source) in argon atmosphere (pressure was 6.65×10^{-2} Pa, ion acceleration voltage—3 kW, ion source current—100 mA) during 5 min. Then, the substrates were placed in a chamber pumped to a residual vacuum of less than 10^{-3} Pa. The ion cleaning was performed (Hall type ion source) in argon atmosphere (pressure was 6.65×10^{-2} Pa, ion acceleration voltage—3 kW, ion source current—100 mA) during 5 min. Then the deposition of the coatings of Ta, Ta_2O_5, TaN, TaON of about 1 μm thickness was conducted. The tantalum target with a diameter of 170 mm was used. The magnetron discharge power was 4–5 kW. The distance between magnetron and samples was about 30 cm. The feature of this system was the additional oxygen activation by discharge plasma source induction. The oxygen was supplied through a plasma source for activation. The deposition parameters of the coatings are summarized in Table 1. These deposition modes were selected after optimization of the spraying technology carried out in [42,43].

Table 1. The parameters of the magnetron sputtering for the coatings of tantalum and its compounds.

Type of Coating	Gas Pressure P [Pa]	Magnetron Voltage U_m [V]	Magnetron Current I_m [A]	Time of Deposition [min]	Gas Mass Flow Rate Q [cm^3/min]
Ta	1×10^{-1} (Ar)	495	6.6	30	-
Ta_2O_5	1.3×10^{-1} (general)	500	6.4	20	25 (O_2)
TaN	1.2×10^{-1} (N_2)	800	3.4	60	95 (N_2)
TaON	1.5×10^{-1} (general)	620	4.0	30	10 (O_2) 45 (N_2)

The morphology of the coatings was evaluated by the AFM Dimension FastScan (Bruker, Santa Barbara, CA, USA) in PeakForce QNM (Quantitative NanoMechanics, Bruker, Santa Barbara, CA, USA) regime with CSG10_SS (Micromasch, Tallinn, Estonia) cantilevers. The study of the microstructure and elemental composition of the samples was carried out on the JSM7001F (JEOL, Tokyo, Japan) scanning electron microscope (SEM) equipped with the X-ray energy dispersive microanalysis probe system INCA ENERGY 350 (Oxford Instruments, Abingdon, Oxfordshire, UK) at x10,000 magnification. The operating voltage and probe current were 20 KV and 5 nA, respectively. The working distance was 10 mm. The microstructure was analyzed in the secondary electron mode (SEI mode).

The phase compositions were studied by X-ray phase analysis (XRD) on a DRON-4-07 (LNPO "Burevestnik", Saint-Petersburg, Russia) unit in copper radiation. To analyze the amorphous and crystal structure formation, coatings were annealed at a temperature of 700 °C for 15 min and one hour in air in a Nabertherm GmbH L5 /13/ B180 furnace.

The method for the investigation of the friction coefficient (CoF) using AFM is based on measuring the twist angle of the silicon cantilever of the probe around its axis under the action of friction between the surface and the tip. It is described in detail in the paper by Chizhik et al. [44]. In the present research, we used NCS 11A silicon cantilevers (Mikromasch, Tallinn, Estonia) with the stiffness 3 N/m to determine the CoF using AFM

NT-206 in a "sliding" mode. During the tests on the five different 20 × 4 μm areas and 256 × 50 points, the probe load was 0.005 μN, and the friction speed was 4.9 μm/s. The radius of curvature of the cantilever was increased to 100 nm by scanning at high loads on the silicon surface in order to prevent these changes during research.

The CoF in a "plowing" mode and the wear of the coatings were studied using a Dimension FastScan AFM in the Contact Mode using a diamond probe on silicon console of D300 type (SCDprobes, Tallinn, Estonia) with an initial tip radius of curvature of 33 nm The stiffness of cantilevers was 13.84 $N \cdot m^{-1}$. During the tests, the normal load of 1.164 μN (calculated = 0.6 V) per probe was controlled. The process parameters, which were kept constant, were as follows: the scanning area 20 × 4 μm, 100 cycles, 256 × 256 points, friction speed 2.0 μm/s. The movement of the probe on the surface was reciprocating. Thus, the use of silicon and diamond probes with different loads allowed exploring the surface of tantalum coatings by various friction mechanisms (Figure 1). At the "sliding" mode, the action of adhesive forces was influenced on the friction coefficient, and the "plowing" mode characterized the strength properties of the material during friction.

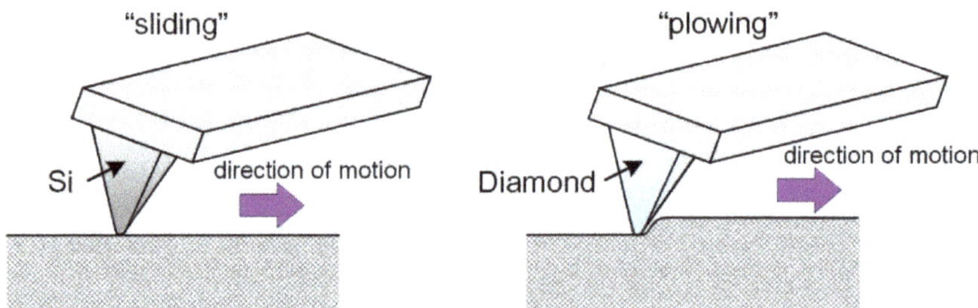

Figure 1. The principle scheme of friction mechanisms: "sliding" and "plowing".

The friction force (F) was recorded separately in the forward and in the reverse scanning. In the processing program, the image of the reverse scanning was subtracted from the obtained image of the forward scanning, and thus the average value of the friction force was determined. The mechanical stresses in the contact zone of the AFM probe with the coating surface (contact pressure) were determined using the AMES (Advanced Mechanical Engineering Solutions) contact stress calculator [45,46], setting the values of the radius of curvature of the probe, the elastic modulus of the coatings, and the probe. The value of the specific volumetric wear k_v was calculated as the ratio of the volume of worn material (V) to the normal load (L) and the indenter path length along the sample (S) and expressed in $m^3/N \cdot m$:

$$k_v = V/(L \cdot S) \quad (1)$$

The volume of wear V was estimated by the cross-sectional area and the perimeter of the wear track [47].

The thickness of the coatings was determined via AFM scanning of cross sections of coatings–substrate obtained after cooling samples into the fluid nitrogen during 10 min and fracture of cooled samples.

The microhardness (H) and the elastic modulus (E) were measured with using Hysitron 750 Ubi nanoindentation device (Bruker, Minneapolis, MN, USA). The radius of curvature of the diamond Berkovich indenter was 150 nm. For each sample, 9 curves were obtained at the load of 2000 μN. The indentation depth (h) into the coatings was 50–80 nm. The radius of the tip was calibrated by a set of indentations with increasing load into a surface of standard fused silica sample.

3. Results and Discussion

3.1. Elemental Analysis of Coatings

The EDX spectra (Figure 2) confirm the presence of basic characteristic elements in the films, such as tantalum, oxygen, and nitrogen [48]. The amount of tantalum is close to its atomic content in the compounds Ta_2O_5, TaN, and TaON. Except for the main lines of tantalum, the peak at 2.2 KeV corresponds to a secondary line of Ta (Figure 2b–d). Probably, the peak intensity is related to the texture of different Ta-based coatings. Deviations from the stoichiometric composition of coatings can be associated with some method accuracy. Non-uniformity of a scan coating surface and the presence of a small amount of other elements in the spectrum, for example Ar, may affect the error in the results normalizing. Previously carried out research [49] demonstrated the stoichiometric composition of the tested coatings. X-ray photoelectron spectroscopy was carried out using ESCALAB MkII (VG Scientific, East Grinstead, UK) with 1486.6 eV Al Kα radiation. Detailed scans were detected for the C1s, O1s, N1s, and Ta4f regions (Figure A1). An error on the binding energy (BE) values was obtained by standard deviation about 0.2 eV. Data analysis was made with a Shirley-type background subtraction, non-linear least-squares curve fitting with Gaussian-Lorentzian peak shapes. The atomic compositions were calculated using peak areas. The compositional analysis of oxide Ta_2O_5, oxynitride TaON, and nitride TaN coatings by X-ray photoelectron spectroscopy was performed. The photoelectron spectra of Ta4f, O1s, and N1s coatings were observed [49]. The spectrum included the photoelectron lines for Ta (4f7/2) and Ta (4f5/2). The Ta^{+5} signals were detected at binding energies 26.8 eV and 28.7 eV. The O1s high-resolution spectra demonstrated the peak at binding energy position E_b = 530.9 eV, associated with Ta-O chemical bond. The N1s peak was detected at binding energy E_b = 396.2 eV, associated with Ta-N chemical bond. This peak is generally considered to be the evidence for replacement of O atoms by N atoms in Ta_2O_5 crystal lattices [50]. In addition, a slight N1s peak at binding energy E_b = 398.0 eV was corresponded to Ta-N-O chemical bonds [51]. All binding energies of the high resolution spectra were calibrated with the C1s binding energy of 285.0 eV.

XRD spectra of as-deposited and annealed Ta_2O_5 and TaON coatings were analyzed (Figure A2). According to the XRD data, the amorphous nature of the as deposited Ta_2O_5 coatings was confirmed. Changes were detected for the Ta_20_5 coatings after the treatment at 700 °C for 15 min, as confirmed by the XRD pattern peaks which became sharper and more intense (Figure A2). The increase in the crystallinity of the coatings as a function of thermal treatment temperature was detected. After 15 min annealing at a temperature of 700 °C, the peaks typical for the formation of the crystal structure of Ta_2O_5 (001), (110), (111), as well as the peaks typical for Ta (200), were clearly identified. Further heating for 1 h led to an increase in the intensity of main peaks and the appearance of a new one (020) in the angular range of 24–72 degrees (2θ). In the case of TaON, characteristic peaks of oxynitride at angles of 27°, 33°, 36°, 38°, as well as spectra associated with the formation of TaON structure at angles in the range of 61–63 degrees (2θ) were detected after 15 min of annealing. In addition, some characteristic peaks of the nitride structure (110), (111), (220) were revealed. Subsequent annealing for an hour was resulted in the further formation of the oxynitride structure and increase in the characteristic peaks at the angles of 23°, 37°, 47°, and 67°.

3.2. The Thickness of the Coatings and Fracture Microstructure

AFM images of fractures of the investigated coatings of tantalum compounds on steel and their surface profiles are demonstrated on the Figure 3. According to these profiles, the values of the thickness for the coatings were as follows: Ta_2O_5 and TaON—1500 nm, TaN—800 nm, and Ta—500 nm, which is confirmed by the SEM data (Figure A3). In addition, the fracture sites of coatings allow a qualitative assessment of the brittleness of coatings. Thus, the fracture sites of Ta_2O_5 and TaN coatings have in their microstructure significant fragments of a "columnar" texture (arrows in Figure 3a,c), which are typical for more brittle fracture. The microstructure of TaON coatings shows strips with a thickness of 20–40 nm (arrows in Figure 3b). The layered Ta and fracture sites coatings consist of nanosized grains of 40 nm in diameter (arrows in Figure 3d). In the case of fracture, the grains are arranged in rows under the action of deformation. These structures are typical for the more plastic materials, according to the plasticity values determined by NI: the highest value for Ta (59.2%), the high for TaON (52.3%), the middle for TaN (42.3%), and Ta_2O_5 has the lowest η (27.2%).

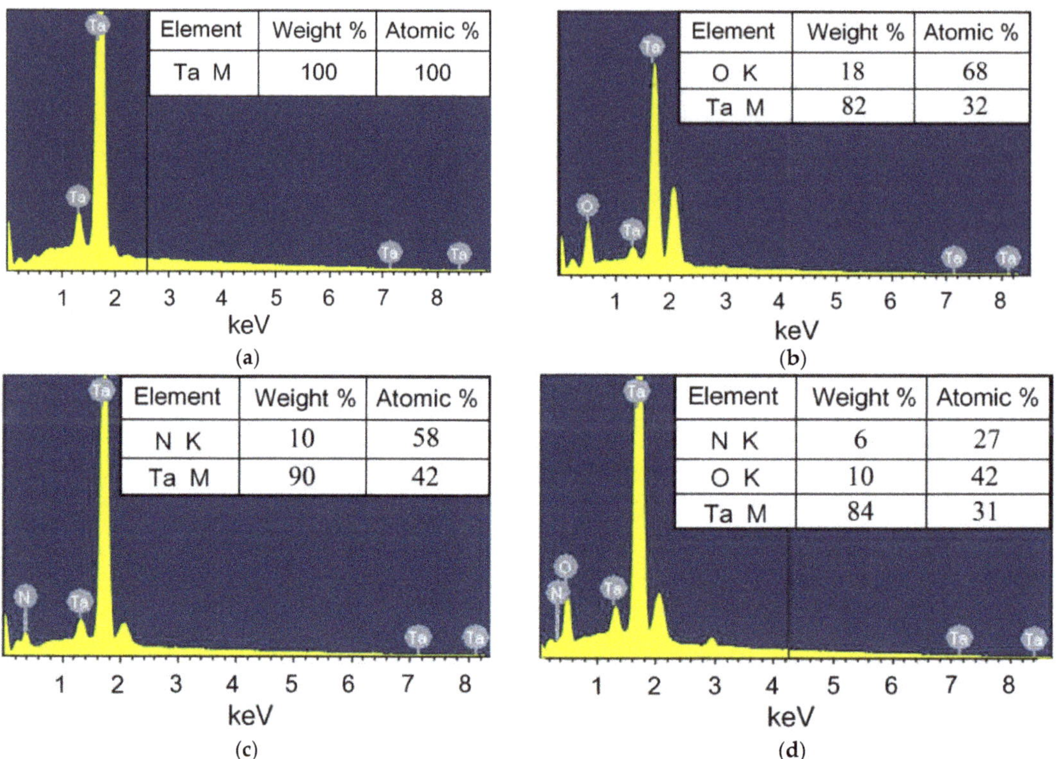

Figure 2. EDX spectra of nanostructured coatings: tantalum (**a**), tantalum oxide (**b**), tantalum nitride (**c**), and tantalum oxynitride (**d**).

3.3. *The Surface Microstructure of Coatings*

In the process of magnetron sputtering, the mechanism of growth is determined by the balance of the energy of the substrate surface, the deposited material, the energy of the material-substrate interface, and the energy of elastic stresses in the growing film. High-resolution AFM is needed to reveal the surface morphology and roughness of smooth amorphous coatings on polished substrates. According to AFM-images, the polished surface of the stainless steel has a microstructure with irregularities and protruding particles of alloying phases of 20–200 nm in diameter (arrows Figure 4a). The arithmetic mean roughness (R_a) for the steel surface on the area of 4 × 4 μm was 3.8 nm.

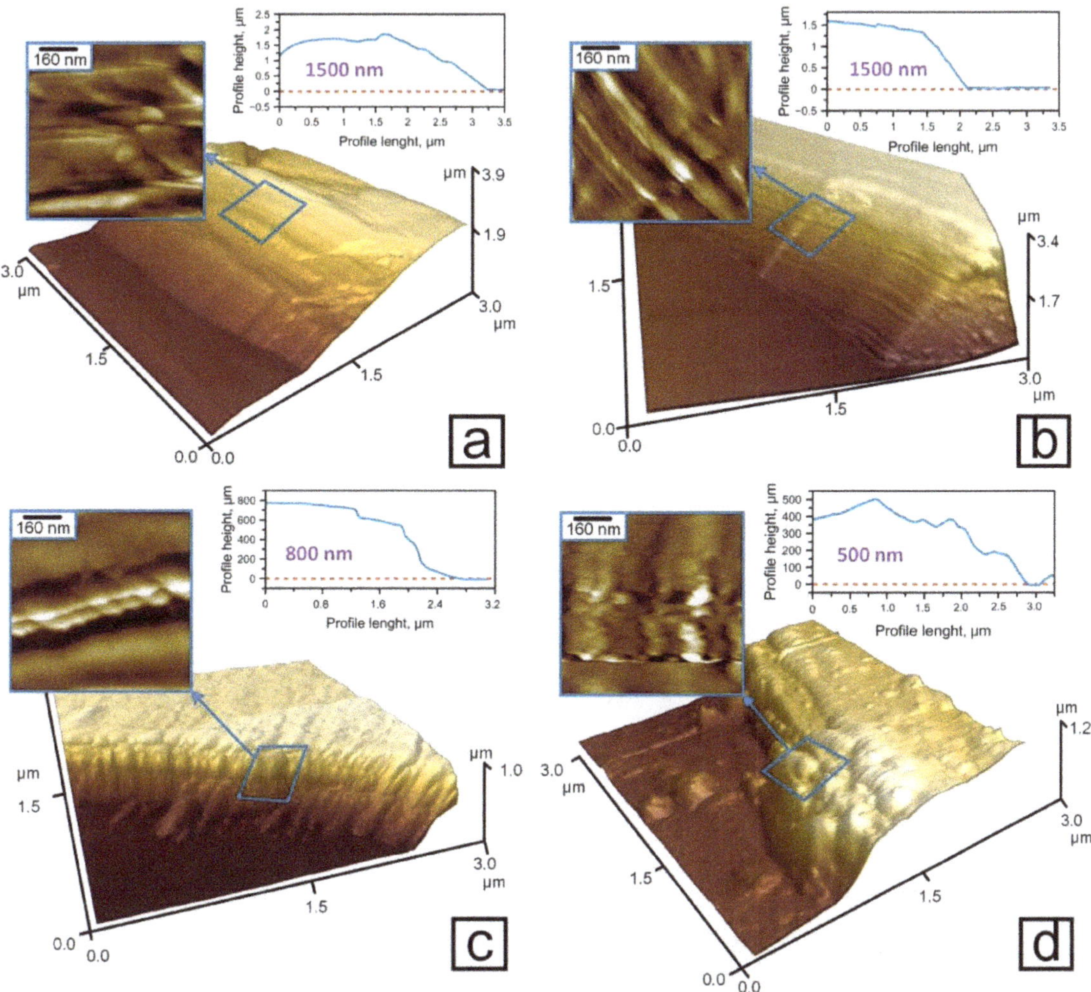

Figure 3. AFM-images of nanostructured coatings fractures: tantalum oxide (**a**), tantalum oxynitride (**b**), tantalum nitride (**c**), and tantalum (**d**).

Surface uniformity was increased after tantalum was deposited on the steel surface. On the area of 1 × 1 µm, Ta coatings have the cellular surface with ribbings that are limitative for the cells (arrows Figure 4b). The ribbings have granular microstructure with the diameter of grains of 20 nm that are shown on the area of 60 × 60 nm scanning field (inset Figure 4b). R_a of Ta coatings on the area 4 × 4 µm was 4.8 nm.

Flat islets with the height of 4–6 nm and the size of 100–500 nm appeared on the Ta_2O_5 coatings in the process of the sputtering (arrows Figure 4c). These flat islets indicate that a growth mechanism of tantalum oxide is different compared to the one of tantalum films. Ta coatings are characterized by polycrystalline growth, while tantalum oxide is characterized by layer-by-layer growth. On the area of 60 × 60 nm (inset Figure 4c), Ta_2O_5 coatings have granular microstructure with the diameter of grains of 5–20 nm. R_a of Ta_2O_5 coatings on the area 4 × 4 µm was 4.2 nm.

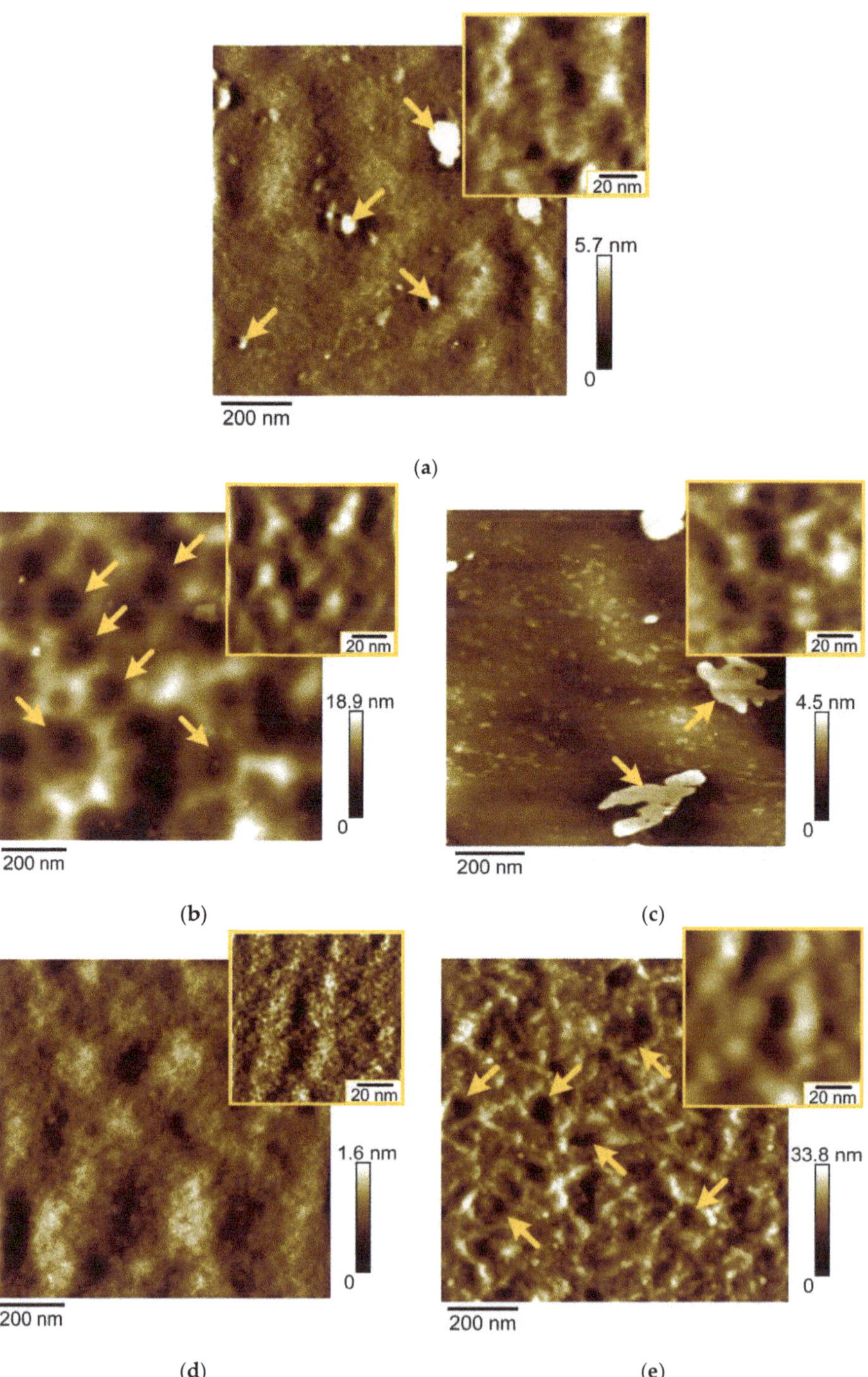

Figure 4. AFM-images of stainless steel (**a**) and the coatings: tantalum (**b**), tantalum oxide (**c**), tantalum oxynitride (**d**), and tantalum nitride (**e**).

The surface of TaON coatings is characterized by a granular structure with the small crystallites of 5 nm in diameter (inset Figure 4d). The surface of TaN coatings consists of grains of 20 nm in diameter (inset Figure 4e). These grains assembled into chains, which formed the edges of cells 200 nm in size (arrows Figure 4e). R_a values of TaN and TaON coatings on the area 4 × 4 μm were 9.5 and 5.6 nm, respectively.

The obtained morphology of the coatings of tantalum and its compounds makes it possible to explain the numerical values of the roughness.

The obtained results on AFM microstructure of the coatings of tantalum compounds are consistent with the previously published studies of Ta [52] and TaN [53] coatings which had a granular structure. The grain size is significantly smaller and is close to calculations in the research of Alishahi et al. [54].

3.4. The Mechanical Properties of Coatings

The diagrams, showing dependence of the indentation depth on indentation load for the coatings and the steel are demonstrated in the Figure 5. The shape of the curves and their position according to the Y axis show the distribution of coatings by H: the closer the curve is to the Y axis, the harder the material is. The area bounded by the curves of the approach–retraction characterizes the plastic deformation. According to the curves the significant difference in mechanical properties of coatings and substrate is visible.

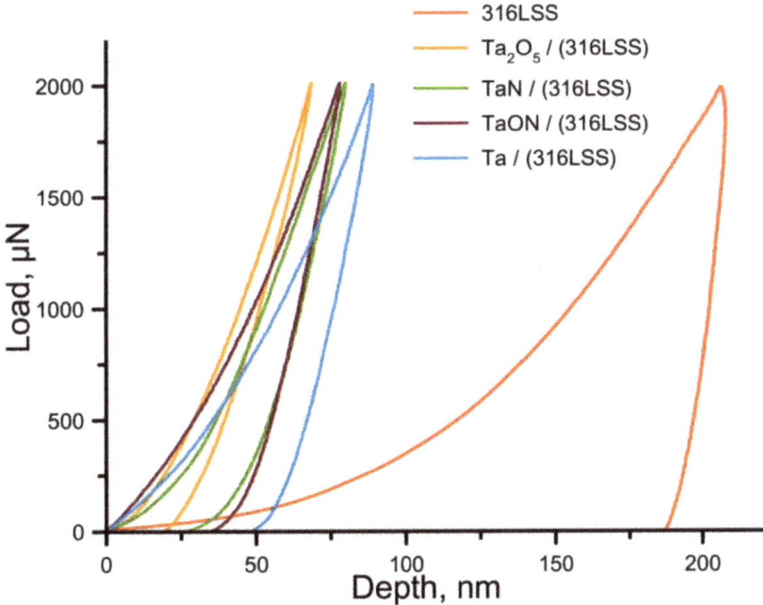

Figure 5. The dependence curves of load on the indentation depth h.

It was shown that the values of H and E of stainless steel 316 L SS were determined as 1.7 ± 0.2 and 150.0 ± 10.0 GPa, respectively. Plasticity according to the indentation curve area was the heighest—91.4% and H/E—the lowest—was 0.01. Metallic Ta coatings are characterized by the lowest H of 8.3 ± 0.2 GPa. Nonmetallic addition of O and N increased the value of H to 10.0 ± 0.3 GPa for TaN, 13.3 ± 0.6 GPa for TaON, and 16.0 ± 3.5 GPa for Ta_2O_5. The values of the elastic moduli of four studied coatings on the stainless steel substrates are about 158.0 GPa. These values are close to E of 156 GPa and H of 10 GPa for bulk pure Ta, obtained and described by Kommel et al. [55]. After the coatings' deposition, the strength of the steel substrate increased and H/E allows to estimate by how much:

0.05 for Ta, 0.06 for TaN, 0.08 for TaON, and 0.10 for Ta_2O_5. The coatings on stainless steel substrates (304 SS), formed via reactive magnetron sputtering by the authors of [56], showed lower mechanical properties for coatings of Ta compounds: for Ta_2O_5—H of 5.8 GPa, E of 135 GPa, and H/E of 0.049 and for TaON—H of 7.5 GPa, E of 119 GPa, and H/E of 0.056. The reason of the underestimated values of H and E in [56] may be the lower content of Ta in the coatings in comparison to those studied in this work. In [57], for coatings of 700–100 nm thickness, the values of E were 127 GPa for Ta and 108 GPa for Ta_2O_5, and the values of H—6.8 GPa for Ta and 8.4 GPa for Ta_2O_5. The somewhat underestimated values of the characteristics can be explained by the excessive indentation depth and the influence of the substrate.

The dependences of the mechanical properties of investigated coatings on the atomic content of Ta are shown in Figure 6. The higher oxygen and nitrogen content in the coatings based on Ta, the greater the microhardness of the surface. This tendency is associated with the introduction of oxygen and nitrogen atoms into the tantalum crystal structure, what leads to compressive residual stress in the coatings [58]. In addition, the authors of work [59] found that the higher the oxygen content, the higher the average hardness.

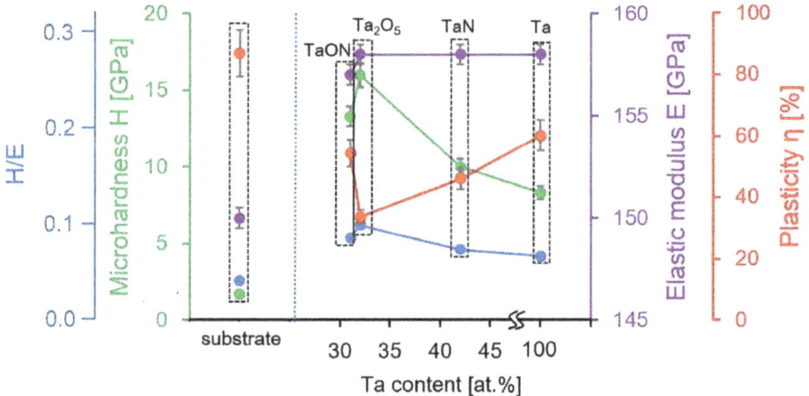

Figure 6. The dependences of H/E, H, E and η of Ta based coatings on the Ta atomic content.

On the base of the dependence of E and H on the atomic content of Ta coatings on the steel (Figure 6), the Ta content in the coatings does not affect the elastic modulus but affect the hardness. Such effect was described in [55] where two existing polytypes of tantalum α-Ta and β-Ta showed the same E of 188 GPa and the different H: 10 GPa for α-Ta and 18 GPa for β-Ta. The discrepancy in the values of E and H can be explained by the film thickness (of 100 and 300 nm), which is usually higher for thin films.

Taking into account the whole set of physical and mechanical characteristics of Ta coatings and its compounds, TaON is the optimal coating, which simultaneously has sufficient hardness and high plasticity.

3.5. The Tribological Characteristics

The results of friction and wear tests in the "plowing" regime are shown in Figure 7. Wear tracks on TaON and Ta coatings are visible only in the Friction regime. These tracks show the efficacy of Ta based coatings of wear protection of steel. Wear tracks on TaN (Figure 7d) and Ta (Figure 7e) coatings are visible only in the PeakForce Error data type. Since PeakForce QNM uses Peak Force as the feedback signal, the PeakForce Error data type is essentially the Peak Force Setpoint with the error. It is recorded simultaneously with the topography. In this mode, the boundaries of the wear mark are better visible. We had to use the error signal as the wear on these materials was very low. The topography mode did not allow it to be visualized against the background of scratches and protruding microparticles. Moreover, the error signal showed the contours of the worn material at the boundaries of the wear marks located across the scanning direction due to a significant change in the Peak Force Setpoint in this scanning area.

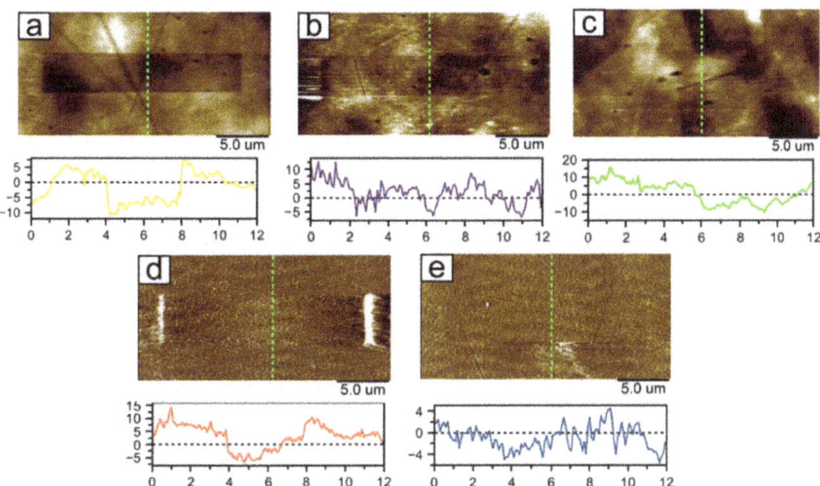

Figure 7. AFM-images of wear test results at loading of 1.164 μN, speed of 2 μm/s, and 100 cycles on the steel substrate and the tantalum coatings: steel (**a**), Ta_2O_5 (**b**), TaON (**c**), TaN (**d**), Ta (**e**).

The dependences of the CoF and *F* of the coatings during the wear in the "plowing" mode to the number of friction cycles are presented in Figure 8. Each point in the obtained diagrams is averaged over 50 scanning lines. "Teeth" in values of CoF are explained by position of the probe in each cycle with respect to the surface and AFM- photodetector: «upper–down» or «down–upper». The higher the value of CoF, the larger the "part" of probe twisting per when changing the position of the probe at reciprocating motion. CoF values were decreased after 15 cycles; it is explained by «breaking-in» to expressed in the change in the subnanometer layer of the material under the tribological load from the start of scanning and its uniform distribution over the surface. CoF in steady-state for steel was 0.448. The coatings TaON, Ta, and TaN allow decreasing CoF of the surface down to 0.444, 0.336, and 0.308, respectively (Figure 8). The deposition of Ta_2O_5 coatings on the steel substrate increases CoF to 0.780.

Specific volumetric wear allows comparing coatings of tantalum and its compounds with other materials quantitatively. All the tantalum coatings under research are capable of reducing the wear of steel (25.4×10^{-13} $m^3/N{\cdot}m$) more than twice (Figure 9). The minimum value of specific volumetric wear were recorded for the Ta coatings— 2.1×10^{-13} $m^3/N{\cdot}m$. TaN and TaON showed the middle values—4.2×10^{-13} and 6.1×10^{-13} $m^3/N{\cdot}m$. The minimum wear were determined for Ta_2O_5 coatings—11.6×10^{-13} $m^3/N{\cdot}m$. The dependences of specific volumetric wear on Ta atomic content in coatings are in good agreement with their CoF (Figure 9). The "plowing" mode allows assessing the real strength properties of the material during the friction.

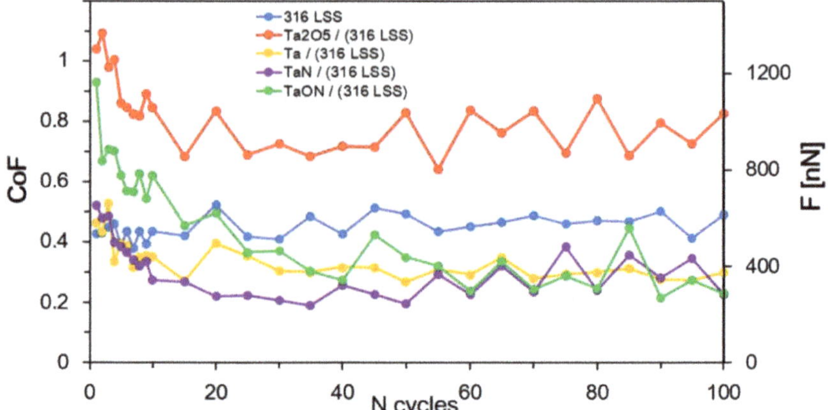

Figure 8. Dependence of the obtained friction coefficients on the number of friction cycles of the studied coatings on the steel substrates.

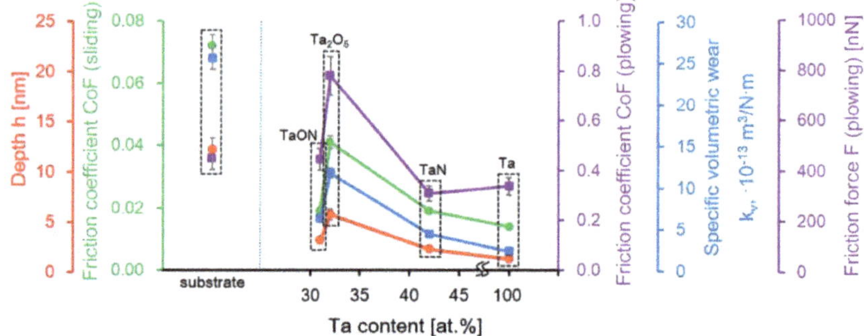

Figure 9. The dependences of h, CoF (sliding), CoF (plowing), k_v, and F of Ta based coatings on the Ta atomic content.

The influence of the adhesive forces is better characterized by the "sliding" mode [60,61]. The determined CoF for stainless steel in the "sliding" mode is 0.072 (Figure 9). After the deposition of nanostructured tantalum coatings, the CoF decreases to 0.014 (Ta) and 0.019 (TaN and TaON). CoF of Ta_2O_5 coatings in the "sliding" mode is 0.041. The dependences of CoF in the "sliding" and "plowing" modes on Ta atomic content in coatings have a similar behavior (Figure 9).

The better tribological properties of Ta coatings can be explained by its microstructure and plasticity of the coating. The low CoF values of the oxynitride film can be explained by the high roughness, and the high values of tantalum oxide, by the low roughness. Microhardness is often considered as the main characteristic to predict the wear. The Ta_2O_5 coatings with the highest H of 18 GPa and R_a of 4.2 nm showed the weakest nanotribological properties. This result can be explained by the significantly different mechanism of wear in the environment of nanofriction contact from usual classical mechanism of macrocontact.

The authors of [35] obtained TaNx coatings by the high-frequency magnetron sputtering method. Tribological researches were carried out by the nanoindentation method using a Berkovich diamond tip at a load of 5 µN. The average friction coefficient was ~0.18. The significant difference between CoF in [35] and those obtained by us for TaN (0.30) is explained by the low applied load during tribological tests of AFM. In [36], tantalum nitride films were synthesized on silicon using magnetron sputtering. A pin-on-disk (alumina ball) tribometer at 1 N load was used to obtain the coefficient of friction and the wear rate of these films. It was found that 13% N_2 films have the highest hardness, a low coefficient of friction (~0.6), and a reduced wear rate (7.09×10^{-15} $m^3/N{\cdot}m$). The discrepancy in the values of CoF is associated with an increased load and a significantly larger contact area in [35]. Wear during AFM tests is overestimated due to high mechanical stresses in the contact associated with the small radius of the tip, wherein AFM-wear makes it possible to accurately determine the difference between all compared coatings.

The used normal forces about 1 μN and a nanoscale radius of the probe (8–76 nm) lead to the creation on the surface of coatings of significant contour pressure (or mechanical stress in contact, contact pressure) of 15–65 GPa. This value is many times higher than the level of contact stresses at macro-tribotests of about 1.3–1.6 GPa. The localization of maximal mechanical stresses at nanometers depth and displacement of the material from the surface by atomic layers gives a different mechanism of the friction process under conditions of microcontact. The high η of Ta coatings allow light moving of atomic layer of materials under the probe acting.

The TaON coatings were chosen according to the strength properties because of the rather high values of H and simultaneously high η and H/E value of 0.08, then according to tribological characteristics—Ta, TaN, and TaON.

4. Conclusions

Nanostructured films of tantalum compounds were deposited on the stainless steel substrates by magnetron sputtering. It was determined that the microstructure of coatings depends on the elemental composition. All tantalum based coatings are characterized by a granular structure. Depending on the composition of the coatings, the grains vary in size from 5 to 20 nm. In some cases (Ta and TaN), the grains associate into cells. All obtained coatings have low roughness values.

The best combination of properties among the studied coatings have TaN (H of 10.0 GPa, E of 158.0 GPa and $H/E = 0.06$) and TaON (H of 13.3 GPa, E of 157.0 GPa and $H/E = 0.08$).

The tribological characteristics of obtained coatings were: TaN—CoF of 0.019 in the "sliding" mode and 0.308 in the "plowing" mode, specific volumetric wear of 4.2×10^{-13} $m^3/N{\cdot}m$, TaON—CoF of 0.019 in the "sliding" mode and 0.444 in the "plowing" mode, and specific volumetric wear of 6.1×10^{-13} $m^3/N{\cdot}m$. Thus, deposition of TaN decreases specific volumetric wear of steel by more than 6 times, TaON—by more than 4 times. These tribological characteristics allow reducing of platelets on the surface of stents, and thereby, preventing the formation of a blood clot.

Thus, TaN and TaON coatings, which have a special complex of mechanical and tribological properties, can be used as upper layer for stainless steel stents.

Author Contributions: Conceptualization, G.M. and S.C.; methodology, G.M. and A.Z.; validation, S.C.; investigation, T.K., V.L., A.P. and A.Z.; resources, V.S. and S.Y.; writing—original draft preparation, G.M. and A.P.; writing—review and editing, T.K., W.S. and E.S.; visualization, V.L.; supervision, S.A. and E.S.; project administration, T.K.; funding acquisition, T.K., A.P., V.S., S.A. and E.S. All authors have read and agreed to the published version of the manuscript.

Funding: This research was supported by the grant of the Belarusian Republican Foundation for Fundamental Research (BRFFR) No. T20UKA-030 and the grant of the National Academy of Sciences of Ukraine No. 08-03-20, and by the World Federation of Scientists. E.S. and S.A. were supported by the grant of the Government of the Russian Federation, grant number 14.Z50.31.0046.

Conflicts of Interest: The authors declare no conflict of interest.

Appendix A

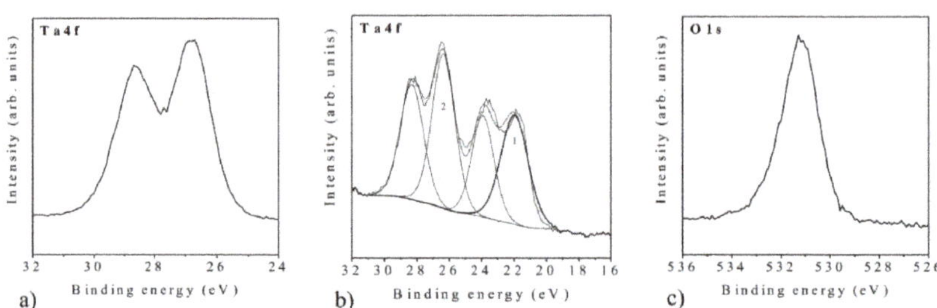

Figure A1. XPS spectra of Ta-based coatings deposited on the steel substrates: Ta^{+5} for Ta_2O_5 (**a**), Ta^{+5} and Ta^0 for Ta/Ta_2O_5 (**b**), O1s for Ta_2O_5 (**c**) [49].

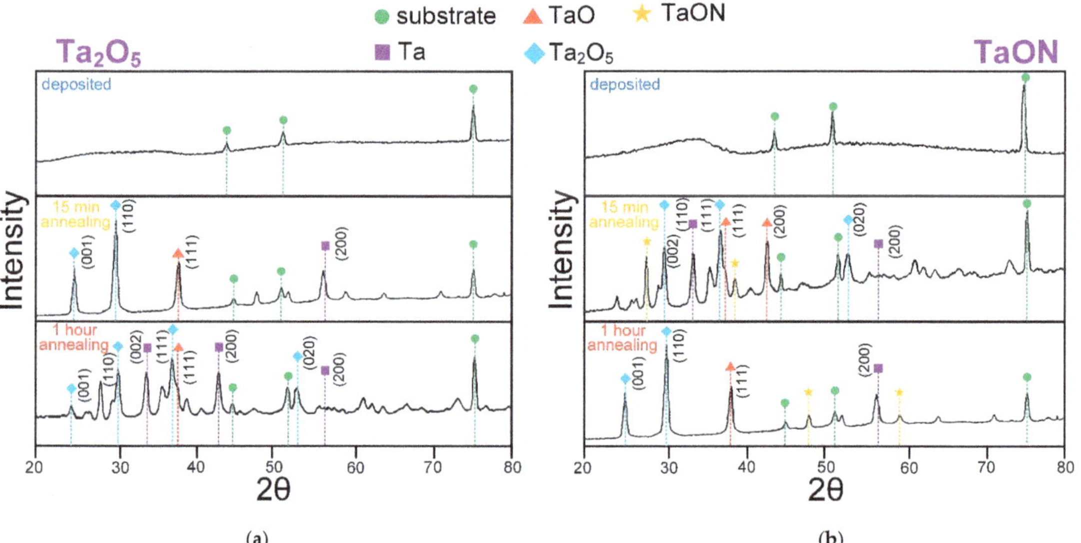

Figure A2. Diffraction X-ray profiles of the Ta_2O_5 (**a**) and TaON (**b**) coatings after application and annealing at a temperature of 973 K for 15 min and 1 h.

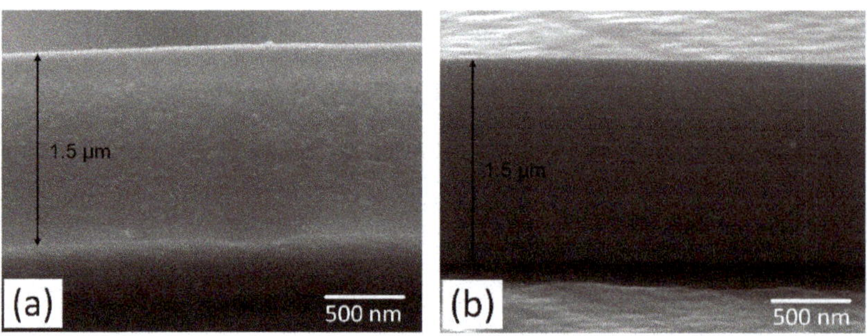

Figure A3. *Cont.*

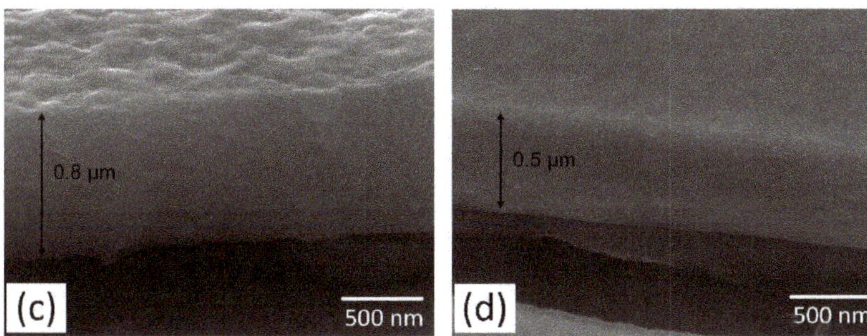

Figure A3. SEM-images of nanostructured coatings cross-sections: tantalum oxide (**a**), tantalum oxynitride (**b**), tantalum nitride (**c**), tantalum (**d**).

References

1. Mani, G.; Feldman, M.D.; Patel, D.; Agrawal, C.M. Coronary stents: A materials perspective. *Biomaterials* **2007**, *28*, 1689–1710. [CrossRef]
2. Papirov, I.I. *Materials of Medical Stents: Review*; NSC KIPT: Kharkov, Ukraine, 2010. (In Russian)
3. Rudnev, V.S.; Medkov, M.A.; Kilin, K.N.; Ustinov, A.Y.; Belobeletskaya, M.V.; Steblevskaya, N.I.; Mutylina, I.N.; Zherebtsov, T.O. Plasma-electrolytic formation of Ta-containing oxide coatings on titanium. Their composition and properties. *Prot. Met. Phys. Chem. Surf.* **2013**, *49*, 717–723. [CrossRef]
4. Matsuno, H. Biocompatibility and osteogenesis of refractory metal implants, titanium, hafnium, niobium, tantalum and rhenium. *Biomaterials* **2001**, *22*, 1253–1262. [CrossRef]
5. Leng, Y.; Chen, J.; Yang, P.; Sun, H.; Wang, J.; Huang, N. The biocompatibility of the tantalum and tantalum oxide films synthesized by pulse metal vacuum arc source deposition. *Nucl. Instrum. Methods Phys. Res. Sect. B* **2006**, *242*, 30–32. [CrossRef]
6. Matsumoto, A.H.; Teitelbaum, G.P.; Barth, K.H.; Carvlin, M.J.; Savin, M.A.; Strecker, E.P. Tantalum vascular stents: In Vivo evaluation with MR imaging. *Radiology* **1989**, *170*, 753–755. [CrossRef]
7. Wang, C.; Li, X.; Tong, C.; Cai, A.; Guo, H.; Yin, H. Preparation, In Vitro bioactivity and osteoblast cell response of Ca-Ta_2O_5 nanorods on tantalum. *Surf. Coat. Technol.* **2020**, *391*, 125701. [CrossRef]
8. Cardonne, S.M.M.; Kumar, P.; Michaluk, C.A.A.; Schwartz, H.D.D. Tantalum and its alloys. *Int. J. Refract. Met. Hard Mater.* **1995**, *13*, 187–194. [CrossRef]
9. Catania, P.; Doyle, J.P.; Cuomo, J.J. Low resistivity body-centered cubic tantalum thin films as diffusion barriers between copper and silicon. *J. Vac. Sci. Technol. A* **1992**, *10*, 3318–3321. [CrossRef]
10. Kim, S.; Cha, B. Deposition of tantalum nitride thin films by D.C. magnetron sputtering. *Thin Solid Films* **2005**, *475*, 202–207. [CrossRef]
11. Gladczuk, L.; Patel, A.; Paur, C.S.; Sosnowski, M. Tantalum films for protective coatings of steel. *Thin Solid Films* **2004**, *467*, 150–157. [CrossRef]
12. Natishan, P.; McCafferty, E.; Puckett, P.; Michel, S. Ion beam assisted deposited tantalum oxide coatings on aluminum. *Corros. Sci.* **1996**, *38*, 1043–1049. [CrossRef]
13. Hino, T. Effect of Droplets on Corrosion Resistance of Tantalum Oxide Films Fabricated by PLD. *J. Laser Micro Nanoeng.* **2011**, *6*, 10–14. [CrossRef]
14. Khemasiri, N.; Jessadaluk, S.; Chananonnawathorn, C.; Vuttivong, S.; Lertvanithphol, T.; Horprathum, M.; Eiamchai, P.; Patthanasettakul, V.; Klamchuen, A.; Pankiew, A.; et al. Optical band engineering of metal-oxynitride based on tantalum oxide thin film fabricated via reactive gas-timing RF magnetron sputtering. *Surf. Coat. Technol.* **2016**, *306*, 346–350. [CrossRef]
15. Bah, S.T.; Ba, C.O.; D'Auteuil, M.; Ashrit, P.; Sorelli, L.; Vallée, R.; Soreli, L. Fabrication of TaOxNy thin films by reactive ion beam-assisted ac double magnetron sputtering for optical applications. *Thin Solid Films* **2016**, *615*, 351–357. [CrossRef]
16. Venkataraj, S.; Kittur, H.; Drese, R.; Wuttig, M. Multi-technique characterization of tantalum oxynitride films prepared by reactive direct current magnetron sputtering. *Thin Solid Films* **2006**, *514*, 1–9. [CrossRef]
17. Jong, C.-A.; Chin, T.S. Optical characteristics of sputtered tantalum oxynitride Ta(N,O) films. *Mater. Chem. Phys.* **2002**, *74*, 201–209. [CrossRef]
18. Navid, A.; Hodge, A. Nanostructured alpha and beta tantalum formation—Relationship between plasma parameters and microstructure. *Mater. Sci. Eng. A* **2012**, *536*, 49–56. [CrossRef]
19. Shiri, S.; Zhang, C.; Odeshi, A.; Yang, Q. Growth and characterization of tantalum multilayer thin films on CoCrMo alloy for orthopedic implant applications. *Thin Solid Films* **2018**, *645*, 405–408. [CrossRef]
20. Liu, L.L.; Xu, J.; Lu, X.; Munroe, P.; Xie, Z.-H. Electrochemical Corrosion Behavior of Nanocrystalline β-Ta Coating for Biomedical Applications. *ACS Biomater. Sci. Eng.* **2016**, *2*, 579–594. [CrossRef]

21. Pham, V.-H.; Lee, S.-H.; Li, Y.; Kim, H.-E.; Shin, K.-H.; Koh, Y.-H. Utility of tantalum (Ta) coating to improve surface hardness In Vitro bioactivity and biocompatibility of Co–Cr. *Thin Solid Films* **2013**, *536*, 269–274. [CrossRef]
22. Li, X.; Wang, L.; Yu, X.; Feng, Y.; Wang, C.; Yang, K.; Su, D. Tantalum coating on porous Ti6Al4V scaffold using chemical vapor deposition and preliminary biological evaluation. *Mater. Sci. Eng. C* **2013**, *33*, 2987–2994. [CrossRef]
23. Moreira, H.; Barbosa, A.A.C.; Marques, S.M.; Sampaio, P.; Carvalho, S. Evaluation of cell activation promoted by tantalum and tantalum oxide coatings deposited by reactive DC magnetron sputtering. *Surf. Coat. Technol.* **2017**, *330*, 260–269. [CrossRef]
24. Hee, A.C.; Cao, H.; Zhao, Y.; Jamali, S.S.; Bendavid, A.; Martin, P.J. Cytocompatible tantalum films on Ti6Al4V substrate by filtered cathodic vacuum arc deposition. *Bioelectrochemistry* **2018**, *122*, 32–39. [CrossRef] [PubMed]
25. Gruen, T.A.; Poggie, R.A.; Lewallen, D.G.; Hanssen, A.D.; Lewis, R.J.; O'Keefe, T.J.; Stulberg, S.D.; Sutherland, C.J. Radiographic evaluation of a monoblock acetabular component: A multicenter study with 2-to 5-year results. *J. Arthroplast.* **2005**, *20*, 369–378. [CrossRef]
26. Maccauro, G.; Iommetti, P.R.; Muratori, F.; Raffaelli, L.; Manicone, P.F.; Fabbrii, C. An overview about biomedical applications of micron and nano size tantalum. *Recent Pat. Biotechnol.* **2009**, *3*, 157–165. [CrossRef]
27. Levine, B.R.; Sporer, S.; Poggie, R.A.; Della Valle, C.J.; Jacobs, J.J. Experimental and clinical performance of porous tantalum in orthopedic surgery. *Biomaterials* **2006**, *27*, 4671–4681. [CrossRef]
28. Khan, F.A.; Rose, P.S.; Yanagisawa, M.; Lewallen, D.G.; Sim, F.H. Surgical Technique: Porous Tantalum Reconstruction for Destructive Nonprimary Periacetabular Tumors. *Clin. Orthop. Relat. Res.* **2012**, *470*, 594–601. [CrossRef] [PubMed]
29. Yang, C.; Li, J.; Zhu, C.; Zhang, Q.; Yu, J.; Wang, J.; Wang, Q.; Tang, J.; Zhou, H.; Shen, H. Advanced antibacterial activity of biocompatible tantalum nanofilm via enhanced local innate immunity. *Acta Biomater.* **2019**, *89*, 403–418. [CrossRef] [PubMed]
30. Silva, R.A.; Walls, M.; Rondot, B.; Belo, M.D.C.; Guidoin, R. Electrochemical and microstructural studies of tantalum and its oxide films for biomedical applications in endovascular surgery. *J. Mater. Sci. Mater. Electron.* **2002**, *13*, 495–500. [CrossRef]
31. Papirov, I.I.; Tikhonovsky, M.A.; Shokurov, V.S.; Pikalov, A.I.; Sivtsov, V.S.; Storozhilov, G.E.; Emljaninova, T.G.; Mazin, A.I.; Shkuropatenko, V.A. Reception of fine-grained tantalum. *Bull. Kharkov Univ.* **2005**, *664*, 99–102.
32. Liu, L.; Xu, J.; Jiang, S. Nanocrystalline β-Ta Coating Enhances the Longevity and Bioactivity of Medical Titanium Alloys. *Metals* **2016**, *6*, 221. [CrossRef]
33. Ding, M.; Wang, B.; Li, L.; Zheng, Y. Preparation and characterization of TaCxN1−x coatings on biomedical 316L stainless steel. *Surf. Coat. Technol.* **2010**, *204*, 2519–2526. [CrossRef]
34. Rodionov, I.V.; Proskuryakov, V.I.; Koshuro, V.A. Morphology of oxide films on the tantal after thermal air treatment. *Mod. Mater. Technol. Technol.* **2018**, *17*, 58–65.
35. Dai, W.; Shi, Y. Effect of Bias Voltage on Microstructure and Properties of Tantalum Nitride Coatings Deposited by RF Magnetron Sputtering. *Coatings* **2021**, *11*, 911. [CrossRef]
36. Zaman, A.; Shen, Y.; Meletis, E.I. Microstructure and Mechanical Property Investigation of TaSiN Thin Films Deposited by Reactive Magnetron Sputtering. *Coatings* **2019**, *9*, 338. [CrossRef]
37. Kumar, M.; Kumari, N.; Kumar, V.P.; Karar, V.; Sharma, A.L. Determination of optical constants of tantalum oxide thin film deposited by electron beam evaporation. *Mater. Today Proc.* **2018**, *5*, 3764–3769. [CrossRef]
38. Cheviot, M.; Gouné, M.; Poulon-Quintin, A. Monitoring tantalum nitride thin film structure by reactive RF magnetron sputtering: Influence of processing parameters. *Surf. Coat. Technol.* **2015**, *284*, 192–197. [CrossRef]
39. Riekkinen, T.; Molarius, J.; Laurila, T.; Nurmela, A.; Suni, I.; Kivilahti, J. Reactive sputter deposition and properties of TaxN thin films. *Microelectron. Eng.* **2002**, *64*, 289–297. [CrossRef]
40. Petrovskaya, A.S.; Lapitskaya, V.; Melnikova, G.; Kuznetsova, T.A.; Chizhik, S.A.; Zykova, A.V.; Safonov, V.I. Hydrophilic properties of surface of nanostructured tantalum films and its oxynitride compounds. *J. Phys. Conf. Ser.* **2019**, *1281*, 012061. [CrossRef]
41. Yakovin, S.; Dudin, S.; Zykov, A.; Farenik, V. Integral cluster set-up for complex compound composites synthesis. *Probl. At. Sci. Technol. Ser. Plasma Phys.* **2011**, *1*, 152–154.
42. Yakovin, S.D.; Dudin, S.V.; Zykov, A.V.; Farenik, V.I.; Karazin, V.N. Synthesis of thin-film Ta_2O_5 coatings by reactive magnetron sputtering. *Probl. At. Sci. Technol. Ser. Plasma Phys.* **2016**, *6*, 248.
43. Dudin, S.; Yakovin, S.; Zykov, A.; Yefymenko, N. Optical and mass spectra from reactive plasma at magnetron deposition of tantalum oxynitride. *Probl. At. Sci. Technol. Ser. Plasma Phys.* **2021**, *1*, 122.
44. Chizhik, S.A.; Rymuza, Z.; Chikunov, V.V.; Kuznetsova, T.A.; Jarzabek, D. Micro-and nanoscale testing of tribomechanical properties of surfaces. In *Recent Advances in Mechatronics, Proceedings of the 7th International Conference Mechatronics 2007, Warsaw University of Technology, Warsaw, Poland, 19–21 September 2007*; Jablonski, R., Turkowski, M., Szewczyk, R., Eds.; Springer: Berlin/Heidelberg, Germany, 2007; pp. 541–545.
45. Hertzian Contact Stress Calculator. 2013. Available online: http://www.amesweb.info/HertzianContact/HertzianContact.aspx/ (accessed on 9 October 2018).
46. Budynas, R.G.; Nisbett, J.K. *Shigley's Mechanical Engineering Design*, 10th ed.; McGraw-Hill: New York, NY, USA, 2014.
47. Kuznetsova, T.; Lapitskaya, V.; Chizhik, S.; Kuprin, A.; Tolmachova, G.; Ovcharenko, V.; Gilewicz, A.; Lupicka, O.; Warcholinski, B. Friction and Wear of Cr-O-N Coatings Characterized by Atomic Force Microscopy. *Tribol. Ind.* **2019**, *41*, 274–285. [CrossRef]

48. Zykova, A.; Safonov, V.; Yakovin, S.; Dudin, S.; Melnikova, G.; Petrovskaya, A.; Tolstaya, T.; Kuznetsova, T.; Chizhik, S.A.; Donkov, N. Comparative analysis of platelets adhesion to the surface of Ta-based ceramic coatings deposited by magnetron sputtering. *J. Phys. Conf. Ser.* **2020**, *1492*. [CrossRef]
49. Donkov, N.; Walkowicz, J.; Zavaleyev, V.; Zykova, A.; Safonov, V.; Dudin, S.; Yakovin, S. Mechanical properties of tantalum-based ceramic coatings for biomedical applications. *J. Phys. Conf. Ser.* **2018**, *992*, 012034. [CrossRef]
50. Chun, W.-J.; Ishikawa, A.; Fujisawa, H.; Takata, T.; Kondo, J.N.; Hara, M.; Kawai, M.; Matsumoto, Y.; Domen, K. Conduction and Valence Band Positions of Ta_2O_5, TaON, and Ta_3N_5 by UPS and Electrochemical Methods. *ChemInform* **2003**, *688*, 137403. [CrossRef]
51. Matizamhuka, W.; Sigalas, I.; Herrmann, M. Synthesis, sintering and characterisation of TaON materials. *Ceram. Int.* **2008**, *34*, 1481–1486. [CrossRef]
52. Abadias, G.; Colin, J.; Tingaud, D.; Djemia, P.; Belliard, L.; Tromas, C. Elastic properties of α- and β-tantalum thin films. *Thin Solid Films* **2019**, *688*, 137403. [CrossRef]
53. Kim, I.-S.; Cho, M.-Y.; Lee, D.-W.; Ko, P.-J.; Shin, W.H.; Park, C.; Oh, J.-M. Degradation behaviors and failure of magnetron sputter deposited tantalum nitride. *Thin Solid Films* **2020**, *697*, 137821. [CrossRef]
54. Alishahi, M.; Mahboubi, F.; Khoie, S.M.M.; Aparicio, M.; Lopez-Elvira, E.; Méndez, J.; Gago, R. Structural properties and corrosion resistance of tantalum nitride coatings produced by reactive DC magnetron sputtering. *RSC Adv.* **2016**, *6*, 89061–89072. [CrossRef]
55. Kommel, L.; Shahreza, B.O.; Mikli, V. Microstructure and physical-mechanical properties evolution of pure tantalum processed with hard cyclic viscoplastic deformation. *Int. J. Refract. Met. Hard Mater.* **2019**, *83*, 104983. [CrossRef]
56. Hirpara, J.; Chawla, V.; Chandra, R. Investigation of tantalum oxynitride for hard and anti-corrosive coating application in diluted hydrochloric acid solutions. *Mater. Today Commun.* **2020**, *23*, 101113. [CrossRef]
57. Liu, Y.; Lin, I.-K.; Zhang, X. Mechanical properties of sputtered silicon oxynitride films by nanoindentation. *Mater. Sci. Eng. A* **2008**, *489*, 294–301. [CrossRef]
58. Marinelli, G.; Martina, F.; Ganguly, S.; Williams, S. Microstructure, hardness and mechanical properties of two different unalloyed tantalum wires deposited via wire + arc additive manufacture. *Int. J. Refract. Met. Hard Mater.* **2019**, *83*, 104974. [CrossRef]
59. Kaliaraj, G.S.; Kumar, N. Oxynitrides decorated 316L SS for potential bioimplant application. *Mater. Res. Express* **2018**, *5*, 036403. [CrossRef]
60. Lapitskaya, V.; Kuznetsova, T.; Chizhik, S.A.; Sudzilouskaya, K.A.; Kotov, D.A.; Nikitiuk, S.A.; Zaparozhchanka, Y.V. Lateral force microscopy as a method of surface control after low-temperature plasma treatment. *IOP Conf. Ser. Mater. Sci. Eng.* **2018**, *443*, 012019. [CrossRef]
61. Grzywacz, H.; Milczarek, M.; Jenczyk, P.; Dera, W.; Michałowski, M.; Jarząbek, D.M. Quantitative measurement of nanofriction between PMMA thin films and various AFM probes. *Measurement* **2021**, *168*, 108267. [CrossRef]

 nanomaterials

MDPI

Article

Fracture and Embedment Behavior of Brittle Submicrometer Spherical Particles Fabricated by Pulsed Laser Melting in Liquid Using a Scanning Electron Microscope Nanoindenter

Daizen Nakamura [1], Naoto Koshizaki [1,*], Nobuyuki Shishido [2], Shoji Kamiya [3] and Yoshie Ishikawa [4,*]

1 Graduate School of Engineering, Hokkaido University, Sapporo 060-8628, Japan; nakamura.daizen@frontier.hokudai.ac.jp
2 Faculty of Science and Engineering, Kindai University, Higashiosaka 577-8502, Japan; shishido@mech.kindai.ac.jp
3 Department of Electrical and Mechanical Engineering, Nagoya Institute of Technology, Nagoya 466-8555, Japan; kamiya.shoji@nitech.ac.jp
4 Research Institute for Advanced Electronics and Photonics, National Institute of Advanced Industrial Science and Technology (AIST), Tsukuba 305-8565, Japan
* Correspondence: koshizaki.naoto@eng.hokudai.ac.jp (N.K.); ishikawa.yoshie@aist.go.jp (Y.I.)

Citation: Nakamura, D.; Koshizaki, N.; Shishido, N.; Kamiya, S.; Ishikawa, Y. Fracture and Embedment Behavior of Brittle Submicrometer Spherical Particles Fabricated by Pulsed Laser Melting in Liquid Using a Scanning Electron Microscope Nanoindenter. *Nanomaterials* **2021**, *11*, 274. https://doi.org/10.3390/nano11092201

Academic Editors: Gregory M. Odegard and Yang-Tse Cheng

Received: 9 July 2021
Accepted: 23 August 2021
Published: 26 August 2021

Publisher's Note: MDPI stays neutral with regard to jurisdictional claims in published maps and institutional affiliations.

Abstract: Generally, hard ceramic carbide particles, such as B_4C and TiC, are angulated, and particle size control below the micrometer scale is difficult owing to their hardness. However, submicrometer particles (SMPs) with spherical shape can be experimentally fabricated, even for hard carbides, via instantaneous pulsed laser heating of raw particles dispersed in a liquid (pulsed laser melting in liquid). The spherical shape of the particles is important for mechanical applications as it can directly transfer the mechanical force without any loss from one side to the other. To evaluate the potential of such particles for mechanical applications, SMPs were compressed on various substrates using a diamond tip in a scanning electron microscope. The mechanical behaviors of SMPs were then examined from the obtained load–displacement curves. Particles were fractured on hard substrates, such as SiC, and fracture strength was estimated to be in the GPa range, which is larger than their corresponding bulk bending strength and is 10–40% of their ideal strength, as calculated using the density-functional theory. Contrarily, particles can be embedded into soft substrates, such as Si and Al, and the local hardness of the substrate can be estimated from the load–displacement curves as a nanoscale Brinell hardness measurement.

Keywords: pulsed laser melting in liquid; spherical submicrometer particles; particle fracture; particle embedment; Brinell hardness; titanium carbide

1. Introduction

Spherical submicrometer particles (SMPs) are larger than well-studied nanoparticles (NPs) but have been an interesting research topic owing to their mechanical applications. The spherical shape of the particles is important, as it can directly transfer the mechanical force without any loss from one side to the other. Submicrometer size is another important factor because suitable fabrication techniques via a top-down approach, such as milling, or a bottom-up approach, such as the nucleation process, are currently unavailable. However, the use of spacers, milling agents, lubricant fillers, etc., has been proposed and is in demand [1,2].

Ductile SMPs made of polymers or glasses are relatively easy to fabricate and are commercially available. Spherical SMPs composed of NP aggregates have also been reported for TiO_2 [3]. However, these SMPs are polycrystalline or porous, and, therefore, their mechanical properties are not good, owing to the high density of grain boundaries or nanopores, which act as defect sources [4].

Our group reported a new fabrication technique for various spherical SMPs, which is called pulsed laser melting in liquid (PLML) [5–9]. In this process, spherical SMPs are synthesized by applying pulsed laser irradiation with relatively weak laser fluence onto raw colloidal NPs dispersed in a liquid medium. Raw aggregated or agglomerated NPs in a liquid medium are selectively heated and melted via unfocused pulsed laser irradiation to form larger molten droplets due to the temperature increase over the melting point. Subsequently, the particles are quenched with the surrounding liquid to produce crystalline non-porous spherical SMPs [10,11].

PLML appears similar to the well-studied pulsed laser ablation in liquid (PLAL) for NP fabrication [12–14], as both are laser processes in liquid medium. However, PLML is not a plasma process like PLAL but a thermal process at high temperatures, ranging from 2000 to 4000 K, due to the difference in the applied laser fluence [15,16]. Therefore, non-porous SMPs of high-temperature materials, such as B, W, TiO_2, and ZnO, can be fabricated via a simple transient melting process, and B_4C SPMs can be reactively fabricated via laser melting of raw B particles in ethanol (as a carbon source) [5,6]. Thus, SMPs obtained by PLML are unique in fabrication as well as possible mechanical applications.

However, only a few studies on mechanical properties of hard, brittle, crystalline SMPs have been reported so far [4,17]. In our previous report, brittle non-porous SMPs of B_4C and TiO_2 were fabricated via PLML, and their mechanical properties were measured using a nanoindenter equipped in a SEM [4]. In that experiment, the particles were placed on a hard SiC substrate and pressed using a diamond indenter. Their fracture strengths were 40–50% of the ideal strength calculated using the density functional theory (DFT) and somewhat stronger than the bending strength of bulk B_4C and TiO_2. Thus, SMPs obtained via PLML were harder compared with those obtained using other techniques, such as chemical methods, and are promising for various mechanical applications.

Here, by extending our previous studies on the mechanical property measurements of SMPs fabricated via PLML [4], indentation tests of various combinations of hard and brittle SMPs fabricated via PLML and single-crystal substrates were conducted using a diamond indenter equipped in an SEM. B_4C, B and newly fabricated TiC were used as hard and brittle SMPs. Not only a hard substrate, such as SiC, but also softer substrates, such as Si and Al, were used to compare the mechanical behaviors using an embedment process at submicrometer scale. In particular, the possibilities of nano-Brinell hardness measurement and local surface enforcement were explored by analyzing the particle embedment process.

2. Materials and Methods

2.1. Particle Fabrication

Brittle SMPs of B_4C, B, and TiC were fabricated via PLML and used for mechanical tests. The fabrication procedures of B_4C and B have been previously reported [5,6]. Briefly, raw B NPs (Sigma-Aldrich, 7440-42-8, nominal size < 100 nm, Tokyo, Japan) were dispersed in deionized water for B SMPs and in ethanol for B_4C SMPs and then irradiated using a Nd:YAG laser (Continuum, Powerlite Precision 8000, pulse width: 7 ns, wavelength: 355 nm, pulse frequency: 10 Hz) with a fluence of 200 mJ pulse^{-1} cm^{-2} for 10 min at room temperature. During laser irradiation, a magnetic stirrer was used to stir the liquid for agitation to disperse the raw particles. For B_4C SMP fabrication, unreacted B was dissolved in nitric acid and removed. Common byproducts of boric acid for both processes were dissolved in water and removed. For TiC SMPs, TiC NPs (Sigma-Aldrich, 636967, <200 nm, Tokyo, Japan) were used as raw particles. A 532 nm Nd:YAG laser was irradiated at 150 mJ pulse^{-1} cm^{-2} in ethanol for 10 min and toluene for 120 min. The particles in toluene required a longer ultrasonication time for pre-dispersion (60 min) compared with those in ethanol.

2.2. Characterization

The morphology of the obtained spherical SMPs following laser irradiation was observed using a field emission scanning electron microscope (FE-SEM; JEOL JSM-6500F,

Akishima, Japan). The crystallinity of the obtained particles was analyzed from X-ray diffraction patterns (XRD; Rigaku SmartLab, Akishima, Japan). The composition profile within individual particles was analyzed via energy-dispersive X-ray analysis (EDX) equipped with an aberration-corrected scanning transmission electron microscope (STEM; FEI Titan Cubed G2 60–300, Tokyo, Japan). The crystallinities of individual particles were measured via the electron diffraction technique using a high-resolution TEM (HR-TEM; JEOL JEM-ARM1300, Akishima, Japan).

2.3. Observation of Fracture and Embedment Behavior of SMPs

An indentation device (Hysitron PI SEM PicoIndenter) installed in a SEM (JEOL JIB-4600 F) was utilized for compressive fracture and embedment tests of the SMPs. The indenter speed for all compression tests was fixed at 12.5 nm s^{-1}. Further detailed experimental conditions and the typical arrangement of the particles, substrate, and indenter tip were the same as those described in a previous report [4]. The diamond indenter tip was flattened to a 1 μm square, using a focused ion beam installed on the SEM, for the facile operation and observation of the particles during mechanical tests. This helped to ensure good reproducibility and repeatability. The substrates used for the mechanical tests of the particles were single crystals of SiC, TiO_2, and $SrTiO_3$, and polycrystals of Al (A5052), which have to be sufficiently flat and electroconductive for SEM observation during indentation. The particles must be sparsely distributed on the substrate to avoid compression of multiple particles with the indenter tip during measurements.

3. Results and Discussion

3.1. Fabrication of TiC SMPs by PLML

Figure 1 shows the SEM images of raw TiC NPs and particles fabricated via PLML in ethanol and toluene. Spherical particles were successfully fabricated in both solvents, forming SMPs with a similar size range from 300 to 500 nm. Figure 2 shows XRD patterns of the particles in Figure 1. Raw particles and particles obtained in toluene were pure TiC, whereas those prepared in ethanol contained TiO_2 rutile phase, which was probably induced by reaction with oxygen in ethanol. Figure 3 shows the compositional line scans of particles prepared in ethanol and toluene obtained via aberration-corrected STEM. The data indicate that the particles fabricated in ethanol were oxidized at the surface and those in toluene were nearly oxygen-free. The electron diffraction technique using high-resolution TEM (not shown here) confirmed that the particles fabricated in toluene were single crystalline and those in ethanol were polycrystalline.

Figure 1. TiC SMP fabrication from TiC NPs via PLML. (**a**) SEM image of raw TiC NPs (<200 nm). SMPs obtained by 532 nm Nd:YAG laser irradiation at 150 mJ pulse $^{-1}$ cm^{-2} (**b**) in ethanol for 10 min and (**c**) in toluene for 120 min.

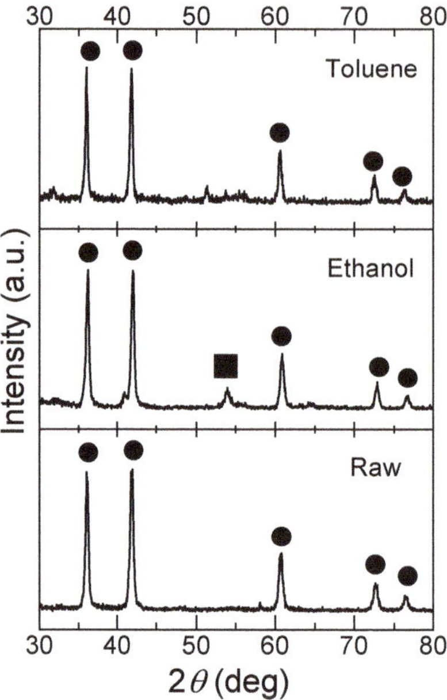

Figure 2. XRD patterns of raw TiC NPs and SMPs obtained via 532 nm Nd:YAG laser irradiation in ethanol and toluene. Circles indicate the TiC peaks, and the square indicates the TiO_2 rutile phase.

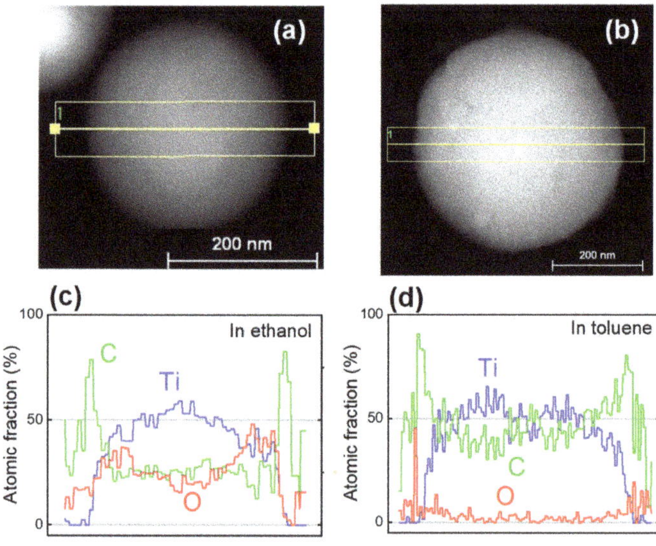

Figure 3. STEM images of TiC SMPs fabricated via PLML (**a**) in ethanol and (**b**) in toluene. The laser irradiation conditions were 532 nm Nd:YAG laser at 150 mJ pulse^{-1} cm^{-2} for 10 min in ethanol and 120 min in toluene. Compositional line scans of corresponding single particles obtained (c) in ethanol and (**d**) in toluene.

3.2. Fracture Strength of TiC SMPs Prepared by PLML

We previously reported fracture tests of B_4C SMPs on SiC substrates using diamond tip indentation [4]. The fracture strength of the B_4C SMPs was calculated to be 6–12 GPa, assuming tensile fracture at the center of a particle. This value was rather large compared with the typical bending strength by tensile fracture of a B_4C sintered body (0.3–0.9 GPa) and was 18–38% of the ideal strength calculated by first-principles DFT. This high strength is due to the less defective nature of SMPs obtained via the PLML process.

Figure 4 shows a typical fracture process of TiC SMPs on a SiC substrate using a diamond indenter with SEM images before and after the test. A clear jump of displacement induced by a fracture is observed in the load–displacement curve. Figure 5 shows the particle size dependence of fracture strength for TiC SMPs obtained using the Hiramatsu–Oka equation [18–20].

$$S_t = 2.8\frac{F}{\pi D^2} \tag{1}$$

where S_t ($N\ m^{-2}$) denotes the fracture strength; F (N), the failure load of a spherical particle; and D (m), the particle diameter. The values indicate that the particles obtained in toluene (red circles, average fracture strength: 7.46 GPa) have higher fracture strength than those obtained in ethanol (black squares, average fracture strength: 2.67 GPa), although the average size obtained in toluene (396 nm) is larger than that in ethanol (301 nm), which is due to the difference in the permittivity of liquid, which controls the aggregation behavior of raw particles [6]. Larger particles generally have lower fracture strength due to the increased possibility of defect inclusion. However, our results indicate that the high fracture strength of particles obtained in toluene is caused by more perfect SMPs with less oxygen inclusion, as suggested in Figures 2 and 3.

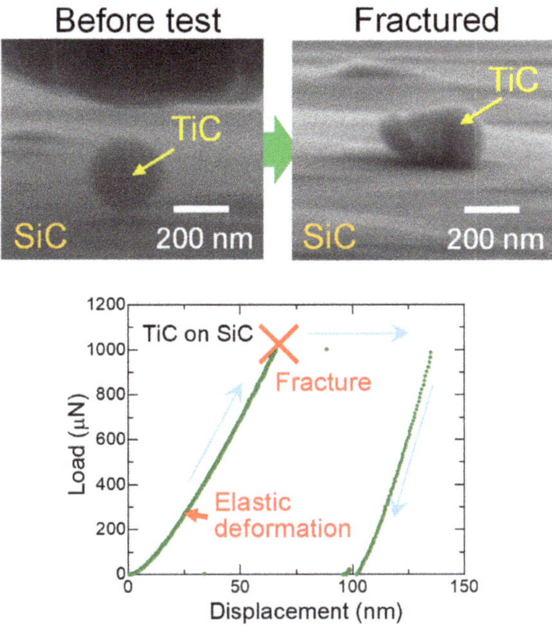

Figure 4. Typical fracture process of a TiC SMP and corresponding load–displacement curve during the compression test.

Figure 5 also shows a comparison of bulk bending strength [21–23], SMP fracture strength, and calculated ideal strength (24.4 GPa) of TiC [24]. The fracture strength of TiC SMPs fabricated via PLML ranged from 3% to 20% of the ideal strength when prepared in ethanol and from 20% to 50% when prepared in toluene. The previously reported fracture strengths of B_4C and TiO_2 SMPs obtained via PLML ranged from 18% to 38% and 10% to 40% of their respective ideal tensile strengths, which is similar to the TiC SMPs obtained in toluene. In relation to bulk fracture strength, TiC SPMs obtained in toluene were greater by one order of magnitude, as in the case of B_4C [4].

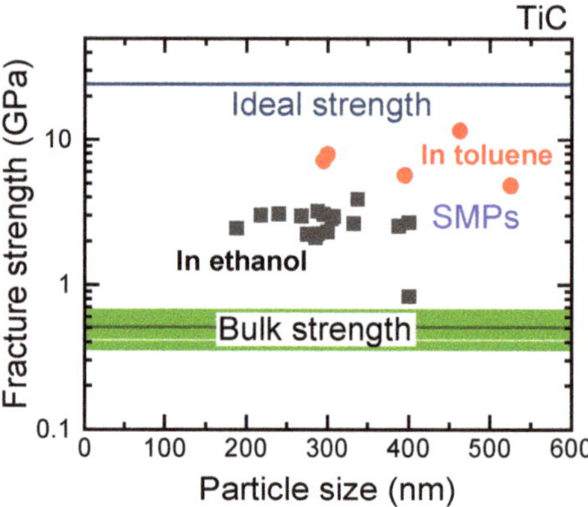

Figure 5. Particle size dependence of the fracture strength of TiC SMPs fabricated in ethanol (black squares) and toluene (red circles). The particles were placed on the SiC substrate and indented by a diamond tip. The calculated ideal tensile strengths and bulk strengths of large compact samples for TiC are also shown.

3.3. *Embedding Process of SMPs Obtained via PLML*

The above data for the SMP fracture strength measurements were obtained using hard SiC substrates, and the particles were interlaid between the substrate and a diamond tip to be fractured. Si substrates can also be used for the indenter test of TiC, as shown in Figure 6. However, most fractured TiC SMPs on Si substrates were larger than 400 nm, which corresponds to the size range wherein phase-separated or polycrystalline SMPs tend to be formed in the PLML process [25]. Thus, the fracture strength gradually degraded with increasing particle size, as in the previous report [4]. In contrast, when smaller TiC SMPs were indented on the Si substrates, the particles were often embedded without appreciable shape change or after slight fracturing. This behavior may be observed on soft substrates and differs from the simple particle fracture process. Therefore, the embedding process of SMPs is systematically studied hereafter.

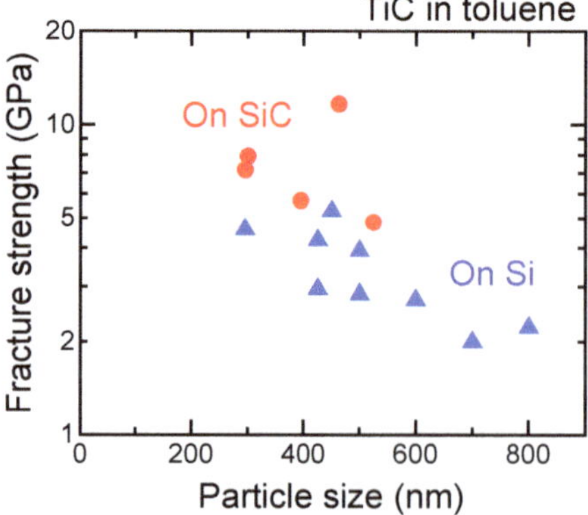

Figure 6. Particle size dependence of fracture strength of TiC SMPs fabricated in toluene. Fracture tests were conducted on SiC and Si substrates indented with a diamond tip.

When soft Al substrates were used, the particles did not fracture but were gradually embedded, as shown in Figure 7. Figure 8 shows SEM images of the Al substrate after embedment of TiC SMPs, indicating complete embedment to the substrate level. Figure 9 shows a typical load–displacement curve wherein a particle is embedded and the corresponding schematic images illustrate how the particle is embedded. In contrast to the particle fracture in Figure 4, the particle is embedded with the load increment at the initial stage (Figure 9a) and then inserted into the substrate without a further load increase, exhibiting a plateau in the load–displacement curve (Figure 9b). This behavior is relevant to the contact area change with embedded depth, as shown in Figure 9c. When the particle is embedded by half, a further push is made only through the contacting hemisphere surface. Further pushing will cause the diamond tip to touch the substrate, resulting in a sharp rise in the load–displacement curve, as demonstrated at the right side of the curve in Figure 9b. Thus, the idealized relation between the contact area and embedded depth is schematically displayed in Figure 9c.

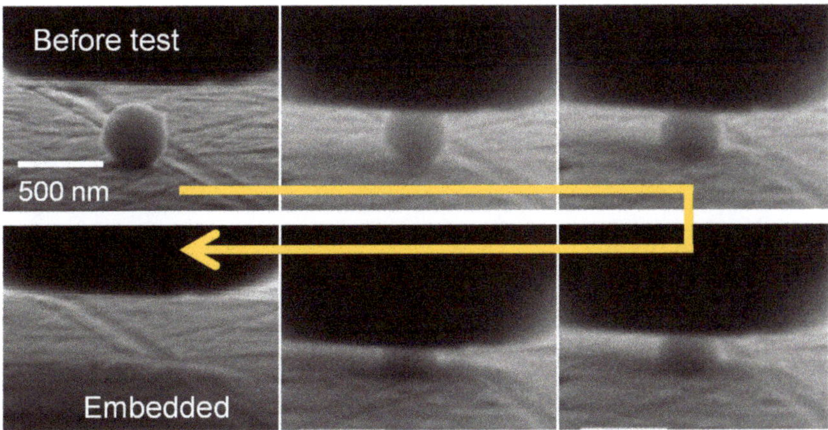

Figure 7. Embedment process of TiC SMPs on an Al substrate.

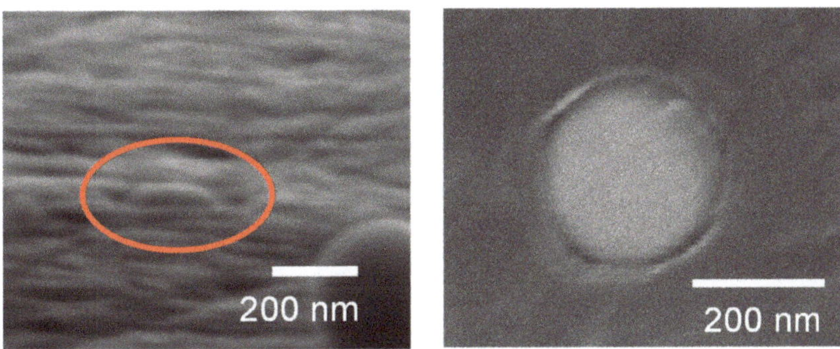

Figure 8. SEM images of the TiC-SMP embedded region on an Al substrate: left: oblique downward view; right: top view.

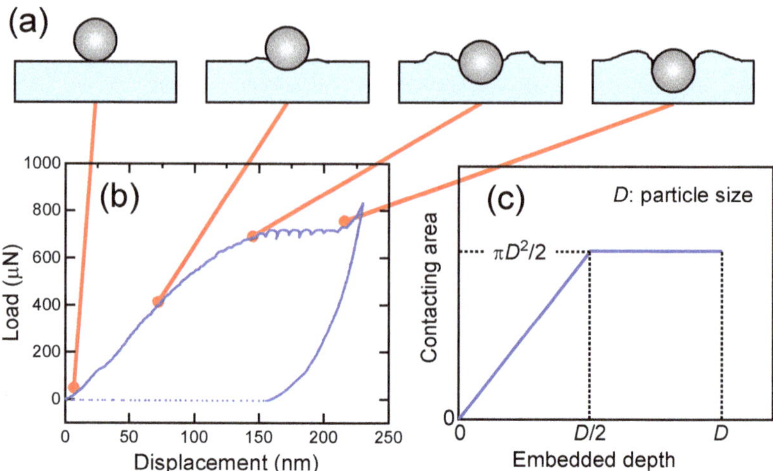

Figure 9. Embedding process of TiC SMPs on an Al substrate. (**a**) Schematic illustration of the particle embedment process. (**b**) Typical load–displacement curve of TiC SMPs on Al substrate. (**c**) Calculated relationship between the particle-surface contact area and embedded depth.

3.4. Hardness Estimation of Substrates via SMP Embedding Process

Figure 10 shows the load–displacement curve of the particle embedding process for 250 and 450 nm B_4C particles into Si single-crystalline substrates with SEM images before and after the load application. Nearly the entire particle was embedded after the load application, and only the top part of the particle can be observed after embedment. Plateau ranges in the load–displacement curves began when the displacement exceeded half of the particle size, suggesting the embedding process described in Figure 9c. The load at the plateau was larger for larger particles. For larger particles, a pop-in process, as in the case of 450 nm indented particles, is sometimes observed, indicating discontinuous embedment by crack formation of the particles [26].

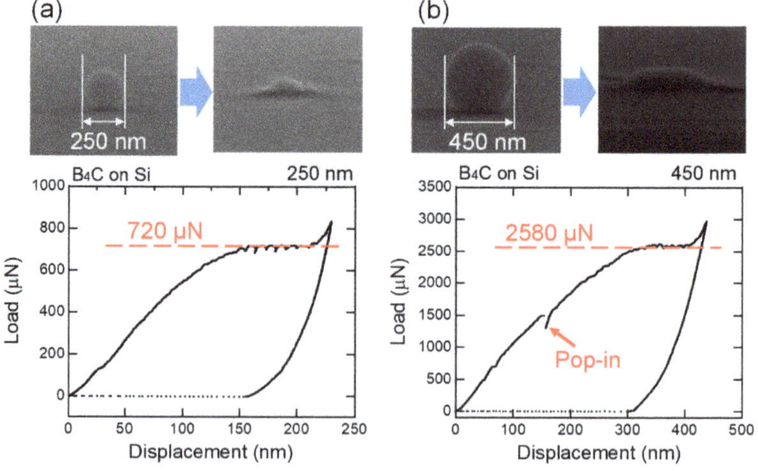

Figure 10. SEM images before and after the load application and load–displacement curves of the particle embedding process for (**a**) 250 and (**b**) 450 nm B_4C particles in Si substrates.

With regard to the tip shape to push the substrate, the experimental arrangement in this study is similar to common Brinell hardness measurements using a macroscopic spherical tip, mostly made of a diamond of millimeter order in size and, recently, with micrometer range [27–34]. However, in our case, the particle indenter size was at the submicrometer scale, which is somewhat smaller than the conventional spherical indenter used for Brinell hardness measurements. When a large spherical tip was pushed onto the specimen, the Brinell hardness, H_B, was calculated from the applied load, F; particle size, D; and measured indentation diameter, d, of the specimen in overhead view after unloading:

$$H_B = \frac{2F}{\pi D\left(D - \sqrt{D^2 - d^2}\right)} \tag{2}$$

In our experimental setup, a particle was continuously pushed during particle embedment. Thus, the displacement of the tip from the surface, h, can be adopted as the indentation depth, and the Brinell hardness can be calculated as follows:

$$H_B = \frac{F}{\pi D h} = \frac{F}{(\pi D^2)(h/D)} \tag{3}$$

Figure 11 shows the Brinell hardness variations calculated using Equation (3) from the initial stage of particle embedment data in Figure 10. In the embedded fraction range of 0.25–0.50 (d/D range: 0.66–0.87), the calculated Brinell hardness values were close, irrespective of the particle size used to push the Si substrate.

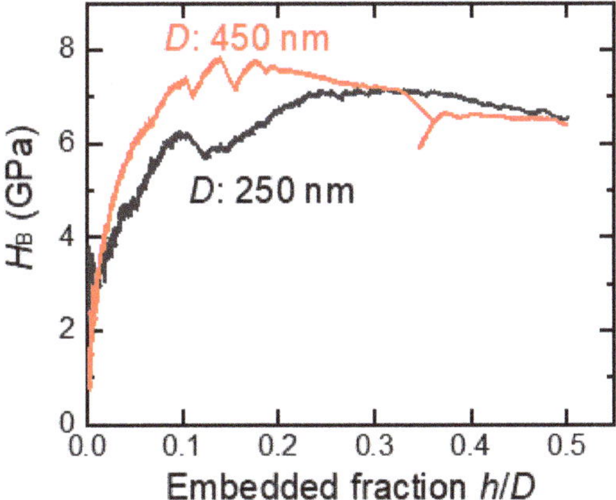

Figure 11. Brinell hardness variations calculated using Equation (3) from the initial stage of particle embedment data in Figure 10.

The load value at the plateau range, F_p, in Figure 10 can be easily determined since the embedded fraction, h/D, is 0.5 and, therefore, Equation (3) can be simplified as follows:

$$F_p = \frac{\pi}{2} H_B D^2 \tag{4}$$

Figure 12 shows the embedding load at the plateau region in the load–displacement curve as a function of particle size for various B_4C SMPs on a Si substrate. The solid black line is a curve fitted with a parabolic curve based on Equation (4) and is well fitted to the experimental data, indicated by red circles, without pop-in. The fit is also good for the data with pop-in behavior (blue triangles)—for particles larger than 450 nm. The Brinell hardness estimated from the fitting parameter in Figure 12 and Equation (4) was 7.01 GPa. Typical hardness values of Si reported were 11.1 (Vickers) [35] and 10.9 GPa

(Berkovich) [36], obtained via microindentation tests, although the values were scattered from 7 to 20 GPa [37], depending on defect concentrations and distributions of the specimens and the difference in size, shape, and material of the indenter.

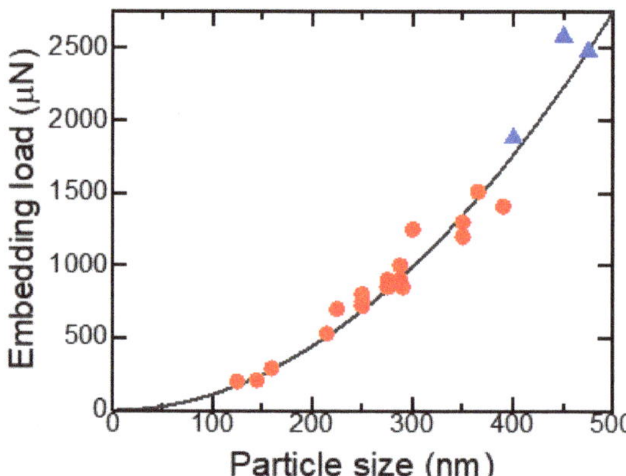

Figure 12. Particle size dependence of embedding load at the plateau region for B_4C particles on Si substrate. Red circles and blue triangles are experimental data without and with pop-in behavior during indentation. The black curve is a fitted curve.

Nanoindentation techniques for hardness estimation have been developed to measure the mechanical properties of nano-objects. However, most nanoindentation devices use Vickers tips based on a pyramidal shape. Nanosized, sphere-shaped tips have seldom been utilized, probably owing to the unavailability of hard and small spherical particles. Spherical SMPs are a possible candidate for nano-Brinell tips to evaluate the hardness of nanosized objects.

3.5. Substrate Effects on the SMP Embedding Process

Another possible application of hard and brittle SMPs is the site-selective surface post-enforcement of soft materials via particle embedment. For such enforcement to be sufficiently effective, particles should not be easily detached after embedment and, therefore, more than half of the particles have to be embedded. Particle embedment is completed when the indenter starts to push the substrate. The tip displacement on this occasion can be estimated from the right end of the load–displacement curve in Figure 9b. From the tip displacement at the embedment completion, h_c; the particle size, D; embedded volume, V_h; and particle volume, V_D, (as presented in Figure 13), the embedded volume fractions can be calculated as follows:

$$\frac{V_h}{V_D} = 3\left(\frac{h_c}{D}\right)^2 - 2\left(\frac{h_c}{D}\right)^3 \tag{5}$$

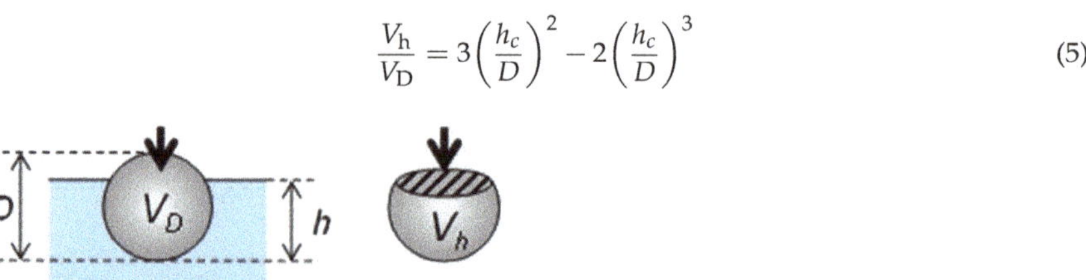

Figure 13. Schematic illustration of particle embedment.

Figure 14 shows the embedded volume fraction as a function of particle size for various SMPs on Si substrates. Various hard brittle particles of TiC, B_4C, and B were embedded in similar volume fractions (50–70%) into Si substrates, irrespective of the particle size.

The average volume fractions, indicated by lines with corresponding colors, were nearly the same.

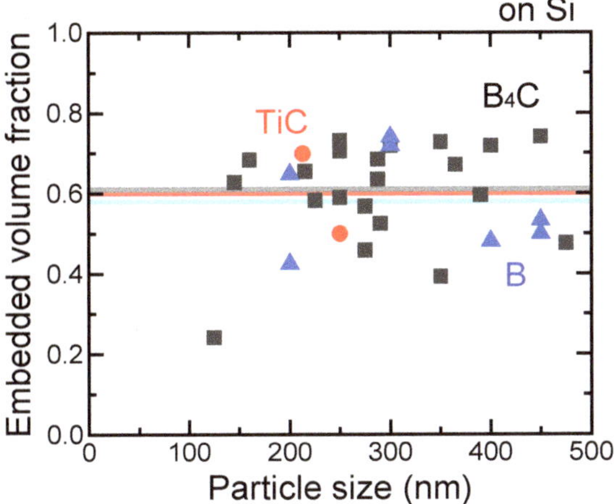

Figure 14. Particle size dependence of embedded volume fraction of TiC (red circles), B_4C (black squares), and B (blue triangles) on Si substrates. The horizontal lines indicate the average values of the embedded volume fraction of particles with corresponding colors.

Figure 15 shows the embedded volume fraction as a function of particle size for various SMPs on TiO_2, Si, $SrTiO_3$, and Al substrates. The embedded volume fractions do not have a clear size dependence in the case of Si substrates, though the embedded volume fraction increased with the decrease in Vickers hardness of the substrates, as presented in Figure 16. When more than half of the B_4C, B, or TiC particles must be embedded, substrates with a Vickers hardness smaller than 12 GPa have to be used. Thus, site-selective surface enforcement is considered to be possible simply by spreading mechanically hard SPMs on a soft material surface and applying local force.

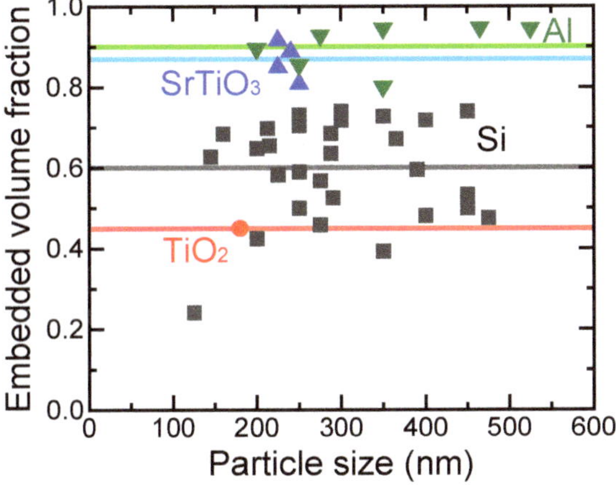

Figure 15. Particle size dependence of the embedded volume fraction for SMPs on various substrates of TiO_2 (red circle), Si (black squares), $SrTiO_3$ (blue triangles), and Al (green inverted triangles). The horizontal lines indicate the average values of the embedded volume fraction of particles with corresponding colors.

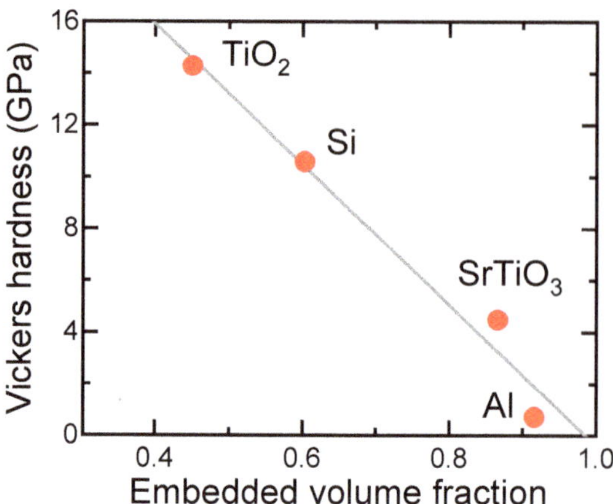

Figure 16. Relationship between Vickers hardness of substrates and average embedded volume fraction of SMPs.

4. Conclusions

Spherical non-porous SMPs of hard and brittle ceramics, such as B_4C, TiC, and B, were experimentally fabricated via the PLML process, which employs instantaneous pulsed laser heating of raw particles dispersed in a liquid. To explore and evaluate their potential for mechanical applications, these hard SMPs were compressed using a diamond tip on various substrates. The hard particles were fractured on hard substrates, such as SiC, and embedded into soft substrates, such as Si or Al. From the fracture process, particle strength can be estimated to be 5–12 GPa, which is larger by one order of magnitude than the bulk bending strength and 10–40% of the ideal strength of 24.4 GPa from the DFT calculations. From the embedment process, the submicrometer-scale Brinell hardness of 7.01 GPa of the Si substrate can be estimated by analyzing the load–displacement curves. The relationship of particle embedding depth and the hardness of substrates was elucidated from various combinations of SMPs and substrates.

Author Contributions: Conceptualization, D.N., N.K., and Y.I.; methodology, N.S. and S.K.; software, N.S. and N.K.; validation, D.N., N.K., N.S., S.K., and Y.I.; formal analysis, D.N. and N.K.; investigation, D.N., N.K., and Y.I.; resources, N.K., N.S., and S.K.; data curation, D.N. and N.K.; writing—original draft preparation, N.K. and Y.I.; writing—review and editing, D.N., N.S., and S.K.; visualization, N.K. and Y.I.; supervision, N.K.; project administration, N.K.; funding acquisition, N.K. All authors have read and agreed to the published version of the manuscript.

Funding: This research was partially funded by JSPS KAKENHI, grant numbers 26289266 and 26870908.

Data Availability Statement: The data are included in the main text.

Acknowledgments: This work was conducted at "Joint-use Facilities: Laboratory of Nano-Micro Material Analysis", Hokkaido University, supported by the "Nanotechnology Platform" Program of the Ministry of Education, Culture, Sports, Science and Technology (MEXT), Japan.

Conflicts of Interest: The authors declare no conflict of interest.

References

1. Hu, X.; Gong, H.; Wang, Y.; Chen, Q.; Zhang, J.; Zheng, S.; Yang, S.; Cao, B. Laser-induced reshaping of particles aiming at energy-saving applications. *J. Mater. Chem.* **2012**, *22*, 15947–15952. [CrossRef]
2. Jendrzej, S.; Gondecki, L.; Debus, J.; Moldenhauer, H.; Tenberge, P.; Barcikowski, S.; Gökce, B. Tribological properties of laser-generated hard ceramic particles in a gear drive contact. *Appl. Surf. Sci.* **2019**, *467–468*, 811–818. [CrossRef]
3. Murugavel, P.; Kalaiselvam, M.; Raju, A.R.; Rao, C.N.R. Sub-micrometre spherical particles of TiO_2, ZrO_2 and PZT by nebulized spray pyrolysis of metal–organic precursors. *J. Mater. Chem.* **1997**, *7*, 1433–1438. [CrossRef]

4. Kondo, M.; Shishido, N.; Kamiya, S.; Kubo, A.; Umeno, Y.; Ishikawa, Y.; Koshizaki, N. High-Strength Sub-Micrometer Spherical Particles Fabricated by Pulsed Laser Melting in Liquid. *Part. Part. Syst. Char.* **2018**, *35*, 1800061. [CrossRef]
5. Ishikawa, Y.; Shimizu, Y.; Sasaki, T.; Koshizaki, N. Boron carbide spherical particles encapsulated in graphite prepared by pulsed laser irradiation of boron in liquid medium. *Appl. Phys. Lett.* **2007**, *91*, 161110. [CrossRef]
6. Ishikawa, Y.; Feng, Q.; Koshizaki, N. Growth fusion of submicron spherical boron carbide particles by repetitive pulsed laser irradiation in liquid media. *Appl. Phys. A* **2010**, *99*, 797–803. [CrossRef]
7. Wang, H.; Pyatenko, A.; Kawaguchi, K.; Li, X.; Swiatkowska-Warkocka, Z.; Koshizaki, N. Selective pulsed heating for the synthesis of semiconductor and metal submicrometer spheres. *Angew. Chem. Int. Ed. Engl.* **2010**, *49*, 6361–6364. [CrossRef] [PubMed]
8. Wang, H.; Miyauchi, M.; Ishikawa, Y.; Pyatenko, A.; Koshizaki, N.; Li, Y.; Li, L.; Li, X.; Bando, Y.; Golberg, D. Single-crystalline rutile TiO_2 hollow spheres: Room-temperature synthesis, tailored visible-light-extinction, and effective scattering layer for quantum dot-sensitized solar cells. *J. Am. Chem. Soc.* **2011**, *133*, 19102–19109. [CrossRef] [PubMed]
9. Wang, H.; Koshizaki, N.; Li, L.; Jia, L.; Kawaguchi, K.; Li, X.; Pyatenko, A.; Swiatkowska-Warkocka, Z.; Bando, Y.; Golberg, D. Size-tailored ZnO submicrometer spheres: Bottom-up construction, size-related optical extinction, and selective aniline trapping. *Adv. Mater.* **2011**, *23*, 1865–1870. [CrossRef] [PubMed]
10. Pyatenko, A.; Wang, H.; Koshizaki, N.; Tsuji, T. Mechanism of pulse laser interaction with colloidal nanoparticles. *Laser Photonics Rev.* **2013**, *7*, 596–604. [CrossRef]
11. Ishikawa, Y.; Koshizaki, N.; Pyatenko, A.; Saitoh, N.; Yoshizawa, N.; Shimizu, Y. Nano- and Submicrometer-Sized Spherical Particle Fabrication Using a Submicroscopic Droplet Formed Using Selective Laser Heating. *J. Phys. Chem. C* **2016**, *120*, 2439–2446. [CrossRef]
12. Zhang, D.; Gökce, B.; Barcikowski, S. Laser Synthesis and Processing of Colloids: Fundamentals and Applications. *Chem. Rev.* **2017**, *117*, 3990–4103. [CrossRef] [PubMed]
13. Xiao, J.; Liu, P.; Wang, C.X.; Yang, G.W. External field-assisted laser ablation in liquid: An efficient strategy for nanocrystal synthesis and nanostructure assembly. *Prog. Mater Sci.* **2017**, *87*, 140–220. [CrossRef]
14. Amendola, V.; Amans, D.; Ishikawa, Y.; Koshizaki, N.; Scirè, S.; Compagnini, G.; Reichenberger, S.; Barcikowski, S. Room-Temperature Laser Synthesis in Liquid of Oxide, Metal-Oxide Core-Shells, and Doped Oxide Nanoparticles. *Chem. Eur. J.* **2020**, *26*, 9206–9242. [CrossRef] [PubMed]
15. Sakaki, S.; Ikenoue, H.; Tsuji, T.; Ishikawa, Y.; Koshizaki, N. Pulse-Width Dependence of the Cooling Effect on Sub-Micrometer ZnO Spherical Particle Formation by Pulsed-Laser Melting in a Liquid. *ChemPhysChem* **2017**, *18*, 1101–1107. [CrossRef] [PubMed]
16. Suehara, K.; Takai, R.; Ishikawa, Y.; Koshizaki, N.; Omura, K.; Nagata, H.; Yamauchi, Y. Reduction Mechanism of Transition Metal Oxide Particles in Thermally Induced Nanobubbles during Pulsed Laser Melting in Ethanol. *ChemPhysChem* **2021**, *22*, 675–683. [CrossRef]
17. Yoshida, M.; Ogiso, H.; Nakano, S.; Akedo, J. Compression test system for a single submicrometer particle. *Rev. Sci. Instrum.* **2005**, *76*, 093905. [CrossRef]
18. Hiramatsu, Y.; Oka, Y.; Kiyama, H. Rapid Determination of the Tensile Strength of Rocks with Irregular Test Pieces (in Japanese). *J. Mining Metall. Inst. Jpn.* **1965**, *81*, 1024–1030. [CrossRef]
19. Hiramatsu, Y.; Oka, Y. Determination of the tensile strength of rock by a compression test of an irregular test piece. *Int. J. Rock Mech. Min. Sci. Geomech. Abstr.* **1966**, *3*, 89–90. [CrossRef]
20. Hiramatsu, Y.; Oka, Y.; Kiyama, H. Investigations on the Disc Test, Ring Test and Indentation Test for Rocks (in Japanese). *J. Mining Metall. Inst. Jpn.* **1969**, *85*, 8–14. [CrossRef]
21. Fu, Z.; Koc, R. Pressureless sintering of submicron titanium carbide powders. *Ceram. Int.* **2017**, *43*, 17233–17237. [CrossRef]
22. Iseki, T.; Yano, T.; Chung, Y.-S. Wetting and Properties of Reaction Products in Active Metal Brazing of SiC. *J. Ceram. Soc. Jpn.* **1989**, *97*, 710–714. [CrossRef]
23. Ono, T.; Ueki, M.; Shimizu, M. Fabrication and Characterization of Titanium Carbide-Graphite Composite Materials (1st Report), Mechanical Properties and Microstructure of Titanium Carbide and Its Compsoites with Addition of a Small Amount of Graphite (in Japanese). *Trans. JSME (A)* **1993**, *59*, 1978–1984. [CrossRef]
24. Tang, B.; Shen, Y.; An, Q. Shear-induced brittle failure of titanium carbide from quantum mechanics simulations. *J. Am. Ceram. Soc.* **2018**, *101*, 4184–4192. [CrossRef]
25. Fuse, H.; Koshizaki, N.; Ishikawa, Y.; Swiatkowska-Warkocka, Z. Determining the Composite Structure of Au-Fe-Based Submicrometre Spherical Particles Fabricated by Pulsed-Laser Melting in Liquid. *Nanomaterials* **2019**, *9*, 198. [CrossRef]
26. Sato, Y.; Shinzato, S.; Ohmura, T.; Hatano, T.; Ogata, S. Unique universal scaling in nanoindentation pop-ins. *Nat. Commun.* **2020**, *11*, 4177. [CrossRef]
27. Liu, M.; Lin, J.-y.; Lu, C.; Tieu, K.A.; Zhou, K.; Koseki, T. Progress in Indentation Study of Materials via Both Experimental and Numerical Methods. *Crystals* **2017**, *7*, 258. [CrossRef]
28. Broitman, E. Indentation Hardness Measurements at Macro-, Micro-, and Nanoscale: A Critical Overview. *Tribol. Lett.* **2016**, *65*, 23. [CrossRef]
29. Hess, D.R.; Doty, H.W. Brinell hardness testing methods and their applicability. In Proceedings of the AFS 124th Metalcasting Congress, Cleveland, OH, USA, 21–23 April 2020; p. 12.
30. Pathak, S.; Kalidindi, S.R. Spherical nanoindentation stress–strain curves. *Mater. Sci. Eng. R Rep.* **2015**, *91*, 1–36. [CrossRef]

31. Chudoba, T.; Schwarzer, N.; Richter, F. Determination of elastic properties of thin films by indentation measurements with a spherical indenter. *Surf. Coat. Technol.* **2000**, *127*, 9–17. [CrossRef]
32. Bei, H.; Lu, Z.P.; George, E.P. Theoretical Strength and the Onset of Plasticity in Bulk Metallic Glasses Investigated by Nanoindentation with a Spherical Indenter. *Phys. Rev. Lett.* **2004**, *93*, 125504. [CrossRef] [PubMed]
33. Martinez, R.; Xu, L.R. Comparison of the Young's moduli of polymers measured from nanoindentation and bending experiments. *MRS Commun.* **2014**, *4*, 89–93. [CrossRef]
34. Kim, M.; Marimuthu, K.P.; Lee, J.H.; Lee, H. Spherical indentation method to evaluate material properties of high-strength materials. *Int. J. Mech. Sci.* **2016**, *106*, 117–127. [CrossRef]
35. Walls, M.G.; Chaudhri, M.M.; Tang, T.B. STM profilometry of low-load Vickers indentations in a silicon crystal. *J. Phys. D Appl. Phys.* **1992**, *25*, 500–507. [CrossRef]
36. Budnitzki, M.; Kuna, M. Experimental and numerical investigations on stress induced phase transitions in silicon. *Int. J. Solids Struct.* **2017**, *106–107*, 294–304. [CrossRef]
37. INSPEC. *Properties of Silicon*; INSPEC, Institution of Electrical Engineers: New York, NY, USA; London, UK, 1988; Volume 4.

MDPI

Article

Properties of $CrSi_2$ Layers Obtained by Rapid Heat Treatment of Cr Film on Silicon

Tatyana Kuznetsova [1], Vasilina Lapitskaya [1], Jaroslav Solovjov [2], Sergei Chizhik [1], Vladimir Pilipenko [2] and Sergei Aizikovich [3,*]

1 Nanoprocesses and Technology Laboratory, A.V. Luikov Institute of Heat and Mass Transfer of National Academy of Science of Belarus, 15, P.Brovki str, 220072 Minsk, Belarus; t_kuzn@hmti.ac.by (T.K.); lapitskayava@hmti.ac.by (V.L.); chizhik_sa@tut.by (S.C.)

2 JSC "INTEGRAL"—"INTEGRAL" Holding Managing Company, 121 A Kazintsa, 220108 Minsk, Belarus; jsolovjov@integral.by (J.S.); office@bms.by (V.P.)

3 Research and Education Center "Materials", Don State Technical University, 344000 Rostov-on-Don, Russia

* Correspondence: s.aizikovich@sci.donstu.ru; Tel.: +7-863-238-1558

Abstract: The changes in the morphology and the electrophysical properties of the Cr/n-Si (111) structure depending on the rapid thermal treatment were considered in this study. The chromium films of about 30 nm thickness were deposited via magnetron sputtering. The rapid thermal treatment was performed by the irradiation of the substrate's back side with the incoherent light flux of the quartz halogen lamps in nitrogen medium up to 200–550 °C. The surface morphology was investigated, including the grain size, the roughness parameters and the specific surface energy using atomic force microscopy. The resistivity value of the chromium films on silicon was determined by means of the four-probe method. It was established that at the temperatures of the rapid thermal treatment up to 350 °C one can observe re-crystallization of the chromium films with preservation of the fine grain morphology of the surface, accompanied by a reduction in the grain sizes, specific surface energy and the value of specific resistivity. At the temperatures of the rapid thermal treatment from 400 to 550 °C there originates the diffusion synthesis of the chromium disilicide $CrSi_2$ with the wave-like surface morphology, followed by an increase in the grain sizes, roughness parameters, the specific surface energy and the specific resistivity value.

Keywords: thin films; chromium; chromium disilicide; silicon substrate; rapid thermal treatment; roughness

Citation: Kuznetsova, T.; Lapitskaya, V.; Solovjov, J.; Chizhik, S.; Pilipenko, V.; Aizikovich, S. Properties of $CrSi_2$ Layers Obtained by Rapid Heat Treatment of Cr Film on Silicon. *Nanomaterials* **2021**, *11*, 274. https://doi.org/10.3390/nano11071734

Academic Editor: Victor A. Eremeyev

Received: 21 May 2021
Accepted: 28 June 2021
Published: 30 June 2021

Publisher's Note: MDPI stays neutral with regard to jurisdictional claims in published maps and institutional affiliations.

1. Introduction

The chromium films represent an important and multifunctional material in the various contemporary branches of industry owing to their exceptional properties (enhanced mechanical, corrosion, conductive, optical and catalytic properties) [1–3]. The chromium thin films (thick up to 200 nm) are widely applied as durable conductive coatings and contact pads in microelectronics and sensors [4,5]. Chromium can be applied as the underlayer material for the multipurpose coatings in the microelectronic devices and microsensor products on silicon and glass [6–8], where its morphology is important, as it determines the structure and properties of the subsequent functional layer [9]. In this case, chromium promotes the crystallization extent of the upper layer, increases the grain size and increases the Hall ratio and the concentration of carriers owing to the suppression of Na atom diffusion into the films (for glass) and reduces resistance. Meanwhile, the increase in roughness that takes place in the upper films with deposition on the underlayer of Cr enhances the functional parameters of the upper layer [6].

The electric conductivity of the chromium films, applied on the silicon substrate, substantially changes during the formation of the chromium disilicide ($CrSi_2$) [10–12]. Such properties of $CrSi_2$ as the high temperature of melting, the resistance to oxidation and the capability of standing a considerable deformation make it a prospective material under the conditions of the energy influences [11,13]. Earlier, $CrSi_2$ layers were mainly applied as the barriers of Schottky diodes [3–5,11,14,15]. At present, the $CrSi_2$ layers are also used as a joint between silicon and the working element in the integrated circuits and the sensors owing to the low transient resistance and the semiconductor properties [2,11]. This peculiarity in combination with a good compatibility with the regular silicon technologies makes it possible to successfully apply $CrSi_2$ in the thermoelectric and photoelectric devices [1,2,16]. The narrow prohibited zone [17] makes it suitable for application in the converters and sensors in micro- and nanoelectronics. Meanwhile, the electrical properties of $CrSi_2$ substantially depend on the method of formation and the microstructure.

The chromium disilicide layers may be formed by direct deposition of the given phase on the silicon substrate [12] and by means of modification of the chromium films at the expense of diffusion of silicon atoms into chromium [18]. This diffusion may progress even under low (<300 °C) temperatures [18]. The conventional heating (furnace annealing) is performed at the temperature of 450 °C and lasts for 30 min [16].

The standard temperature of transition of the nanocrystalline film of $CrSi_2$ to the semiconductor state is the temperature of 1200 °C [19]. In the case of $CrSi_2$ layers synthesized at 450 °C and possessing metallic conductivity, they can be subjected to annealing for 300 s. The progressive method of the rapid thermal treatment (RTT) of the silicon wafer with the applied metal layer makes it possible to obtain the required semiconductor layers at a substantially lower temperature, and the process parameters make it possible to control the electrical properties of the synthesized coatings [20]. An abrupt change in the electrical properties of the metallic films at the RTT temperature of 400 °C is related to the phase transitions [21]. The rapid thermal treatment of the silicon wafer with the applied Cr layer makes it possible to obtain the $CrSi_2$ layers in several seconds and at a temperature of 400–600 °C [22].

The scientific literature presents no dependences relating the RTT temperature of the Cr films to their parameters of roughness, grain size, specific surface energies and resistivity. The purpose of the given paper is to study the influence of the RTT temperature of the chromium thin films, applied on the silicon substrate by means of the magnetron sputtering, on the parameters of the surface roughness, grain size, specific surface energy and resistivity.

2. Materials and Methods

2.1. Coating Deposition

The chromium films about 30 nm thick were applied on the silicon substrates by means of the magnetron sputtering of the chromium target with the purity of 99.5% in the argon medium with the purity of 99.993% under the Ar pressure of 0.35 Pa, at the Ar flow rate of 180 ccm, the power of discharge of 5.1 kW and the discharge current of 7.5 A (the power density constituted about 5.85 W/cm^2 with the discharge voltage of 680 V) on the unit SNT (StratNanoTech) "Sigma" (StratNanoTech Invest, Minsk, Belarus). The silicon substrates were essentially the epitaxial layers of phosphorus-doped silicon with the resistivity of 0.58–0.63 Ω·cm and the thickness of 5.3–5.8 μm, formed on the substrates of p-type monocrystalline silicon with the resistivity of 10 Ω·cm and the orientation of (111).

Further on, the substrates were subjected to the rapid thermal treatment in the mode of the heat balance by means of irradiation of the reverse side of the substrates with the incoherent light flux of the quartz halogen lamps of the constant power in the nitrogen medium for 7 s to attain the temperature from 200 to 550 °C at atmospheric pressure on the unit JetFirst 100 (Jipelec Qualiflow, Montpellier, France). The heating rate was 30–80°/s. The temperature control of the working side of the substrate was performed by means of the thermal couple with the precision of ±0.5 °C.

2.2. Coating Characterization

The surface roughness parameters after the rapid thermal treatment of the Cr/Si structure were determined by means of AFM [23]. All studies of the surface morphology, roughness and adhesion forces (F_{ad}) values were carried out using AFM Dimension FastScan (Bruker, Santa Barbara, CA, USA) in the PeakForce QNM (Quantitative Nanoscale Mechanical Mapping, Bruker, Santa Barbara, CA, USA) mode with standard silicon cantilevers of the CSG10_SS type (TipsNano, Moscow, Russia) with a tip radius of 3.2 nm and console stiffness of 0.26 N/m. The probe performs an "approach–retraction" motion to the sample surface with the record of the forces curves at each point of the image. The forces between the AFM tip and the sample can be precisely controlled. Obtaining and recording entry of such curves is the basis of the PeakForce QNM mode, based on which recomputation of the values of the adhesion forces is automatically performed with consideration of the specific characteristics of the used probe. For each sample, 220–250 grains were considered to determine the grain size of Cr and $CrSi_2$ films. The "Particle Analysis" function in the "NanoScope analysis" processing program was applied for this purpose.

The specific surface energy (γ) (the work of adhesion) was determined by means of the expression $\gamma = F_{ad}/(2\pi R)$ [24], where F_{ad} is the force of adhesive interaction between the AFM probe tip and the surface, N, and R is the radius of the probe tip, m. *Ra*, *Rq* and *Rz* parameters were measured on 1×1 μm^2 areas; the grain diameters were measured on 250×250 nm^2 areas.

The surface resistance (*Rs*) of the Cr/Si and $CrSi_2$/Si layers was measured by means of the four-probe method with the unit RS-30 (KLA Tencor, Milpitas, CA, USA). The error of the surface resistance determination was less than 5%. Meanwhile, the thickness of the layers (t) was determined by means of the scanning electron microscope S-4800 (Hitachi, Tokyo, Japan). To measure the thickness of the films by the SEM (scanning electron microscopy) method, the silicon wafer with the film was vertically cleaved. This cross-sectional cleavage was examined by SEM to determine the thickness (Figure 1). The resistivity value (ρ) was determined by means of the expression $\rho = Rs \times t$ [21].

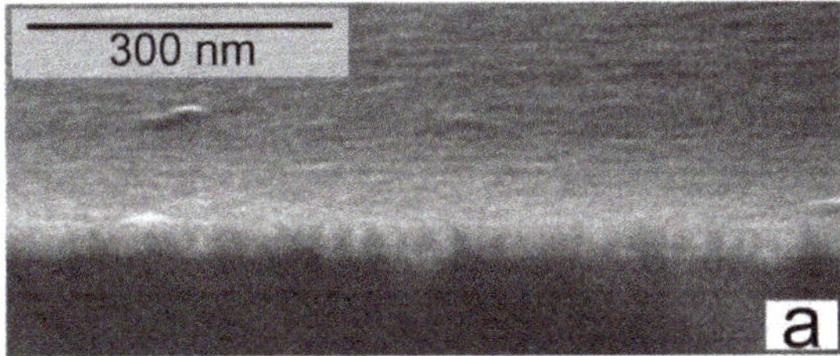

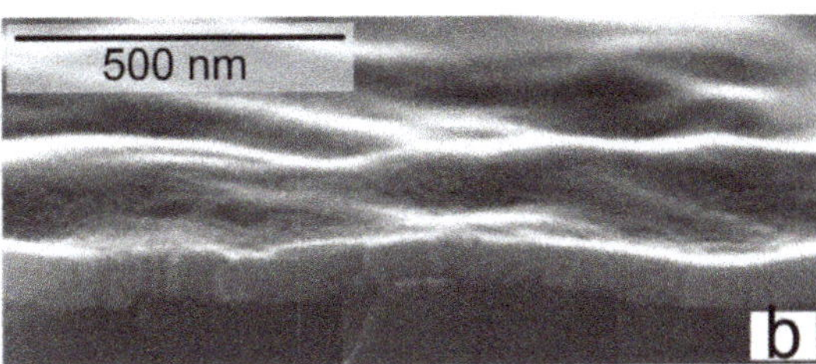

Figure 1. The initial Cr film (**a**) and $CrSi_2$ film (**b**) obtained by RTT Cr at 400 °C.

3. Results and Discussion

Structural properties of the deposited Cr film and the Cr film after RTT measured by XRD are shown in Figure 2. The main phase in the initial film and after RTT at 300 °C and 350 °C is the Cr phase. The main phase after RTT at 400 °C and 450 °C is the $CrSi_2$ phase.

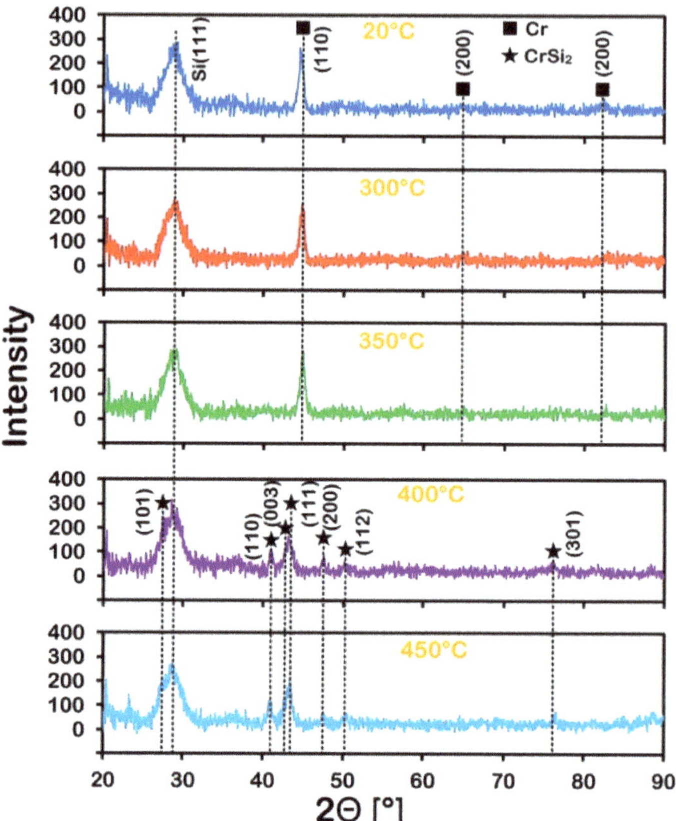

Figure 2. X-ray diffraction (XRD) patterns of Cr coatings in the initial state (20 °C) and after RTT with Cr lines (300 °C, 350 °C) and $CrSi_2$ lines (400 °C, 450 °C).

The results of the investigations of the surface morphology of the obtained coating are represented in Figure 3. In the initial microstructure of Cr/Si, as well as after its rapid thermal treatment at the temperatures from 200 to 350 °C, one observes the smooth surface with the fine grain morphology (Figure 3a,b). At the RTT temperature of 400–550 °C, there arises the "wave-like" morphology of the Cr/Si structure surface (Figure 3c,d). The given changes are explained by the phase transition of the chromium film into the chromium silicide in the temperature range from 350 to 400 °C, which is well matched with the results of the earlier academic papers [10], where the phase transition of Cr to $CrSi_2$ was observed at the temperature of 400 °C. The presence of the "wave-like" morphology in the structures of $CrSi_2$/Si obtained by diffusion synthesis conforms well with the results of the literature [15,25] and is explained by the mismatch of the crystal lattice parameters of $CrSi_2$ and Cr and the prevailing diffusion of silicon from the substrate to the boundary of Cr/Si. It is known that the parameter of the crystal lattice of $CrSi_2$ by far exceeds the parameter of the crystal lattice of Cr (4.405 and 2.885 Å, respectively) [14,21]. Formation of $CrSi_2$ starts at the boundary of Cr/Si under the layer of Cr and is realized by diffusion of Si from the substrate to the separation boundary and by embedding into the lattice of Cr with its subsequent transformation into $CrSi_2$.

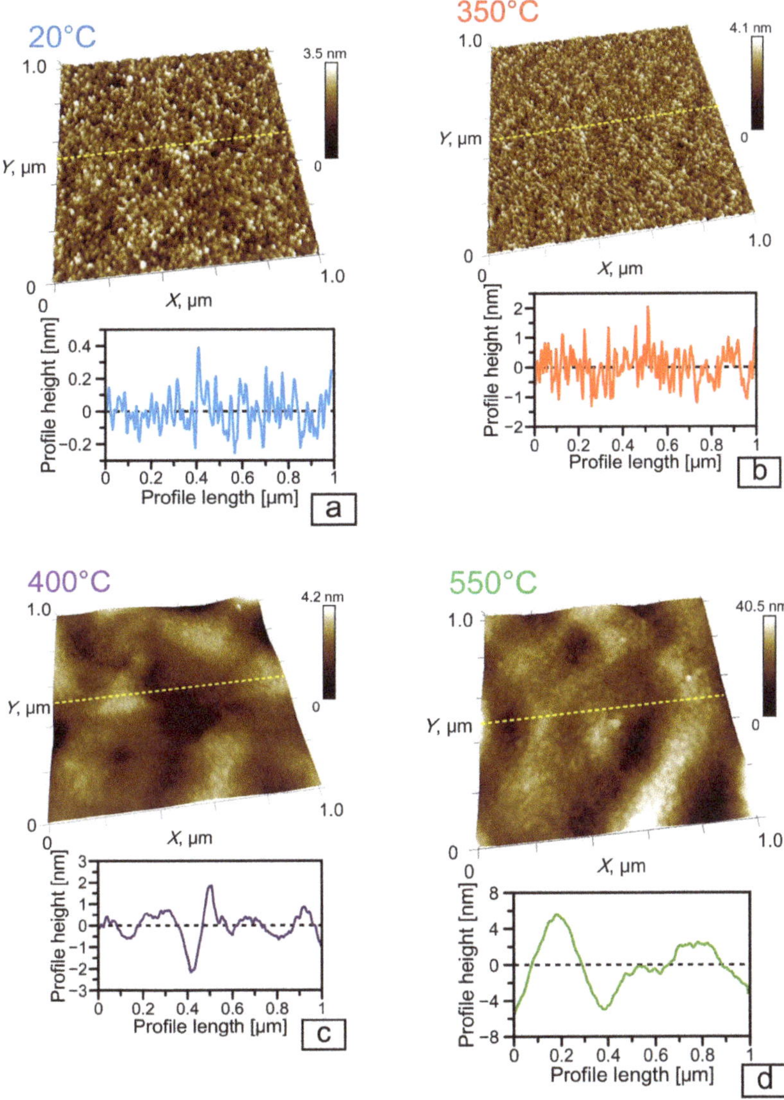

Figure 3. Morphology and the surface nanoprofiles of the Cr/Si structure after deposition and rapid thermal treatment: (**a**) initial film; (**b**) at the temperature of 350 °C; (**c**) at the temperature of 400 °C; (**d**) at the temperature of 550 °C.

Table 1 shows the difference in the characteristics of the Cr films after the conventional heating and RTT of the same film at the same temperature. The complete absence of Cr lines in the XRD analysis of the Cr/Si structure after RTT treatment at 450 °C suggests that the top layer is $CrSi_2$ (Figure 2). The thickness of the $CrSi_2$ layer obtained by RTT is almost twice that after conventional heating. The difference in Cr thickness after conventional heating and RTT heat treatment is explained by the difference in the activation energy of the process. During RTT, additional stimulating factors act on the process of $CrSi_2$ formation, reducing the activation energy of the process. Such factors can be the breaking of silicon–silicon bonds and the excitation of electrons in the silicon wafer under the action of photon flux.

Table 1. Comparison of the $CrSi_2$ layer properties obtained by RTT and conventional heating.

Treatment	T (°C)	Time	Film Thickness (nm)	Ra (nm)	Rq (nm)	Rz (nm)	Grain Size (nm)	γ (N/m)	Specific Resistivity (10^{-4} Ω·cm)
RTT	450	7 s	60.0 *	3.33	4.25	3.1	18.1	0.28	31.1
Conventional heating		30 min	35.7	1.59	2.17	1.7	16.8	0.28	1.2

* The thickness of film after RTT; the thickness before RTT was about 30 nm.

At the time of the rapid thermal treatment (for several seconds) the entire layer of Cr is replaced with $CrSi_2$, and as its volume is greater, distortion of the straight line of the boundary between the silicon wafer and the film occurs with distortion of the coating surface. Formation of $CrSi_2$ in the system of Cr–Si is a most probable outcome on the assumption of the lowest standard enthalpy of formation (ΔH°) of an individual phase of $CrSi_2$ amongst other probable phases [25]. The paper [26] indicates that chromium disilicide is formed at the expense of silicon atom diffusion and not because of chromium, as for metals a higher energy of activation is required, 1.4 times higher as a minimum.

The surface morphological change during the phase transition of Cr → $CrSi_2$ also determines the appropriate changes of its nanoprofiles (compare Figure 3a,b and Figure 3c,d). The comparative analysis of the samples' nanoprofiles demonstrates an abrupt increase in both the average values of nanoasperities at the RTT temperature of 400 °C and of the incidences of their scatter (Figure 4). Thus, in the initial structure of Cr/Si and after its rapid thermal treatment at the temperature from 200 to 350 °C, the mean values of nanoasperities are within the limits from 2 to 7 nm. An increase in the RTT temperature from 400 to 550 °C results in the shift of the distribution maximum and in the scatter of the average values of nanoasperities from 10–35 nm to 30–60 nm appropriately. Such an evolution of the surface nanoprofiles is determined not only by the samples' surface morphology transformation but also by the size change of the crystal grains.

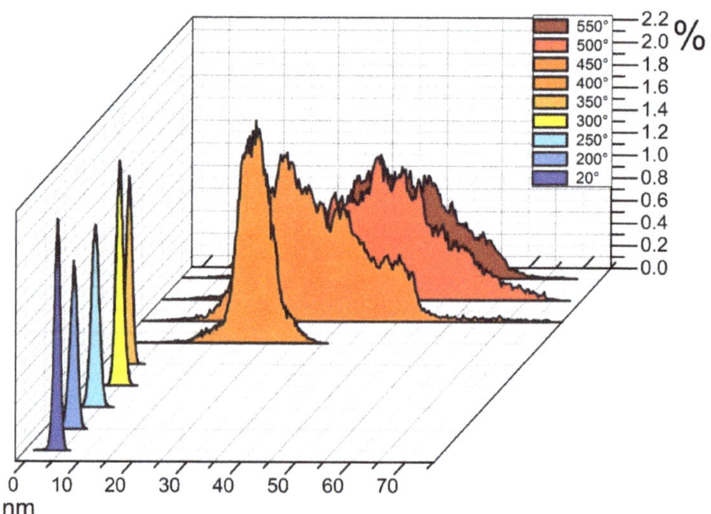

Figure 4. Histograms of the distribution densities of the height values of the nanoasperities on the surface of the structure with the size of 1 × 1 μm^2 after deposition and rapid thermal treatment.

Analysis of the changes in the grain sizes in the Cr/Si structure, obtained by means of AFM on the fields with the size of 100 × 100 nm^2, with the increase in the RTT temperature confirms the given conclusion (Figure 5). The mean grain size in the initial film of Cr is about 15 nm (Figure 5a). At the RTT temperature of the Cr/Si structure from 200 to 350 °C, it is reduced to 10–12 nm (Figure 5b). This somewhat contradicts the classic presumptions in compliance with which the process of heating increases the grain size as during the prolonged time the collective recrystallization manages to occur [27]. However, in our case, this does not happen due to the high rate of heating and absence of the prolonged retention at the temperature.

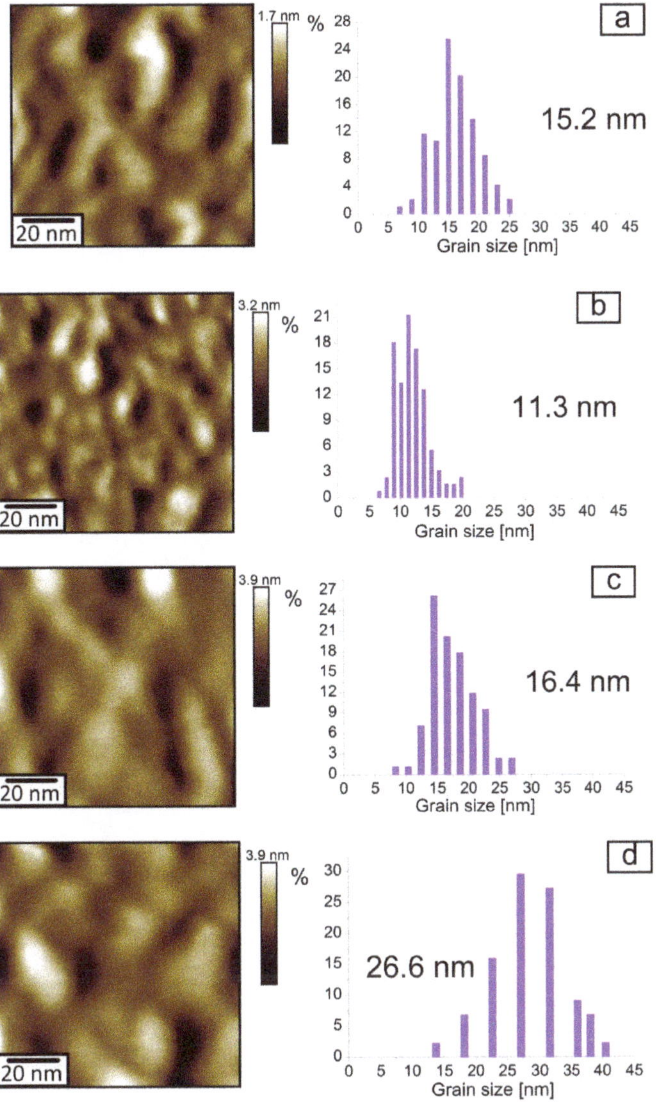

Figure 5. Atomic force microscope image (2D view) of the surface and the histograms of the distribution densities of the grain sizes of the Cr/Si structure after deposition and rapid thermal treatment: (**a**) initial film; (**b**) at the temperature of 350 °C; (**c**) at the temperature of 400 °C; (**d**) at the temperature of 550 °C.

Nonetheless, the given treatment results in recrystallization of the Cr films, accompanied by the density reduction of the structural and admixture defects, which enhances their density. At the RTT temperature of the Cr/Si structure from 400 to 550 °C, the average grain size increases from 16 to 26 nm, which is determined by the increase in the rate of the Si diffusion into the Cr film with the RTT temperature increase.

In all the applications of the thin films of Cr and $CrSi_2$, the important parameters determining the level of their functional properties are the grain size and roughness of the surface. In some cases, the best option is low roughness, while in others, the best option is high roughness [6]. The values of the roughness parameters determined with the application of the atomic force microscope (AFM) can be found in academic papers on the electronic properties of the thin films of Cr and $CrSi_2$ [1,2,5,6,28–30]. However, grain size investigation results are far more rarely found, although the atomic force microscope is a well-suited tool for the evaluation of the grain size in polycrystalline films [31,32]. For example, in [2], the grain size of the Cr film was close to that obtained in this work and was 13 and 14 nm at the negative bias voltage on the substrate of 50 and 250 V. At the negative bias on the substrate of 450 V, the grain size was 20 nm [2]. The roughness was also close and amounted to 0.57–1.49 nm.

Analyses of dependences the roughness parameters *Ra*, *Rq* and *Rz* on the RTT temperature (Figure 6) also show their relation, primarily, with the phase changes in the structure of Cr/Si. Roughness parameters *Ra* and *Rq* of the initial structure and after its rapid thermal treatment at the temperature from 200 to 350 °C are at the level of 0.5 nm and do not experience substantial changes. The *Rz* roughness parameter within the given RTT temperature range is at the level of 1.0–2.0 nm. The considerable changes in the values of the surface roughness parameters of the Cr/Si structure are observed in the result of the phase transition of Cr $\rightarrow$ $CrSi_2$ with the RTT temperature increase from 400 to 550 °C. Thus, the *Ra* surface roughness parameter in the given RTT temperature range increases from 1.5 to 5.5 nm, the *Rq* parameter increases from 2.0 to 7.0 nm and the *Rz* parameter increases from 1.5 to 11.0 nm, reflecting the changes in the phase composition and the crystal structure.

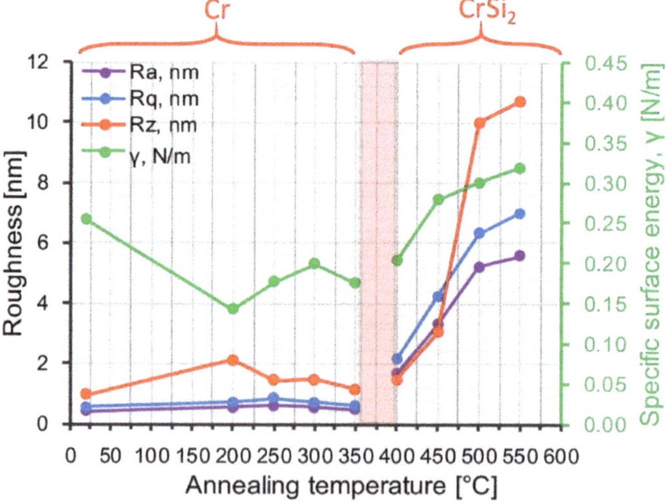

Figure 6. Dependences of the roughness parameters *Ra*, *Rq* and *Rz* and the specific surface energy of the Cr/Si structure on the rapid thermal treatment temperature.

The value of the specific surface energy in the samples with the initial structure of Cr/Si is at the level of 0.26 N/m and after the rapid thermal treatment at the temperature of 200 °C is reduced almost 2 times to values of the order of 0.14 N/m. The given behavior of the γ parameter serves as evidence of the unbalanced state of the initial Cr film, obtained immediately after the magnetron sputtering and containing the excessive mechanical stresses, structural defects and argon admixture atoms. Thus, the RTT of the Cr/Si structure results in the density reduction of the structural and admixture defects in the Cr film, as well as in the relaxation of the film's mechanical stresses. The insignificant growth of values of the γ parameter up to 0.18–0.20 N/m with the RTT temperature increase to 250–350 °C is caused by the growth in the number of unbalanced states, related to the appropriate increase in the heating and cooling rate of the structure. The unbalanced states of some atoms are associated with the action of photon energy and additional breaking of Si–Si bonds. The changes of the γ parameter at the RTT temperatures from 400 to 550 °C, apparently, are determined both by the structural-phase transformations in the Cr/Si structure and the by the changing morphology of its surface, as well as by the cooling rate of the structure after the rapid thermal treatment.

The specific resistivity value of the initial Cr thin films on silicon is about 1.0×10^{-4} Ω·cm, and after the rapid thermal treatment at the temperatures from 200 to 350 °C, it is reduced to 8.0×10^{-5} Ω·cm (Figure 5), which confirms the earlier assumptions about the recrystallization processes resulting in the density reduction of the admixture and structural defects and in the higher density of grains. The rapid thermal treatment at the temperature of 400 °C results in the avalanche growth of specific resistivity to values of the order of 1.2×10^{-3} Ω·cm, which is definitely determined by the transition of the Cr film to the $CrSi_2$ phase. Within the RTT temperature range from 450 to 550 °C, one observes the growth in the specific resistivity from 3.1×10^{-3} to 4.0×10^{-3} Ω·cm. Comparison of the dependences of the specific resistivity and the grain sizes on the RTT temperature of the Cr/Si structure (Figure 7) makes it possible to draw a conclusion that the ϱ value is directly influenced specifically by the grain size of the $CrSi_2$ phase. The significant changes in all parameters (roughness, specific surface energy, grain size, thickness, surface resistivity) of the thin Cr and $CrSi_2$ films are indicators of the phase transition in the films. The increased diffusion rate at 400–550 °C RTT is associated with the breaking of Si–Si bonds under the action of photon energy. The increased diffusion rate and the difference in the parameters of the crystal lattices of Cr and $CrSi_2$ cause an increased surface roughness in the $CrSi_2$ layers. A high correlation has been found between roughness and surface resistance.

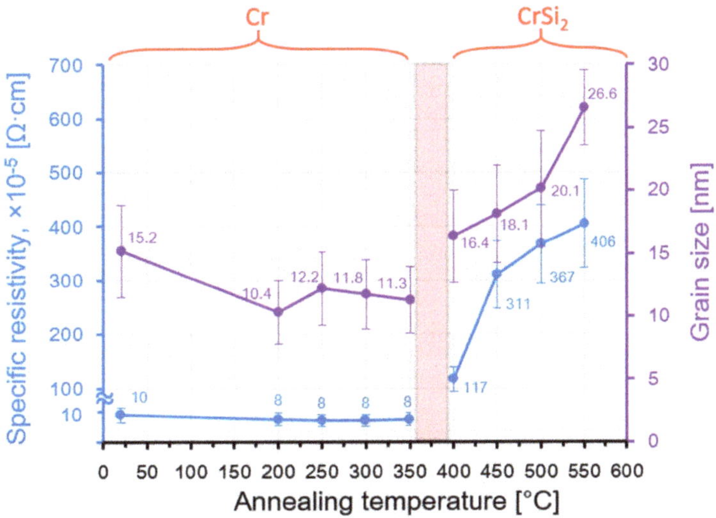

Figure 7. Dependences of resistivity and the mean grain size in the Cr film on the temperature of the rapid thermal treatment of the Cr/Si structure.

4. Conclusions

By means of the atomic force microscope and the electrophysical measurements, the influence of the rapid thermal treatment temperature of the Cr/n-Si structure (111) with incoherent light flux irradiation at constant power on the back side of the silicon wafer for 7 s in the nitrogen medium on the surface morphology and the specific resistivity of the Cr films was studied.

It has been established that at the rapid thermal treatment temperature of the Cr/Si structure from 200 to 350 °C, the fine grain morphology of the initial Cr film is retained, as are the values of the surface roughness parameters *Ra*, *Rq* and *Rz*. Meanwhile, one observes an insignificant reduction in the grain size, specific surface energy and resistivity of the chromium films, which is determined by recrystallization of the chromium films, accompanied by the density reduction of the structural and admixture defects.

Within the rapid thermal treatment temperature range of the Cr/Si structure from 400 to 550 °C, the surface acquires the wave-like morphology determined by the phase transition of the Cr film to the $CrSi_2$ layer with a greater parameter of the crystal lattice and is formed at the expense of silicon atom diffusion from the substrate to the layer in growth. The given phase changes are accompanied by the growth of the crystal grain size and determine the increase in the surface roughness parameters *Ra*, *Rq* and *Rz* and the specific surface energy and resistivity of the $CrSi_2$ layers.

Author Contributions: Conceptualization, T.K., S.C., V.P. and S.A.; methodology, T.K. and V.L.; software, T.K. and V.L.; validation, T.K. and J.S.; formal analysis, V.L.; investigation, T.K., J.S. and V.L.; resources, J.S. and S.A.; data curation, T.K., S.C., V.P. and S.A.; writing—original draft preparation, T.K.; writing—review and editing, J.S.; visualization, V.L.; supervision, S.C., V.P. and S.A.; project administration, T.K.; funding acquisition, S.A. All authors have read and agreed to the published version of the manuscript.

Funding: This research was partly financed by the grant of the Belarusian Republican Foundation for Fundamental Research BRFFR No. F20M-083 and by the Government of Russia (grant No. 14.Z50.31.0046).

Conflicts of Interest: The authors declare no conflict of interest.

References

1. Rawal, S.K.; Kumar, A.; Jayaganthan, R.; Chandra, R. Applied Surface Science Optical and hydrophobic properties of co-sputtered chromium and titanium oxynitride films. *Appl. Surf. Sci.* **2011**, *257*, 8755–8761. [CrossRef]
2. Yate, A.L.; Mart, L.; Esteve, J.; Lousa, A. Ultra Low Nanowear in Novel Chromium/Amorphous Chromium Carbide. *Appl. Surf. Sci.* **2017**. [CrossRef]
3. Danilov, F.I.; Protsenko, V.S.; Gordiienko, V.O.; Kwon, S.C.; Lee, J.Y.; Kim, M. Applied Surface Science Nanocrystalline hard chromium electrodeposition from trivalent chromium bath containing carbamide and formic acid: Structure, composition, electrochemical corrosion behavior, hardness and wear characteristics of deposits. *Appl. Surf. Sci.* **2011**, *257*, 8048–8053. [CrossRef]
4. Santoni, F.; Giovine, E.; Torrioli, G.; Chiarello, F.; Castellano, M.G. Study of the Fabrication Process for a Dual Mass Tuning Fork Gyro. *Procedia Eng.* **2014**, *87*, 991–994. [CrossRef]
5. Ghoranneviss, M.; Sari, A.H.; Esmaeelpour, M. Nitrogen implantation and heat treatment effect on the hardness improvement of the chromium film surface deposited on Si (1 1 1) substrate. *Appl. Surf. Sci.* **2004**, *237*, 326–331. [CrossRef]
6. Lin, Y.C.; Hsu, C.Y.; Hung, S.K.; Chang, C.H.; Wen, D.C. The structural and optoelectronic properties of Ti-doped ZnO thin films prepared by introducing a Cr buffer layer and post-annealing. *Appl. Surf. Sci.* **2012**, *258*, 9891–9895. [CrossRef]
7. Wang, X.; Chen, F.; Liu, H.; Liang, W.; Yang, Q.; Si, J.; Hou, X. Fabrication of micro-gratings on Au-Cr thin film by femtosecond laser interference with different pulse durations. *Appl. Surf. Sci.* **2009**, *255*, 8483–8487. [CrossRef]
8. Shi, F.; Wang, Z.; Xue, C. Synthesis and characterization of GaN nanowires by ammoniating Ga_2O_3/Cr thin films deposited on Si(1 1 1) substrates. *Appl. Surf. Sci.* **2010**, *256*, 4883–4887. [CrossRef]
9. Balu, R.; Raju, A.R.; Lakshminarayanan, V.; Mohan, S. Investigations on the influence of process parameters on the structural evolution of ion beam sputter deposited chromium thin films. *Mater. Sci. Eng. B Solid-State Mater. Adv. Technol.* **2005**, *123*, 7–12. [CrossRef]
10. Martinez, A.; Esteve, D.; Guivarc'h, A.; Auvray, P.; Henoc, P.; Pelous, G. Metallurgical and electrical properties of chromium silicon interfaces. *Solid State Electron.* **1980**, *23*, 55–64. [CrossRef]
11. Deneb Menda, U.; Özdemir, O.; Tatar, B.; Ürgen, M.; Kutlu, K. Transport and storage properties of $CrSi_2$/Si junctions made using the CAPVD technique. *Mater. Sci. Semicond. Process.* **2010**, *13*, 257–266. [CrossRef]

12. Sobe, G.; Schreiber, H.; Weise, G.; Voigtmann, R.; Zies, G.; Sonntag, J.; Grotzschel, R. Deposition Of Cr-Si Thin Films By Reactive Plasmatron-Magnetron Sputtering. *Thin Solid Film* **1985**, *128*, 149–159. [CrossRef]
13. Tam, P.L.; Cao, Y.; Nyborg, L. Thin film characterisation of chromium disilicide. *Surf. Sci.* **2013**, *609*, 152–156. [CrossRef]
14. Galkin, N.G.; Dózsa, L.; Chusovitin, E.A.; Dotsenko, S.A.; Pécz, B.; Dobos, L. Influence of CrSi2 nanocrystals on the electrical properties of Au/Si-p/CrSi2 NCs/Si(111)-n mesa-diodes. *Phys. Procedia* **2011**, *11*, 35–38. [CrossRef]
15. Gül, F. Addressing the sneak-path problem in crossbar RRAM devices using memristor-based one Schottky diode-one resistor array. *Results Phys.* **2019**, *12*, 1091–1096. [CrossRef]
16. Henrion, W.; Lange, H.; Jahne, E.; Giehler, M. Optical properties of chromium and iron disilicide layers. *Appl. Surf. Sci.* **1993**, *70–71*, 569–572. [CrossRef]
17. Galkin, N.G.; Konchenko, A.V.; Vavanova, S.V.; Maslov, A.M.; Talanov, A.O. Transport, optical and thermoelectrical properties of Cr and Fe disilicides and their alloys on Si(1 1 1). *Appl. Surf. Sci.* **2001**, *175–176*, 299–305. [CrossRef]
18. Kovsarian, A.; Shannon, J.M.; Cristiano, F. Comparison of amorphous Mo and Cr disilicides in hydrogenated amorphous silicon. *J. Non. Cryst. Solids* **2000**, *276*, 40–45. [CrossRef]
19. Zhu, J.; Barbier, D.; Mayet, L.; Gavand, M.; Chaussemy, G. Interstitial chromium behaviour in silicon during rapid thermal annealing. *Appl. Surf. Sci.* **1989**, *36*, 413–420. [CrossRef]
20. D'Anna, E.; Leggieri, G.; Luches, A.; Majni, G.; Ottaviani, G. Chromium silicide formation under pulsed heat flow. *Thin Solid Films* **1986**, *136*, 93–104. [CrossRef]
21. Solovjov, J.A.; Pilipenko, V.A. Effect of rapid thermal treatment conditions on electrophysical properties of chromium thin films on silicon. *Doklady Bguir.* **2019**, *7–8*, 157–164. [CrossRef]
22. Filonenko, O.; Mogilatenko, A.; Hortenbach, H.; Allenstein, F.; Beddies, G.; Hinneberg, H.J. Influence of ultrathin templates on the epitaxial growth of CrSi 2 on Si(0 0 1). *Microelectron. Eng.* **2004**, *76*, 324–330. [CrossRef]
23. Newton, P.; Houzé, F.; Guessab, S.; Noël, S.; Boyer, L.; Lécayon, G.; Viel, P. Atomic force microscopy study of the topographic evolution of polyacrylonitrile thin films submitted to a rapid thermal treatment. *Thin Solid Films* **1997**, *303*, 200–206. [CrossRef]
24. Lamprou, D.A.; Smith, J.R.; Nevell, T.G.; Barbu, E.; Stone, C.; Willis, C.R.; Tsibouklis, J. A comparative study of surface energy data from atomic force microscopy and from contact angle goniometry. *Appl. Surf. Sci.* **2010**, *256*, 5082–5087. [CrossRef]
25. Galkin, N.G.; Astashynski, V.M.; Chusovitin, E.A.; Galkin, K.N.; Dergacheva, T.A.; Kuzmitski, A.M.; Kostyukevich, E.A. Ultra high vacuum growth of $CrSi_2$ and β-$FeSi_2$ nanoislands and Si top layers on the plasma modified monocrystalline silicon surfaces. *Phys. Procedia.* **2011**, *11*, 39–42. [CrossRef]
26. Finstad, T.L.; Thomas, O.; D'heurle, F.M. Bilayers with Chromium Dishlicide: Chromium-Vanadium. *Appl. Surf. Sci.* **1989**, *38*, 106–116. [CrossRef]
27. Chaplanov, A. Pulsed annealing of metal films and metal-semiconductor systems. *Vacuum* **1993**, *44*, 1085–1088. [CrossRef]
28. Höflich, A.; Bradley, N.; Hall, C.; Evans, D.; Murphy, P.; Charrault, E. Packing density/surface morphology relationship in thin sputtered chromium films. *Surf. Coatings Technol.* **2016**, *291*, 286–291. [CrossRef]
29. Lintymer, J.; Martin, N.; Chappe, J.M.; Takadoum, J. Glancing angle deposition to control microstructure and roughness of chromium thin films. *Wear* **2008**, *264*, 444–449. [CrossRef]
30. Patnaik, L.; Maity, S.R.; Kumar, S. Comprehensive structural, nanomechanical and tribological evaluation of silver doped DLC thin film coating with chromium interlayer (Ag-DLC/Cr) for biomedical application. *Ceram. Int.* **2020**, *46*, 22805–22818. [CrossRef]
31. Kuznetsova, T.; Lapitskaya, V.; Khabarava, A.; Chizhik, S.; Warcholinski, B.; Gilewicz, A. The influence of nitrogen on the morphology of ZrN coatings deposited by magnetron sputtering. *Appl. Surf. Sci.* **2020**, *522*, 146508. [CrossRef]
32. Anishchik, V.M.; Uglov, V.V.; Kuleshov, A.K.; Filipp, A.R.; Rusalsky, D.P.; Astashynskaya, M.V.; Samtsov, M.P.; Kuznetsova, T.A.; Thiery, F.; Pauleau, Y. Electron field emission and surface morphology of a-C and a-C:H thin films. *Thin Solid Films* **2005**, *482*, 248–252. [CrossRef]

nanomaterials

MDPI

Article

Non-Isothermal Decomposition as Efficient and Simple Synthesis Method of NiO/C Nanoparticles for Asymmetric Supercapacitors

Daria Chernysheva [1], Ludmila Pudova [1], Yuri Popov [2], Nina Smirnova [1], Olga Maslova [3], Mathieu Allix [4], Aydar Rakhmatullin [4], Nikolay Leontyev [5], Andrey Nikolaev [6] and Igor Leontyev [2,*]

1 Platov South-Russian State Polytechnic University (NPI), 346428 Novocherkassk, Russia; da.leontyva@mail.ru (D.C.); pudowa.lyuda@yandex.ru (L.P.); smirnova_nv@mail.ru (N.S.)
2 Physics Department, Southern Federal University, 344090 Rostov-on-Don, Russia; popov@sfedu.ru
3 Institute of Strength Physics and Materials Science, Siberian Branch, Russian Academy of Sciences, 346428 Tomsk, Russia; o_maslova@rambler.ru
4 CNRS, CEMHTI UPR3079, Univ. Orléans, F-45071 Orléans, France; mathieu.allix@cnrs-orleans.fr (M.A.); aydar.rakhmatullin@cnrs-orleans.fr (A.R.)
5 Azov-Black Sea Engineering Institute, Don State Agrarian University, Rostov region, 347740 Zernograd, Russia; lng48@mail.ru
6 Research and Education Center "Materials", Don State Technical University, 344000 Rostov-on-Don, Russia; andreynicolaev@eurosites.ru
* Correspondence: i.leontiev@rambler.ru; Tel.: +7-918-552-4024

Abstract: A series of NiO/C nanocomposites with NiO concentrations ranging from 10 to 90 wt% was synthesized using a simple and efficient two-step method based on non-isothermal decomposition of Nickel(II) bis(acetylacetonate). X-ray diffraction (XRD) measurements of these NiO/C nanocomposites demonstrate the presence of β-NiO. NiO/C nanocomposites are composed of spherical particles distributed over the carbon support surface. The average diameter of nickel oxide spheres increases with the NiO content and are estimated as 36, 50 and 205 nm for nanocomposites with 10, 50 and 80 wt% NiO concentrations, respectively. In turn, each NiO sphere contains several nickel oxide nanoparticles, whose average sizes are 7–8 nm. According to the tests performed using a three-electrode cell, specific capacitance (SC) of NiO/C nanocomposites increases from 200 to 400 F/g as the NiO content achieves a maximum of 60 wt% concentration, after which the SC decreases. The study of the NiO/C composite showing the highest SC in three- and two-electrode cells reveals that its SC remains almost unchanged while increasing the current density, and the sample demonstrates excellent cycling stability properties. Finally, NiO/C (60% NiO) composites are shown to be promising materials for charging quartz clocks with a power rating of 1.5 V (30 min).

Keywords: supercapacitor; nickel oxide; NiO/C nanocomposite; non-isothermal decomposition

Citation: Chernysheva, D.; Pudova, L.; Popov, Y.; Smirnova, N.; Maslova, O.; Allix, M.; Rakhmatullin, A.; Leontyev, N.; Nikolaev, A.; Leontyev, I. Non-Isothermal Decomposition as Efficient and Simple Synthesis Method of NiO/C Nanoparticles for Asymmetric Supercapacitors. *Nanomaterials* **2021**, *11*, 274. https://doi.org/10.3390/nano11010187

Received: 20 November 2020
Accepted: 27 December 2020
Published: 13 January 2021

Publisher's Note: MDPI stays neutral with regard to jurisdictional claims in published maps and institutional affiliations.

1. Introduction

Asymmetric supercapacitors (SCs) are increasingly attracting attention as promising energy storage devices due to the constantly growing energy and ecological problems that are associated with fossil fuel depletion and global warming [1]. SCs contain transition metal oxides that are able to accumulate large charges through Faraday processes on the electrode surface. Among these oxides, NiO is of great interest [2,3] due to the high estimated capacitance (2584 F/g [4]), good corrosion resistance in alkaline solutions, and environmental safety.

Meanwhile, nickel oxide is known as a broadband p-type semiconductor (3.6–4.0 eV [5]) with a low electron conductivity (0.01–0.32 S/m [6]), which impedes its use in supercapacitor electrodes. This problem can be solved by reinforcing NiO's conductivity with conducting carbon nanomaterials, such as ultradispersed soot, carbon nanotubes, or graphene, that possess highly developed surface, chemical inertness and high electron conductivity [7,8]. In comparison with pure nickel oxides, NiO/carbon composite structures ensure faster transfer of electron and ion charges, reduce the resistance of the rechargeable material and allow efficient application of the latter.

Electrodes of NiO supercapacitors are typically based on various nanostructures. On the one hand, the use of nanocrystalline NiO leads to an increase of the surface area. On the other hand, the size effect causes a decrease of the bandgap and an increase of conductivity [9,10]. The synthesis of nanometer scale NiO with different morphology, i.e., nano-sheets [11], thin film [12], plates [13], nanopyramids [14], and nanoparticles [15], is currently accessible via electrochemical routes [13,16], emulsion nanoreactor method [17], thermal decomposition of $Ni(OH)_2$ at various temperatures [18,19], sol–gel [20], hydrothermal method [11,21], spray pyrolysis [12], etc.

Since the capacitance of oxide-based composites strongly depends on their microstructural characteristics that may differ at various synthesis conditions and component ratios, the development of an efficient and simple synthesis method that would ensure the production of composite electrode materials with optimal specific electrochemical properties is a relevant task. Furthermore, it was previously shown [22,23] that non-isothermal decomposition of Pt acetyl acetonate enables access to Pt/C nanoparticles whose surface area can reach 100 m^2/g. This synthesis method is very simple and requires only two stages; moreover, no washing, filtering or subsequent drying of a specimen are required. Hence, the present study is based on the non-thermal decomposition synthesis for producing NiO/C nanoparticles. This synthesis was performed at constant heating rate, but at different nickel oxide-to-carbon carrier ratios in order to achieve the highest possible capacitance of a composite material.

2. Materials and Methods

Nickel (II) acetylacetonate $Ni(acac)_2$ (95%, Sigma-Aldrich, St. Louis, MO, USA) was used as a precursor for the synthesis of NiO/C nanocomposites, and Vulcan-XC 72 (Cabot, Boston, MA, USA) served as carbon support. At the first stage, nickel acetylacetonate was solved at room-temperature in Tetrahydrofuran solution (99.9%, Panreac Quimica SLU, Barcelona, Spain). The volume of solvent was chosen so that $Ni(acac)_2$ was completely diluted. The carbon support was then added to the mixture using continuous stirring. After that, the blend was dried under continuous stirring using a magnetic stirrer and in an ultrasonic bath at a temperature of 50 °C. The powder was subsequently placed in a Nabertherm P300 high-temperature furnace (Nabertherm, Lilienthal, Germany). The synthesis was conducted under non-isothermal conditions in air at a heating rate of 1 K/min. The final synthesis temperature was determined from thermogravimetry data as 420 °C, which corresponded to complete acetylacetonate decomposition (Figure S1). According to the thermogravimetric data, nickel oxide concentrations in resulting composite materials were respectively 10, 20, 30, 40, 50, 60, 70, 80, and 90 wt%.

The powder X-ray diffraction (XRD) data were collected on ARL X'TRA Thermo Fisher Scientific powder diffractometer (Thermo Electron SA, Ecublens, Switzerland). All XRD powder patterns were refined using the Rietveld method which enabled us to evaluate the average crystalline size D and parameter of the unit cell a. Since the β-form of nickel oxide has a space group Fm3m, the X-ray diffraction patterns were refined similarly to that described in the work [24,25]. The quality of X-ray refinement of NiO/C composite is shown in Figure S2. Scanning electron microscopy (SEM) images were obtained on a ZEISS Supra 25 unit microscope (Carl Zeiss SMT, Oberkochen, Germany) operated at 20 kV. Transmission electron microscopy (TEM) images obtained in a Philips CM 20 TEM microscope (Philips, Amsterdam, The Netherlands) operated at 200 kV. Raman spectra were measured at room temperature through 50× microscope objective using a Renishaw micro-Raman spectrometer (Renishaw plc., Gloucestershire, UK) equipped with argon laser (514.5 nm, max power Pex = 10 mW).

The electrochemical characterization of NiO/C composites as electrode materials for supercapacitors was performed at room temperature in a three-electrode cell using a P-150 potentiostat/galvanostat (Elins STC, Zelenograd, Moscow, Russia). A platinum wire and a saturated Ag/AgCl were used as counter and reference electrodes, respectively. The working electrodes were fabricated in accordance with the following procedure: 15 mg of a NiO/C composite (pure carbon black Vulcan XC-72) was suspended in a mixture of 2-propanol (Sigma-Aldrich) (0.5 mL) and 10 wt% aqueous Nafion^{TM} solution (Sigma-Aldrich) (0.02 mL). The mixture was stirred with a magnetic stirrer for 40 min; the formed ink was then dropped onto a glassy carbon foil electrode (Sigma-Aldrich) and was air-dried. The amount of dried ink in the working electrode was 1.0 mg. A 1 M KOH aqueous solution was used as electrolyte. Cyclic voltammetry measurements (CVs) were conducted in a potential range of −0.1 to 0.65 V at different scan rates. The galvanostatic charge–discharge (GCD) and cycle-life/stability studies were carried out within the potential range of 0 to 0.5 V. Based on the GCD process, the specific capacitance (C_s) was obtained from the following equation [26]:

$$C_s = \frac{I \times \Delta t}{m \times \Delta V}$$

where C_s (F g^{-1}) is the specific capacitance in a three-electrode system; I (A) is the discharging current; Δt (s) is the discharge time; m (g) is the weight of active materials (NiO/C composite or pure carbon black Vulcan XC-72); and ΔV (V) is the voltage charge during the discharge process.

A home-made cell of asymmetrical two-electrode SCs was assembled from polytetrafluoroethylene, where positive and negative current collectors were made of a nickel foil (23 mm in diameter). The separator consisted of a hydrophilic polypropylene membrane, and the electrolyte was a 6 M KOH solution. Both electrodes were prepared by mixing the active electrode material (NiO/C nanocomposite in the case of the positive electrode and carbon black (CB) Timcal Super C65 (Nanografi Nano Technology, Ankara, Turkey) for the negative electrode), the CB conductor (Timcal Super C65) and polyvinylidene difluoride (PVDF) with weight ratio of 7.5:1:1.5. A few drops of N-Methylpyrrolidone (Sigma-Aldrich) were added to form the slurry that was smeared onto the nickel foil current collector and exposed to 6 h of drying at 80 °C.

The electrochemical performance of the two-electrode system was tested via CV and GCD measurements. CV measurements were carried out in a potential range of 0 to 1.65 V at different scan rates. The GCD and cycle-life/stability studies were implemented in the same potential range. In the context the GCD process, the specific capacitance of a two-electrode system was obtained, as follows [27]:

$$C_s = \frac{I \times \Delta t}{M \times \Delta V}$$

where C_s (F g^{-1}) is the specific capacitance of a two-electrode system; I(A) is the discharging current; Δt (s) is the discharge time; M(g) is the total weight of active materials on positive and negative electrodes; and ΔV (V) is the voltage charge during the discharge process.

The mass ratio between the positive and negative electrodes can be found using the charge balance ($q^+ = q^-$) from the equation below [26,27]:

$$\frac{m_+}{m_-} = \frac{C_- \times \Delta V_-}{C_+ \times \Delta V_+}$$

where m is the mass, C is the specific capacitance, and ΔV is the potential window of the electrode. The electrochemical properties of the NiO/C composite and CB in a three-electrode system were examined separately through the CV and GCD measurements. According to the peculiar electrochemical behavior of each component, the optimal loading mass ratio of NiO/C to CB is about 0.045.

3. Results and Discussion

Figure 1 displays the data for all NiO/C composites. The peaks correspond to β-NiO (4.18 Å, *Fm*3*m*). Besides this, the 2θ angle range of NiO-deficient composites exhibits the presence of a broadened peak referred to a Vulcan carbon support. The average NiO nanoparticle size as a function of concentration, determined by XRD data refinement of NiO/C composites, evidences a slight increase in nanoparticle size from 7 to 8 nm while increasing the NiO concentration (Figure 2). The unit cell parameters vary from 4.1830 Å to 4.1792 Å when increasing the NiO concentration from 10 to 90 wt%, which exceeds the unit cell parameter a = 4.1771 Å for bulk NiO (JCPDS card no. 78-0429). A similar effect of the unit cell parameter behaving as an inverse relationship of the average NiO nanoparticle size was also observed in previous works [10].

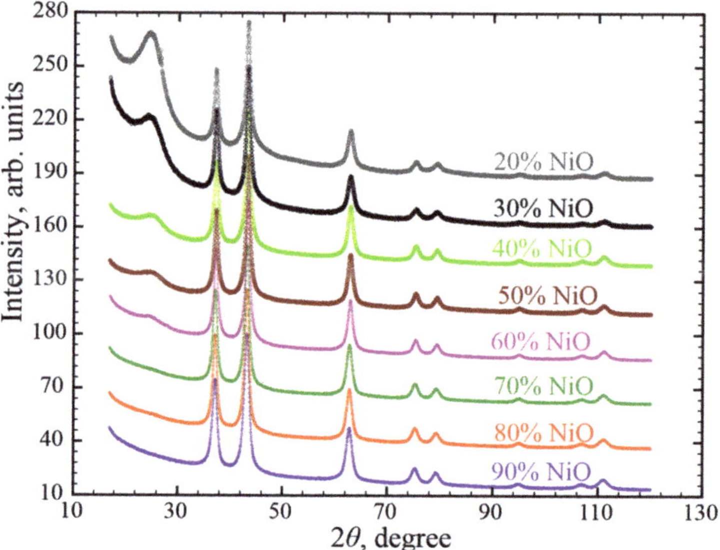

Figure 1. X-ray powder diffraction patterns of freshly prepared NiO/C nanocomposites.

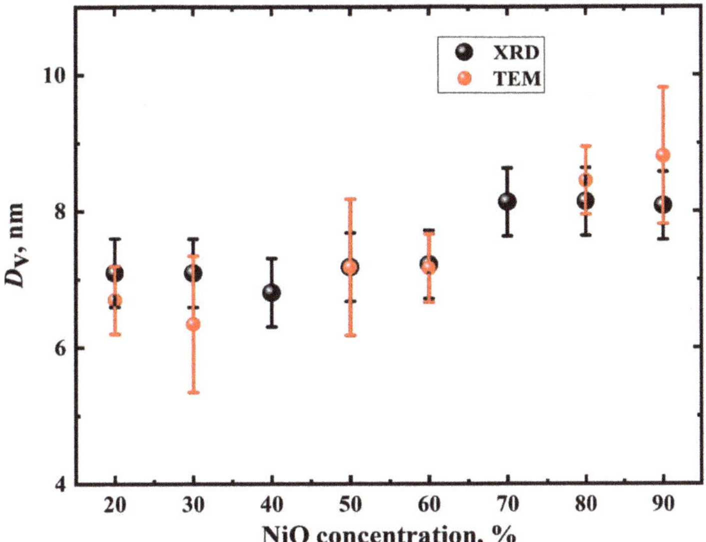

Figure 2. Nanoparticle average sizes evaluated from XRD (black circles) and TEM (red circles) data.

Each spectrum can be roughly divided into three areas: (i) 300 ... 1200 cm^{-1}, region where standard modes of NiO are observed (TO, LO, 2TO, TO+LO, 2LO) (Figure S3); (i) 1200 ... 1800 cm^{-1}, the region where the Vulcan modes are observed, corresponding to the first scattering order and representing a pattern typical for amorphous carbon (bands D and G); in this case, the two-magnon mode of NiO (~1490 cm^{-1}) is not visible (Figures S4 and S5).

Spectra can be conventionally divided into two groups: (i) spectra of samples with NiO concentration of 10%, 20%, 30%, 50% and 80%—in these spectra the Vulcan peaks are most pronounced and the NiO peaks are weakly visible (Figure S4); (ii) spectra of samples with NiO concentration of 40%, 60%, 70%, 90%—in these spectra the NiO peaks are most pronounced and the Vulcan peaks are weakly visible (Figure S5).

Scanning and transmission electron microscopy data of the synthesized NiO/C samples are plotted in Figure 3. As seen from the SEM images (Figure 3a,c), the samples are composed of NiO spherical particles distributed over the carbon support surface. Their sizes increase with NiO concentration in the produced nanocomposites. The average diameters of nickel oxide spheres are found to be 36 nm, 50 nm and 205 nm for nanocomposites with 10, 50 and 80 wt% of NiO concentrations, respectively. Based on these XRPD results, each sphere contains individual nickel oxide nanoparticles. This conclusion is consistent with TEM optical micrographs of NiO/C composites (Figure 3d–f).

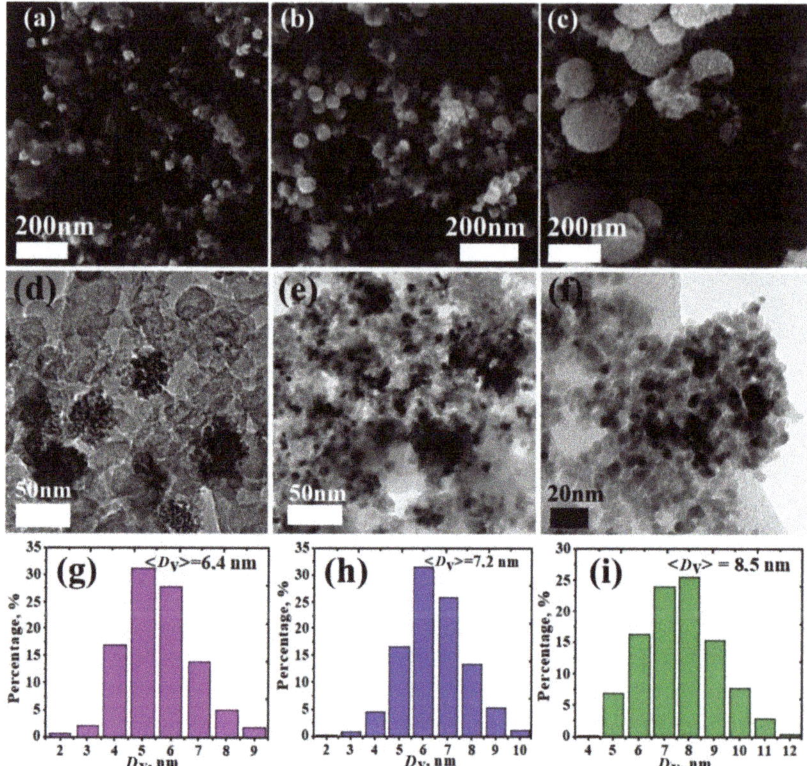

Figure 3. SEM (**a–c**) and TEM (**d–f**) images and grain size distributions (**g–i**) of NiO nanoparticles in NiO/C nanocomposites with different NiO concentrations: (**a**,**d**,**g**) 10 wt%, (**b**,**e**,**h**) 50 wt%, (**c**,**f**,**i**) 90 wt%.

Figure 3g–i illustrate the particle size distributions at various NiO concentrations in composites. The average NiO nanoparticle size as a function of NiO concentration is given in Figure 2. It is evident that the average nanoparticle sizes evaluated from TEM data is commensurate with those estimated via the Rietweld method.

The electrochemical results of all NiO/C composites, obtained in a three-electrode cell, are plotted in Figures 4 and 5. The cyclic voltammograms (CVs) can be divided into two groups with respect to their profiles (Figure 4). The first group comprises the composites with nickel oxide contents ranging from 10 to 60 wt% inclusively (Figure 4a). The current peaks of the higher nickel oxide formation at potentials unfolding towards the anodic range, as well as the appropriate current peaks of the higher nickel oxide reduction upon the potential unfolding towards the cathodic range in a series from 10 wt% to 60 wt%, increase regularly with the NiO amount, and all CVs exhibit an analogous conventional shape. The second group (Figure 4b), including NiO/C composites with NiO concentrations from 70 to 90 wt%, evidences no regularities from the first group. Along with galvanostatic data for all composites, displayed in Figure 5, such a division into two groups can be explained in the context of the percolation theory in strongly inhomogeneous media [28]. NiO/C composites are composed of a low-conductivity semiconducting component of nickel oxide (10^{-9} S cm^{-1} [9]) and an electron conductor of Vulcan carbon soot at their certain ratio, which ensures sufficient conductivity. According to earlier studies [29], starting from the volume soot concentration of 65 vol% (which corresponds to ~40 wt%), the electron conductivity of a composite becomes commensurate to that of pure soot, which allows the electrochemical data to be interpreted in a correct manner.

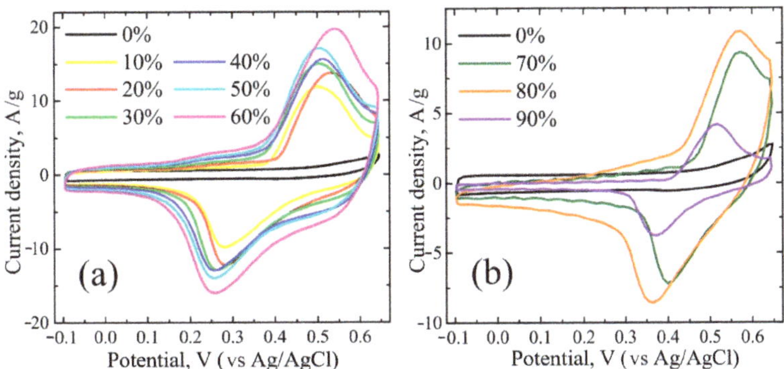

Figure 4. Cyclic voltammograms (CVs) of NiO/C samples, divided into two groups with NiO concentrations of: (**a**): 0, 10, 20, 30, 40, 50, 60 wt%; (**b**) 0, 70, 80, 90 wt%.

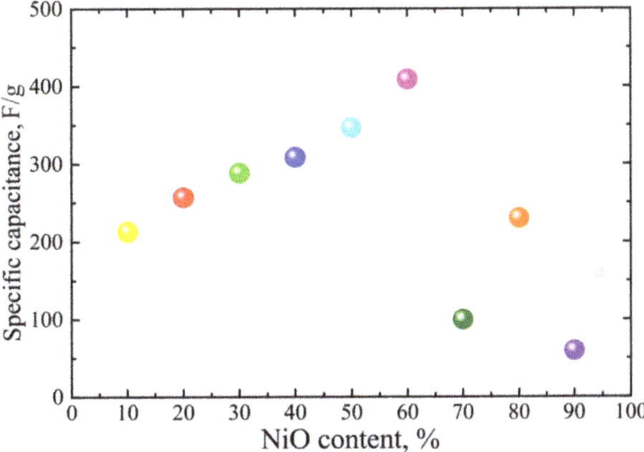

Figure 5. Specific capacitance as a function of NiO content in NiO/C composites at a current density of 0.5 F g^{-1} in a three-electrode cell in a 1 M KOH solution. The results are divided into two groups.

Preliminary examination in a three-electrode cell enables to conclude that a NiO/C composite with a 60 wt% NiO concentration possesses the best electrochemical properties, which makes it a desirable object for more comprehensive investigation of electrochemical characteristics.

The CVs of NiO/C (60 wt%) samples (Figure 6a) exhibit a pronounced anodic peak, associated with a higher oxide formation on the surface, and a cathodic peak of their reduction described by the following Equation (1):

$$NiO + OH^- \longleftrightarrow NiOOH + e^- \quad (1)$$

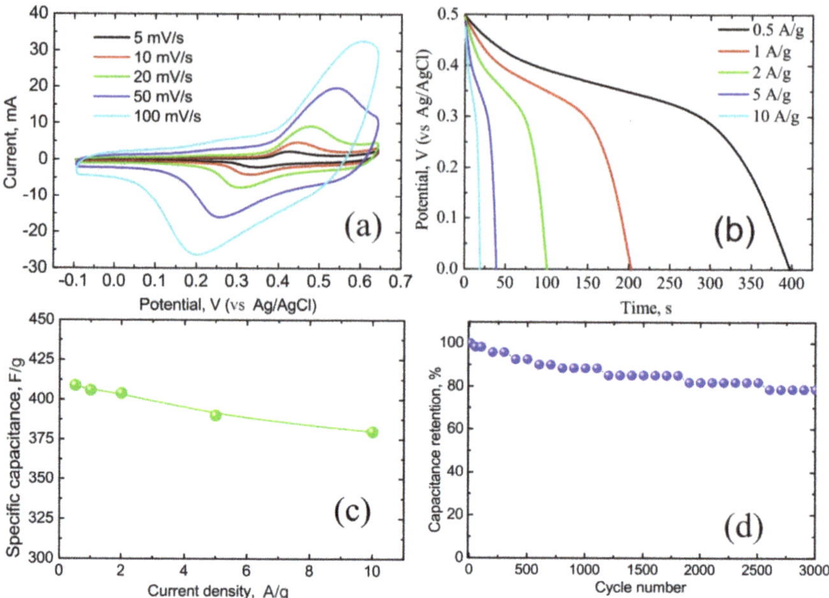

Figure 6. Electrochemical performance of NiO/C nanocomposite (60 wt% NiO) in a three-electrode cell in a 1 M KOH aqueous solution: (**a**) CVs at various scan rates; (**b**) galvanostatic discharge curves at various current densities; (**c**) specific capacitance as a function of current density; (**d**) cycling stability at the current density of 2 A/g.

The maximum current of the anodic peak depends on the potential unfolding rate, meaning that the diffusion limiting process is the proton diffusion in the solid phase.

In order to gather more detailed information on the specific capacitance of NiO/C (60 wt%) composite, the galvanostatic charge/discharge measurements were made at various current densities (Figure 6b). As clearly seen in Figure 6c, the specific capacitance remains almost unchanged while increasing the current density.

In the Table S1 demonstrated a comparation of the electrochemical performance of the obtained NiO/C (60 wt% NiO) and other NiO/C composites with different carbon supports (composite microspheres [30], hollow microspheres [31], CNTs [32], carbon coating on the NiO [33], graphite [34], sulfonated graphene [35], carbon nanospheres [36]), prepared by various methods [30–36]. One can see that the C_s of the NiO/C (60 wt% NiO) at the current density of 1 A g^{-1} is 405 Fg^{-1} (Figure 6c, Table S1) which is nearly the same [35,36] or much higher than C_s [30–34] of other materials.

In order to investigate the electrochemical properties of NiO/C nanocomposite (60 wt% NiO) in terms of its potential applications, an asymmetric two-electrode cell was assembled (Figure 7) with a positive electrode of NiO/C (60 wt% NiO) and a negative electrode of carbon black Timcal Super C65 in a 6 M KOH electrolyte.

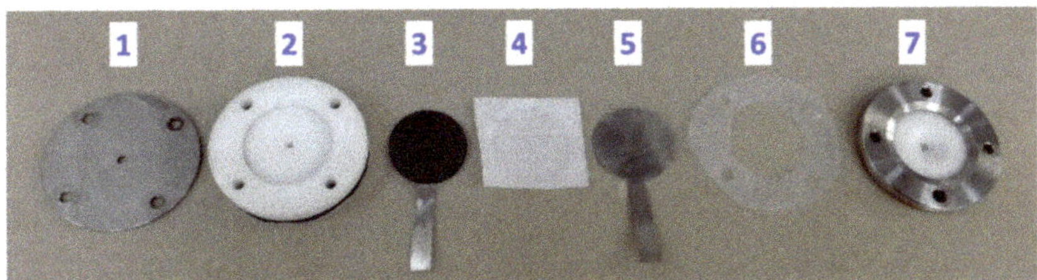

Figure 7. A two-electrode cell: 1—stainless steel plate, 2—polytetrafluoroethylene cell, 3, 5—current collectors (Ni foil), 4—separator (hydrophilic polypropylene membrane), 6—polytetrafluoroethylene isolation gasket, 7—stainless steel cover with polytetrafluoroethylene spacer.

The use of such electrodes conforms to the principle of constructing an asymmetric supercapacitor, because the potential windows for carbon and NiO/C electrodes are −1–0 V and 0–0.65 V versus a saturated Ag/AgCl electrode, respectively (Figure 8), which helps widen the work potentials of the cell.

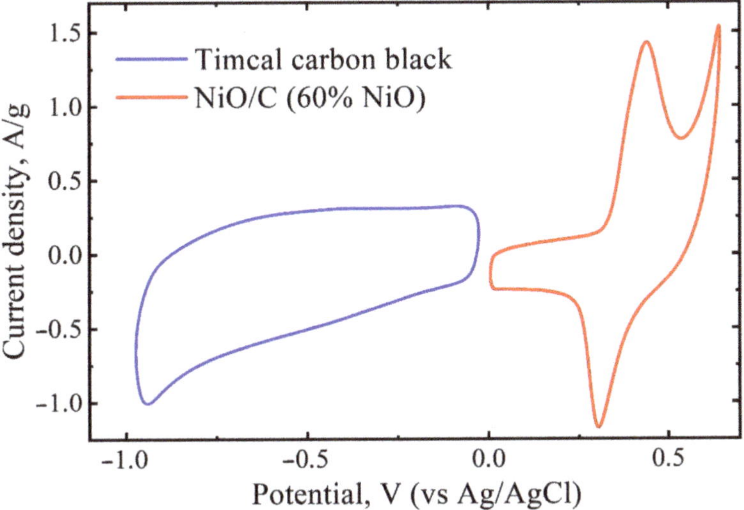

Figure 8. CV curves of carbon black Timcal Super C65 (CB) and NiO/C (60 wt% NiO) in a three-electrode cell in a 6 M KOH solution at a scan rate of 50 mV s^{-1}.

A series of CV curves were measured within different potential ranges at a scan rate of 50 mV/s to estimate the optimal potential window for the asymmetrical two-electrode cell (Figure 9a). When the potential window increased from 1 to 1.65 V, two redox peaks were observed, indicating the pseudocapacitive behavior of the supercapacitor. Based on these results, the CV measurements were performed at different scan rates within a potential window of 1.65 V. As shown in Figure 9b the CV curves of the assembled cell maintained their shape even at high scan rates, evidencing its good rate capability.

The galvanostatic discharge curves exhibit no IR drop due to the internal resistance in the initial range of a discharge curve (Figure 9c). The several-day-long cycling stability tests of the asymmetric two-electrode cell (4000 cycles under a current density of 0.1 A g^{-1}) reveal the excellent cyclic performance and the enduring stability (Figure 9d), which is essential for practical applications.

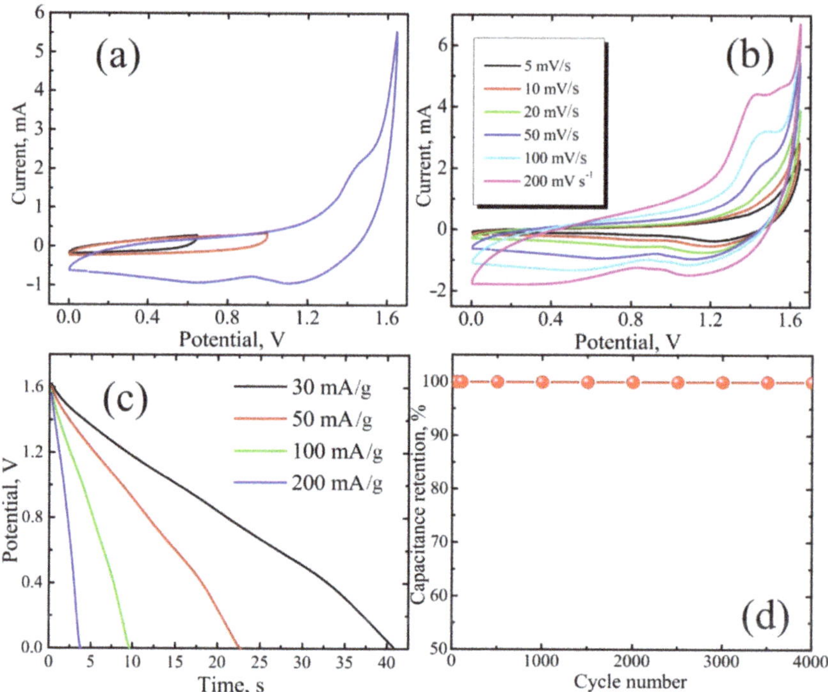

Figure 9. Electrochemical performance of the asymmetrical two-electrode cell with carbon black Timcal Super C65 and NiO/C (60 wt% NiO) as negative and positive electrodes, respectively, in a 6 M KOH solution: (**a**) cyclic voltammogram (CV) curves within different potential ranges up to 0.65, 1 and 1.65 V at 50 mV s^{-1}; (**b**) CV curves at various scan rates; (**c**) galvanostatic discharge curves at various current densities; (**d**) cycling stability at a current density of 0.1 A g^{-1} during 4000 cycles.

The voltage supplied by a single asymmetrical two-electrode cell with a NiO/C positive electrode (60 wt% NiO) and an aqueous electrolyte is sufficient for the use in different micro-devices, for example, in powering quartz clocks with a power rating of 1.5 V for about 30 min (Figure 10).

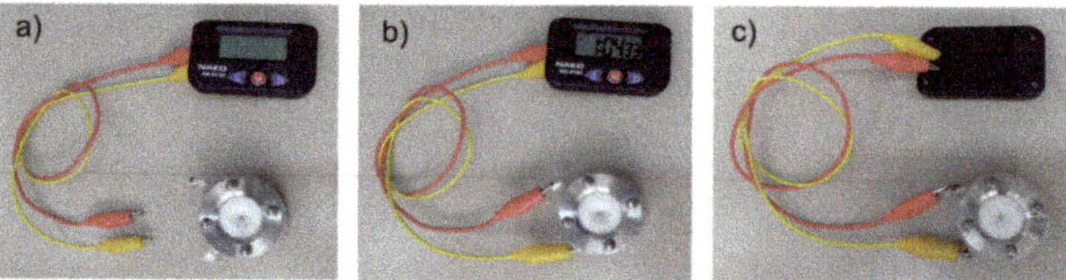

Figure 10. Application of the asymmetrical two-electrode assembled cell (NiO/C (60 wt% NiO) as a positive electrode and CB Timcal Super C65 as a negative electrode) in a 6 M KOH aqueous solution: (**a**) photograph of an asymmetrical two-electrode assembled cell charged at 0.5 mA and a battery-free quartz clock; (**b**) photograph of a quartz clock (front side) in the process of charging; (**c**) photograph of a quartz clock (back side) in the process of charging.

Supplementary Materials: The following are available online at https://www.mdpi.com/2079-4991/11/3/274/s1, Figure S1: TGA curve of Ni(acac)$_2$: thermal decomposition at a heating rate of 1 K/min under air atmosphere., Figure S2: Experimental (red dots) and simulated (black solid line) XRD patterns of NiO/C nanocomposite (80 wt% NiO). The blue line is the difference between experimental and theoretical XRD patterns. The tick marks correspond to the Bragg peak positions of fcc *Fm3m* phase of NiO (Agreement factors were R_p = 4.77%, R_{wp} = 3.82%, and R_{exp} = 2.67%). Figure S3: Decomposition into components of the NiO/Vulcan spectrum (NiO concentration—80%). Empty circles and blue curve correspond to experimental data, green curves—to NiO peaks, gray curves—to Vulcan peaks. Figure S4: The first group of NiO/Vulcan spectra. NiO concentrations: 10%, 20%, 30%, 50%, 80%. Figure S5: The second group of NiO/Vulcan spectra. NiO concentrations: 40%, 60%, 90%, 95%. Table S1: Comparison of electrochemical performances of NiO/C (60 wt% NiO) and various NiO/C composites) in a three-electrode system (alkaline aqueous solution).

Author Contributions: Conceptualization, I.L. and N.S.; methodology, I.L.; investigation, D.C., L.P., Y.P., O.M., M.A., A.R. and I.L.; writing—original draft preparation, I.L., N.L., D.C.; writing—review and editing, A.N., O.M., N.L., M.A. and A.R.; visualization D.C., L.P., I.L. and A.N. All authors have read and agreed to the published version of the manuscript.

Funding: A.N. acknowledge support from the Government of Russia (grant no. 14.Z50.31.0046). All XRD, Raman, electrochemical studies, as well as the assembly of the cells were carried out with funding from grant the Russian Science Foundation (grant no. 20-79-10063).

Data Availability Statement: The data presented in this study are available on request from the corresponding author.

Conflicts of Interest: The authors declare no conflict of interest.

References

1. Lu, X.; Yu, M.; Wang, G.; Tong, Y.; Li, Y. Flexible Solid-State Supercapacitors: Design, Fabrication and Applications. *Energy Environ. Sci.* **2014**, *7*, 2160. [CrossRef]
2. Sk, M.M.; Yue, C.Y.; Ghosh, K.; Jena, R.K. Review on Advances in Porous Nanostructured Nickel Oxides and Their Composite Electrodes for High-Performance Supercapacitors. *J. Power Sources* **2016**, *308*, 121–140. [CrossRef]
3. Paulose, R.; Mohan, R.; Parihar, V. Nanostructured Nickel Oxide and Its Electrochemical Behaviour—A Brief Review. *Nano Struct. Nano Objects* **2017**, *11*, 102–111. [CrossRef]
4. Liu, K.-C.; Anderson, M.A. Porous Nickel Oxide/Nickel Films for Electrochemical Capacitors. *J. Electrochem. Soc.* **1996**, *143*, 124. [CrossRef]
5. Sasi, B.; Gopchandran, K.; Manoj, P.; Koshy, P.; Prabhakara Rao, P.; Vaidyan, V. Preparation of Transparent and Semiconducting NiO Films. *Vacuum* **2002**, *68*, 149–154. [CrossRef]
6. Nandy, S.; Maiti, U.N.; Ghosh, C.K.; Chattopadhyay, K.K. Enhanced P-Type Conductivity and Band Gap Narrowing in Heavily Al Doped NiO Thin Films Deposited by RF Magnetron Sputtering. *J. Phys. Condens. Matter* **2009**, *21*, 115804. [CrossRef]
7. Li, Q.; Li, C.-L.; Li, Y.-L.; Zhou, J.-J.; Chen, C.; Liu, R.; Han, L. Fabrication of Hollow N-Doped Carbon Supported Ultrathin NiO Nanosheets for High-Performance Supercapacitor. *Inorg. Chem. Commun.* **2017**, *86*, 140–144. [CrossRef]
8. Lokhande, V.C.; Lokhande, A.C.; Lokhande, C.D.; Kim, J.H.; Ji, T. Supercapacitive Composite Metal Oxide Electrodes Formed with Carbon, Metal Oxides and Conducting Polymers. *J. Alloys Compd.* **2016**, *682*, 381–403. [CrossRef]
9. Makhlouf, S.A.; Kassem, M.A.; Abdel-Rahim, M.A. Particle Size-Dependent Electrical Properties of Nanocrystalline NiO. *J. Mater. Sci.* **2009**, *44*, 3438–3444. [CrossRef]
10. Yazdani, A.; Zafarkish, H.; Rahimi, K. The Variation of Eg-Shape Dependence of NiO Nanoparticles by the Variation of Annealing Temperature. *Mater. Sci. Semicond. Process.* **2018**, *74*, 225–231. [CrossRef]
11. Bose, P.; Ghosh, S.; Basak, S.; Naskar, M.K. A Facile Synthesis of Mesoporous NiO Nanosheets and Their Application in CO Oxidation. *J. Asian Ceram. Soc.* **2016**, *4*, 1–5. [CrossRef]
12. Ukoba, K.O.; Eloka-Eboka, A.C.; Inambao, F.L. Review of Nanostructured NiO Thin Film Deposition Using the Spray Pyrolysis Technique. *Renew. Sustain. Energy Rev.* **2018**, *82*, 2900–2915. [CrossRef]
13. Aghazadeh, M. Synthesis, Characterization, and Study of the Supercapacitive Performance of NiO Nanoplates Prepared by the Cathodic Electrochemical Deposition-Heat Treatment (CED-HT) Method. *J. Mater. Sci. Mater. Electron.* **2017**, *28*, 3108–3117. [CrossRef]
14. Kumar, A.; Sanger, A.; Kumar, A.; Chandra, R. Single-Step Growth of Pyramidally Textured NiO Nanostructures with Improved Supercapacitive Properties. *Int. J. Hydrogen. Energy* **2017**, *42*, 6080–6087. [CrossRef]
15. Lontio Fomekong, R.; Ngolui Lambi, J.; Ebede, G.R.; Kenfack Tsobnang, P.; Tedjieukeng Kamta, H.M.; Ngnintedem Yonti, C.; Delcorte, A. Effective Reduction in the Nanoparticle Sizes of NiO Obtained via the Pyrolysis of Nickel Malonate Precursor Modified Using Oleylamine Surfactant. *J. Solid State Chem.* **2016**, *241*, 137–142. [CrossRef]

16. Leontyeva, D.V.; Leontyev, I.N.; Avramenko, M.V.; Yuzyuk, Y.I.; Kukushkina, Y.A.; Smirnova, N.V. Electrochemical Dispergation as a Simple and Effective Technique toward Preparation of NiO Based Nanocomposite for Supercapacitor Application. *Electrochim. Acta* **2013**, *114*, 356–362. [CrossRef]
17. Fazlali, F.; Mahjoub, A.; Abazari, R. A New Route for Synthesis of Spherical NiO Nanoparticles via Emulsion Nano-Reactors with Enhanced Photocatalytic Activity. *Solid State Sci.* **2015**, *48*, 263–269. [CrossRef]
18. El-Kemary, M.; Nagy, N.; El-Mehasseb, I. Nickel Oxide Nanoparticles: Synthesis and Spectral Studies of Interactions with Glucose. *Mater. Sci. Semicond. Process.* **2013**, *16*, 1747–1752. [CrossRef]
19. Motahari, F.; Mozdianfard, M.R.; Soofivand, F.; Salavati-Niasari, M. NiO Nanostructures: Synthesis, Characterization and Photocatalyst Application in Dye Wastewater Treatment. *RSC Adv.* **2014**, *4*, 27654. [CrossRef]
20. Danial, A.S.; Saleh, M.M.; Salih, S.A.; Awad, M.I. On the Synthesis of Nickel Oxide Nanoparticles by Sol–Gel Technique and Its Electrocatalytic Oxidation of Glucose. *J. Power Sources* **2015**, *293*, 101–108. [CrossRef]
21. Purushothaman, K.K.; Manohara Babu, I.; Sethuraman, B.; Muralidharan, G. Nanosheet-Assembled NiO Microstructures for High-Performance Supercapacitors. *ACS Appl. Mater. Interfaces* **2013**, *5*, 10767–10773. [CrossRef] [PubMed]
22. Kulbakov, A.A.; Allix, M.; Rakhmatullin, A.; Mikheykin, A.S.; Popov, Y.V.; Smirnova, N.V.; Maslova, O.; Leontyev, I.N. In Situ Investigation of Non-Isothermal Decomposition of Pt Acetylacetonate as One-Step Size-Controlled Synthesis of Pt Nanoparticles. *Phys. Status Solidi* **2018**, 1800488. [CrossRef]
23. Kulbakov, A.A.; Kuriganova, A.B.; Allix, M.; Rakhmatullin, A.; Smirnova, N.; Maslova, O.A.; Leontyev, I.N. Non-Isothermal Decomposition of Platinum Acetylacetonate as a Cost-Efficient and Size-Controlled Synthesis of Pt/C Nanoparticles. *Catal. Commun.* **2018**, *117*, 14–18. [CrossRef]
24. Leontyev, I.N.; Kuriganova, A.B.; Allix, M.; Rakhmatullin, A.; Timoshenko, P.E.; Maslova, O.A.; Mikheykin, A.S.; Smirnova, N.V. On the Evaluation of the Average Crystalline Size and Surface Area of Platinum Catalyst Nanoparticles. *Phys. Status Solidi Basic Res.* **2018**, *255*. [CrossRef]
25. Leontyev, I.N.; Kulbakov, A.A.; Allix, M.; Rakhmatullin, A.; Kuriganova, A.B.; Maslova, O.A.; Smirnova, N.V. Thermal Expansion Coefficient of Carbon-Supported Pt Nanoparticles: In-Situ X-Ray Diffraction Study. *Phys. Status Solidi Basic Res.* **2017**, *254*. [CrossRef]
26. Chernysheva, D.; Vlaic, C.; Leontyev, I.; Pudova, L.; Ivanov, S.; Avramenko, M.; Allix, M.; Rakhmatullin, A.; Maslova, O.; Bund, A.; et al. Synthesis of Co_3O_4/CoOOH via electrochemical dispersion using a pulse alternating current method for lithium-ion batteries and supercapacitors. *Solid State Sci.* **2018**, *86*, 53–59. [CrossRef]
27. Liu, S.; Lee, S.C.; Patil, U.M.; Ray, C.; Sankar, K.V.; Zhang, K.; Kundu, A.; Kang, S.; Park, J.H.; Jun, S.C. Controllable Sulfuration Engineered NiO Nanosheets with Enhanced Capacitance for High Rate Supercapacitors. *J. Mater. Chem. A* **2017**, *5*, 4543–4549. [CrossRef]
28. Shklovskii, B.I.; Éfros, A.L. Percolation Theory and Conductivity of Strongly Inhomogeneous Media. *Uspekhi Fiz. Nauk* **1975**, *117*, 401. [CrossRef]
29. Kuriganova, A.B.; Vlaic, C.A.; Ivanov, S.; Leontyeva, D.V.; Bund, A.; Smirnova, N.V. Electrochemical Dispersion Method for the Synthesis of SnO_2 as Anode Material for Lithium Ion Batteries. *J. Appl. Electrochem.* **2016**, *46*, 527–538. [CrossRef]
30. Wang, Y.; Xing, S.; Zhang, E.; Wei, J.; Suo, H.; Zhao, C.; Zhao, X. One-pot synthesis of nickel oxide–carbon composite microspheres on nickel foam for supercapacitors. *J. Mater. Sci.* **2012**, *47*, 2182–2187. [CrossRef]
31. Liu, J.; Wickramaratne, N.P.; Qiao, S.Z.; Jaroniec, M. Molecular-based design and emerging applications of nanoporous carbon spheres. *Nat. Mater.* **2015**, *14*, 763. [CrossRef] [PubMed]
32. Sannasi, V.; Maheswari, K.U.; Karthikeyan, C.; Karuppuchamy, S. H_2O_2-assisted microwave synthesis of NiO/CNT nanocomposite material for supercapacitor applications. *Ionics* **2020**, *26*, 4067–4079. [CrossRef]
33. Vijayakumar, S.; Nagamuthu, S.; Muralidharan, G. Porous NiO/C Nanocomposites as Electrode Material for Electrochemical Supercapacitors. *ACS Sustain. Chem. Eng.* **2013**, *1*, 1110–1118. [CrossRef]
34. Wu, S.R.; Liu, J.B.; Wang, H.; Yan, H. NiO@graphite carbon nanocomposites derived from Ni-MOFs as supercapacitor electrodes. *Ionics* **2019**, *25*, 1–8. [CrossRef]
35. Wang, L.; Tian, H.; Wang, D.; Qin, X.; Shao, G. Preparation and electrochemical characteristic of porous NiO supported by sulfonated graphene for supercapacitors. *Electrochim. Acta* **2015**, *151*, 407–414. [CrossRef]
36. Liu, M.; Wang, X.; Zhu, D.; Li, L.; Duan, H.; Xu, Z.; Wang, Z.; Gan, L. Encapsulation of NiO nanoparticles in mesoporous carbon nanospheres for advanced energy storage. *Chem. Eng. J.* **2017**, *308*, 240–247. [CrossRef]

MDPI

Article

Critical Temperatures for Vibrations and Buckling of Magneto-Electro-Elastic Nonlocal Strain Gradient Plates

Giovanni Tocci Monaco [1,2], Nicholas Fantuzzi [1,*], Francesco Fabbrocino [3] and Raimondo Luciano [2]

1 Department of Civil, Chemical, Environmental and Materials Engineering, University of Bologna, 40136 Bologna, Italy; giovanni.toccimonaco@studio.unibo.it
2 Engineering Department, Parthenope University, 80133 Naples, Italy; raimondo.luciano@uniparthenope.it
3 Department of Engineering, Telematic University Pegaso, 80132 Naples, Italy; francesco.fabbrocino@unipegaso.it
* Correspondence: nicholas.fantuzzi@unibo.it

Abstract: An analytical method is presented in this work for the linear vibrations and buckling of nano-plates in a hygro-thermal environment. Nonlinear von Kármán terms are included in the plate kinematics in order to consider the instability phenomena. Strain gradient nonlocal theory is considered for its simplicity and applicability with respect to other nonlocal formulations which require more parameters in their analysis. Present nano-plates have a coupled magneto-electro-elastic constitutive equation in a hygro-thermal environment. Nano-scale effects on the vibrations and buckling behavior of magneto-electro-elastic plates is presented and hygro-thermal load outcomes are considered as well. In addition, critical temperatures for vibrations and buckling problems are analyzed and given for several nano-plate configurations.

Keywords: smart nano-plates; semi-analytical solution; critical temperatures; buckling; vibrations

Citation: Tocci Monaco, G.; Fantuzzi, N.; Fabbrocino, F.; Luciano, L. Critical Temperatures for Vibrations and Buckling of Magneto-Electro-Elastic Nonlocal Strain Gradient Plates. *Nanomaterials* **2021**, *11*, 274. https://doi.org/10.3390/nano11010087

Received: 3 December 2020
Accepted: 29 December 2020
Published: 3 January 2021

Publisher's Note: MDPI stays neutral with regard to jurisdictional clai-ms in published maps and institutio-nal affiliations.

1. Introduction

In recent years research has focused heavily on MEMS (Micro-Electro-Mechanical-System) and NEMS (Nano-Electro-Mechanical-System). This interest is mainly due to the wide variety of applications in which these devices could be used [1–4]. These structures, such as nanoplates, nanorods, and nanobeams [5], can be used in medicine [6], electronics [7], aerospace [8] and in civil construction [9], where linear and nonlinear theories are generally needed [10]. The behavior of this type of structures cannot be well described through the classical theories of continuous mechanics, as they are based on the principle of location of stresses. Due to the size of these devices, the effects induced by nanoscales must be taken into account [11,12]. Then to improve the ability of new devices and systems made with these smart materials, it is necessary to accurately investigate the mechanical behavior of these advanced structures [13,14]. Non-local theories have been widely used for the study of nanostructures since Eringen developed his theory of non-local elasticity [15]. These theories consider the nano-scale effects thanks to the introduction of one or more length scale parameters in addition to the well known linear elastic Lamé parameters [16–19]. The classification of nonlocal theories is generally presented as: strain gradient [20–23], stress gradient [24], modified strain gradient [25–27], couple stress [28], modified couple stress [29,30], integral type [31,32] and micropolar [33–35]. Article [36] offers a overview on unified continuous/reduced-order modeling and non-linear dynamic theories for thermomechanical plates. Kim in [37] developed a matrix method for evaluating effective elastic constants of generally anisotropic multilayer composites with various coupled physical effects including piezoelectricity, piezomagnetism and thermoelasticity. In [38–40] a nonlocal nonlinear first-order shear theory is used for investigating the buckling and free vibration of magneto-electro-thermo elastic (METE) nanoplates under magneto-electro-thermo-mechanical loadings. Mota in [41] investigated the influence of shear factor used in the context of the first-order shear deformation theory on functionally graded porous materials. In [42] free and forced vibration of a functionally graded piezoelectric plate with the properties of the material varying along the thickness are investigated. Combined asymptotic-tolerance modelling of dynamic and stability problems for functionally graded shells was given in [43,44]. In [45] are studied the pyroelectric and pyromagnetic effects on multiphase MEE cylindrical shells subjected to a uniform axisymmetric temperature using semi-analytical finite element procedures. Eremeyev et al. in [46,47] investigate the effect of flexoelectricity and flexomagnetic on nanobeams. Using FEM formulation [48,49] analyzed free vibration of orthotropic cross-ply nanoplates and nanowires. Ebrahimi in [50] studied buckling behaviour of magneto-electro-elastic functionally graded nanobeams using higer-order beam theory and Eringen's non-local elasticity. Also in [51] the behaviour of MEE nanobeams is investigated using Eulero-Bernoulli beam theory and including surface effects. In [52] the focus is on free vibration of laminated circular piezoelectric plates and discs using the weak form of the equations of motion. In [53] using the third order shear deformation plate theory, the bending, buckling, free and forced vibration behavior of a nonlocal composite microplate is analyzed. Functionally graded microplates using Kirchhoff plate's theory and straing gradient theory with only one length scale parameter are studied in [54]. In [55] theory of elasticity including surface stresses is used to study behaviour of shells with nano-scaled thickness. In [56] Kirchhoff plate's theory and the modified flexoelectric theory are used to study the nonlinear free vibration of Functionally Graded (FG) flexoelectric nanoplate by taking into account size-dependent effects. In [57] a finite element model based on a higher-order plate's theory is developed to study static and free vibration problem of magneto-electro-elastic plates. In [58] the dynamic problem of thin elastic plates resting on elastic foundation is studied using tolerance averaging method. Vibrations on periodical structures as well as bad gaps problems in dynamics are still an open topic thoroughly discussed by several researchers [59–62]. In [63] the flexural vibration band gaps in periodic beams is investigated using differential quadrature method, moreover the influence of shear deformation on the gaps is analyzed. Similarly, natural frequencies of structures made of period cells was presented for beams in [64,65] and for plates in [66,67]. In [68] the aim is the problem of vibration of band gaps in periodic Mindlin plates and it is solved using spectral element method. Also in [69] is studied the problem of vibration of band gaps in periodic plates, but is solved using differential quadrature element method. Vibration band gap of stiffened thin plate is studied also in [70] and is solved using center finite difference method. In [71,72] the problem of flexural wave propagation of a periodic beam is investigated.

The aim of this paper is the study of buckling and free vibrations of functionally graded nano-plates in hygro-thermal environment. For buckling analysis it will investigate the influence of external applied electric and magnetic potentials on critical load that leads to the instability of the plate, while in the case of dynamic analysis will be studied the behavior of natural vibration frequencies and how they are influenced by external potentials and temperature. Through the graphs will also identify the critical temperature, which corresponds to the temperature at which the vibration frequencies become zero. This paper is structured as described below. After the introduction section, the theoretical background for functionally graded (FG) thin plates in hygro-thermal environment is developed introducing also the non-linear terms of von Kármán that allow to perform the linear analysis of buckling. Using second order strain gradient theory non local effect are take into account. The following is a small paragraph showing how the electrical and magnetic potentials are approximated. Using Hamilton's principle, for the case of METE (magneto-electro-thermo-elastic) materials, the equations of motion are obtained. The analytical solution is obtained using Navier developments in double trigonometric series. Then the results obtained through calculation code implemented in MATLAB for buckling and free vibration are provided. Finally, a conclusion section is reported at the end of this paper.

2. Theoretical Background

As show in Figure 1, consider a METE thin nanoplate with length a, width b and thickness h, in a Cartesian reference system (x, y, z).

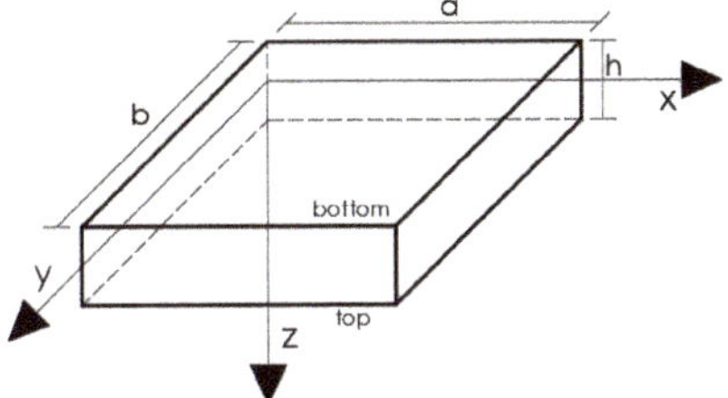

Figure 1. General laminate layout.

The METE nanoplate is in a hygro-thermal environment and is subjected to an electric potential V_0 and to a magnetic potential Ω_0 between the upper and lower surfaces. In this study, classic laminate plate theory is considered. We can define the displacement field of a generic point of the solid by means of the triad of displacement components U, V, W, which are functions of the coordinates (x, y, z) [10].

$$\begin{aligned} U(x,y,z,t) &= u(x,y,t) - z\frac{\partial w}{\partial x} \\ V(x,y,z,t) &= v(x,y,t) - z\frac{\partial w}{\partial y} \\ W(x,y,z,t) &= w(x,y,t) \end{aligned} \tag{1}$$

where u,v and w are the displacements along the x, y and z axis of the point on the middle surface and $\partial w/\partial x$ and $\partial w/\partial y$ are the corresponding rotation. The constitutive equations for a METE material are:

$$\begin{aligned} \sigma &= \mathbf{C}\varepsilon - \mathbf{e}\mathbf{E} - \mathbf{q}\mathbf{H} - \mathbf{C}\alpha\Delta T - \mathbf{C}\beta\Delta H \\ \mathbf{D}_E &= \mathbf{e}^\top\varepsilon + \xi\mathbf{E} + \zeta\mathbf{H} - \mathbf{p}\Delta T - \mathbf{h}\Delta H \\ \mathbf{B}_M &= \mathbf{q}^\top\varepsilon + \zeta\mathbf{E} + \chi\mathbf{H} - \lambda\Delta T - \eta\Delta H \end{aligned} \tag{2}$$

in which σ is the classical stress vector, $\mathbf{D}_E = [D_x,\ D_y,\ D_z]^\top$ and $\mathbf{B}_M = [B_x,\ B_y,\ B_z]^\top$ are respectively the vector of stresses, electrical displacement and magnetic flux. ε, $\mathbf{E}$ and $\mathbf{H}$ are the vector of strain, electric field and magnetic field. $\mathbf{C}$, ξ and χ represent the rigidity matrix, the electrical permittivity matrix and the magnetic permittivity matrix. Finally, $\mathbf{e}$, $\mathbf{q}$, ζ, $\mathbf{p}$, λ, $\mathbf{h}$ and η are respectively the piezo-electric, piezo-magnetic, magneto-electro-elastic (MEE), pyro-electric, pyro-magnetic, hygro-electric and hygro-magnetic coefficients. For the stress plane state ($\sigma_3 = 0$) the matrices can be reduced by carrying out

$$\varepsilon_3 = -\frac{C_{13}}{C_{33}}\varepsilon_1 - \frac{C_{23}}{C_{33}}\varepsilon_2 + \frac{e_{33}}{C_{33}}E_3 + \frac{q_{33}}{C_{33}}H_3 + \frac{C_{13}}{C_{33}}\alpha_1\Delta T + \frac{C_{23}}{C_{33}}\alpha_2\Delta T + \frac{C_{13}}{C_{33}}\beta_1\Delta H + \frac{C_{23}}{C_{33}}\beta_2\Delta H \tag{3}$$

Therefore the constitutive equations can be rewritten as follows

$$\begin{aligned} \sigma_1 = &\left(C_{11} - \frac{C_{13}^2}{C_{33}}\right)\varepsilon_1 + \left(C_{12} - C_{13}\frac{C_{23}}{C_{33}}\right)\varepsilon_2 - \left(e_{31} - C_{13}\frac{e_{33}}{C_{33}}\right)E_3 - \left(q_{31} - C_{13}\frac{q_{33}}{C_{33}}\right)H_3 + \\ &- \left(C_{11} - \frac{C_{13}^2}{C_{33}}\right)\alpha_1\Delta T - \left(C_{12} - C_{13}\frac{C_{23}}{C_{33}}\right)\alpha_2\Delta T - \left(C_{11} - \frac{C_{13}^2}{C_{33}}\right)\beta_1\Delta H - \left(C_{12} - C_{13}\frac{C_{23}}{C_{33}}\right)\beta_2\Delta H \\ = &\, Q_{11}\varepsilon_1 + Q_{12}\varepsilon_2 - \tilde{e}_{31}E_3 - \tilde{q}_{31}H_3 - Q_{11}\alpha_1\Delta T - Q_{12}\alpha_2\Delta T - Q_{11}\beta_1\Delta H - Q_{12}\beta_2\Delta H \end{aligned} \tag{4}$$

similarly for σ_2 it will be

$$\sigma_2 = Q_{12}\varepsilon_1 + Q_{22}\varepsilon_2 - \tilde{e}_{32}E_3 - \tilde{q}_{32}H_3 - Q_{12}\alpha_1\Delta T - Q_{22}\alpha_2\Delta T - Q_{12}\beta_1\Delta H - Q_{22}\beta_2\Delta H \quad (5)$$

finally, D_z can written

$$\begin{aligned} D_z =& \left(e_{31} - e_{33}\frac{C_{13}}{C_{33}}\right)\varepsilon_1 + \left(e_{32} - e_{33}\frac{C_{23}}{C_{33}}\right)\varepsilon_2 + \left(\xi_{33} + \frac{e_{33}^2}{C_{33}}\right)E_3 + \left(\zeta_{33} + e_{33}\frac{q_{33}}{C_{33}}\right)H_3 + \\ &- \left(p_3 - \frac{C_{13}}{C_{33}}\alpha_1 - \frac{C_{23}}{C_{33}}\alpha_2\right)\Delta T - \left(h_3 - \frac{C_{13}}{C_{33}}\beta_1 - \frac{C_{23}}{C_{33}}\beta_2\right)\Delta H \\ =& \tilde{e}_{31}\varepsilon_1 + \tilde{e}_{32}\varepsilon_2 + \tilde{\xi}_{33}E_3 + \tilde{\zeta}_{33}H_3 - \tilde{p}_3\Delta T - \tilde{h}_3\Delta H \end{aligned} \quad (6)$$

and similarly for $B_{M,3}$ it will be

$$B_z = \tilde{q}_{31}\varepsilon_1 + \tilde{q}_{32}\varepsilon_2 + \tilde{\zeta}_{33}E_3 + \tilde{\chi}_{33}H_3 - \tilde{\lambda}_3\Delta T - \tilde{\eta}_3\Delta H \quad (7)$$

So it is possible to write

$$\begin{aligned} &\tilde{\mathbf{e}} = \begin{bmatrix} 0 & 0 & \tilde{e}_{31} \\ 0 & 0 & \tilde{e}_{32} \\ 0 & 0 & 0 \end{bmatrix}, \quad \tilde{\mathbf{q}} = \begin{bmatrix} 0 & 0 & \tilde{q}_{31} \\ 0 & 0 & \tilde{q}_{32} \\ 0 & 0 & 0 \end{bmatrix}, \quad \tilde{\boldsymbol{\xi}} = \begin{bmatrix} \xi_1 & 0 & 0 \\ 0 & \xi_2 & 0 \\ 0 & 0 & \tilde{\xi}_3 \end{bmatrix}, \quad \tilde{\boldsymbol{\chi}} = \begin{bmatrix} \chi_1 & 0 & 0 \\ 0 & \chi_2 & 0 \\ 0 & 0 & \tilde{\chi}_3 \end{bmatrix}, \\ &\tilde{\boldsymbol{\zeta}} = \begin{bmatrix} \zeta_1 & 0 & 0 \\ 0 & \zeta_2 & 0 \\ 0 & 0 & \tilde{\zeta}_3 \end{bmatrix}, \quad \mathbf{p} = \begin{Bmatrix} p_1 \\ p_2 \\ \tilde{p}_3 \end{Bmatrix}, \quad \boldsymbol{\lambda} = \begin{Bmatrix} \lambda_1 \\ \lambda_2 \\ \tilde{\lambda}_3 \end{Bmatrix}, \quad \mathbf{h} = \begin{Bmatrix} h_1 \\ h_2 \\ \tilde{h}_3 \end{Bmatrix}, \quad \boldsymbol{\eta} = \begin{Bmatrix} \eta_1 \\ \eta_2 \\ \tilde{\eta}_3 \end{Bmatrix} \end{aligned} \quad (8)$$

By introducing second order strain gradient theory in the constitutive equations and by considering the mechanical properties variable with respect to the thickness direction we have (the dependency on the time t is omitted for the sake of simplicity)

$$\begin{aligned} \boldsymbol{\sigma}(x,y,z) &= \left(1 - \ell^2\nabla^2\right)\left[\bar{\mathbf{Q}}(z)\boldsymbol{\varepsilon} - \tilde{\mathbf{e}}(z)\mathbf{E} - \tilde{\mathbf{q}}(z)\mathbf{H}\right] - \bar{\mathbf{Q}}(z)\boldsymbol{\alpha}(z)\Delta T - \bar{\mathbf{Q}}(z)\boldsymbol{\beta}(z)\Delta H \\ \mathbf{D}_E(x,y,z) &= \left(1 - \ell^2\nabla^2\right)\left[\tilde{\mathbf{e}}^\top(z)\boldsymbol{\varepsilon} + \tilde{\boldsymbol{\xi}}(z)\mathbf{E} + \tilde{\boldsymbol{\zeta}}(z)\mathbf{H}\right] - \mathbf{p}(z)\Delta T - \mathbf{h}(z)\Delta H \\ \mathbf{B}_M(x,y,z) &= \left(1 - \ell^2\nabla^2\right)\left[\tilde{\mathbf{q}}^\top(z)\boldsymbol{\varepsilon} + \tilde{\boldsymbol{\zeta}}(z)\mathbf{E} + \tilde{\boldsymbol{\chi}}(z)\mathbf{H}\right] - \boldsymbol{\lambda}(z)\Delta T - \boldsymbol{\eta}(z)\Delta H \end{aligned} \quad (9)$$

where ℓ is the nonlocal parameter and the operator $\nabla^2 = \partial^2/\partial x^2 + \partial^2/\partial y^2$ is the second order gradient operator. For the hygro-thermal loads a linear variation is considered along the thickness as

$$\Delta T = T_0 + zT_1/h, \quad \Delta H = H_0 + zH_1/h \quad (10)$$

3. Electric and Magnetic Potentials

To satisfy Maxwell's equations [73] the electrical and magnetic potential are approximated along the thickness with a linear and cosinusoidal combination. The first amends for the open-circuit condition and the latter for the closed-circuit one

$$\begin{aligned} \Phi(x,y,z,t) &= -\cos\frac{\pi z}{h}\phi(x,y,t) + \frac{2z}{h}V_0 \\ \Upsilon(x,y,z,t) &= -\cos\frac{\pi z}{h}\gamma(x,y,t) + \frac{2z}{h}\Omega_0 \end{aligned} \quad (11)$$

in which V_0 represents the difference in electrical potential between the two faces of the plate and Ω_0 represents the difference in magnetic potential. The relationships between electric field and electric potential can be written in accordance with the above

$$\begin{aligned} E_x &= -\frac{\partial \Phi}{\partial x} = \cos\frac{\pi z}{h}\frac{\partial \phi}{\partial x} \\ E_y &= -\frac{\partial \Phi}{\partial y} = \cos\frac{\pi z}{h}\frac{\partial \phi}{\partial y} \\ E_z &= -\frac{\partial \Phi}{\partial z} = -\frac{\pi}{h}\sin\frac{\pi z}{h}\phi - \frac{2}{h}V_0 \end{aligned} \tag{12}$$

that in matrix notation can be rewritten in this form

$$\mathbf{E} = \mathbf{f}_E \mathbb{D}_E \phi + \mathbf{E}_0 \tag{13}$$

with

$$\mathbf{f}_E = \begin{bmatrix} \cos\frac{\pi z}{h} & 0 & 0 \\ 0 & \cos\frac{\pi z}{h} & 0 \\ 0 & 0 & -\frac{\pi}{h}\sin\frac{\pi z}{h} \end{bmatrix}, \quad \mathbb{D}_E = \begin{Bmatrix} \frac{\partial}{\partial x} \\ \frac{\partial}{\partial y} \\ 1 \end{Bmatrix}, \quad \mathbf{E}_0 = \begin{Bmatrix} 0 \\ 0 \\ -\frac{2}{h}V_0 \end{Bmatrix} \tag{14}$$

Similarly for the magnetic field we can write as

$$\begin{aligned} H_x &= -\frac{\partial \Upsilon}{\partial x} = \cos\frac{\pi z}{h}\frac{\partial \gamma}{\partial x} \\ H_y &= -\frac{\partial \Upsilon}{\partial y} = \cos\frac{\pi z}{h}\frac{\partial \gamma}{\partial y} \\ H_z &= -\frac{\partial \Upsilon}{\partial z} = -\frac{\pi}{h}\sin\frac{\pi z}{h}\gamma - \frac{2}{h}\Omega_0 \end{aligned} \tag{15}$$

which in matrix notation becomes

$$\mathbf{H} = \mathbf{f}_H \mathbb{D}_H \gamma + \mathbf{H}_0 \tag{16}$$

with

$$\mathbf{f}_H = \begin{bmatrix} \cos\frac{\pi z}{h} & 0 & 0 \\ 0 & \cos\frac{\pi z}{h} & 0 \\ 0 & 0 & -\frac{\pi}{h}\sin\frac{\pi z}{h} \end{bmatrix}, \quad \mathbb{D}_H = \begin{Bmatrix} \frac{\partial}{\partial x} \\ \frac{\partial}{\partial y} \\ 1 \end{Bmatrix}, \quad \mathbf{H}_0 = \begin{Bmatrix} 0 \\ 0 \\ -\frac{2}{h}\Omega_0 \end{Bmatrix} \tag{17}$$

4. Equations of Motion

The equations of motion are derived through Hamilton's principle

$$\int_{t_1}^{t_2} (\delta H_{ent} + \delta V - \delta K)dt = 0 \tag{18}$$

Writing the variation of enthalpy δH_{ent}

$$\begin{aligned} \delta H_{ent} = \int_{\mathcal{A}} \int_{-\frac{h}{2}}^{\frac{h}{2}} \Big\{ &\sigma_{xx}\delta\varepsilon_{xx} + \sigma_{yy}\delta\varepsilon_{yy} + \sigma_{xy}\delta\gamma_{xy} \\ &- D_x\delta E_x - D_y\delta E_y - D_z\delta E_z - B_x\delta H_x - B_y\delta H_y - B_z\delta H_z \Big\} dz d\mathcal{A} \end{aligned} \tag{19}$$

by introducing the classical stress resultants N_{xx}, N_{yy}, N_{xy} and M_{xx}, M_{yy}, M_{xy} and the piezo and magneto resultants as

$$\begin{Bmatrix} \mathcal{D}_x \\ \mathcal{D}_y \\ \mathcal{D}_z \end{Bmatrix} = \int_{-\frac{h}{2}}^{\frac{h}{2}} \mathbf{f}_E \mathbf{D}_E \, dz, \quad \begin{Bmatrix} \mathcal{B}_x \\ \mathcal{B}_y \\ \mathcal{B}_z \end{Bmatrix} = \int_{-\frac{h}{2}}^{\frac{h}{2}} \mathbf{f}_B \mathbf{B}_M \, dz \tag{20}$$

The definition of the integrated elastic properties are given in the Appendix A in Equations (A1)–(A3). Thus, it is obtained

$$\begin{aligned} \delta H_{ent} = \int_{\mathcal{A}} \Bigg\{ & N_{xx}\left(\frac{\partial \delta u}{\partial x} + \frac{\partial w}{\partial x}\frac{\partial \delta w}{\partial x}\right) + N_{yy}\left(\frac{\partial \delta v}{\partial y} + \frac{\partial w}{\partial y}\frac{\partial \delta w}{\partial y}\right) \\ & + N_{xy}\left(\frac{\partial \delta u}{\partial y} + \frac{\partial \delta v}{\partial x} + \frac{\partial \delta w}{\partial y}\frac{\partial w}{\partial x} + \frac{\partial \delta w}{\partial x}\frac{\partial w}{\partial y}\right) + \\ & + M_{xx}\left(-\frac{\partial^2 \delta w}{\partial x^2}\right) + M_{yy}\left(-\frac{\partial^2 \delta w}{\partial y^2}\right) + M_{xy}\left(-2\frac{\partial^2 \delta w}{\partial x \partial y}\right) + \\ & - \left(\mathcal{D}_x \frac{\partial \delta \phi}{\partial x} + \mathcal{D}_y \frac{\partial \delta \phi}{\partial y} + \mathcal{D}_x \delta \phi + \mathcal{B}_x \frac{\partial \delta \gamma}{\partial x} + \mathcal{B}_y \frac{\partial \delta \gamma}{\partial y} + \mathcal{B}_z \delta \gamma\right) \Bigg\} d\mathcal{A} \end{aligned} \tag{21}$$

Integrating by parts Equation (21) is obtained

$$\begin{aligned} \delta H_{ent} = \int_{\mathcal{A}} \Bigg\{ & \left(\frac{\partial N_{xx}}{\partial x} + \frac{\partial N_{xy}}{\partial y}\right)\delta u + \left(\frac{\partial N_{xy}}{\partial x} + \frac{\partial N_{yy}}{\partial y}\right)\delta v + \left[\frac{\partial}{\partial x}\left(N_{xx}\frac{\partial w}{\partial x} + N_{xy}\frac{\partial w}{\partial y}\right)\right. \\ & \left. + \frac{\partial}{\partial y}\left(N_{xy}\frac{\partial w}{\partial x} + N_{yy}\frac{\partial w}{\partial y}\right) + \frac{\partial^2 M_{xx}}{\partial x^2} + \frac{\partial^2 M_{yy}}{\partial y^2} + 2\frac{\partial^2 M_{xy}}{\partial x \partial y}\right]\delta w \\ & - \left(\frac{\partial \mathcal{D}_x}{\partial x} + \frac{\partial \mathcal{D}_y}{\partial y} + \mathcal{D}_z\right)\delta \phi - \left(\frac{\partial \mathcal{B}_x}{\partial x} + \frac{\partial \mathcal{B}_y}{\partial y} + \mathcal{B}_z\right)\delta \gamma \Bigg\} d\mathcal{A} + \\ - \int_{\Gamma} \Bigg\{ & (N_{xx}\, n_x + N_{xy}\, n_y)\delta u + (N_{xy}\, n_x + N_{yy}\, n_y)\delta v + \left[(N_{xx} n_x + N_{xy} n_y)\frac{\partial w}{\partial x}\right. \\ & \left. + (N_{xy} n_x + N_{yy} n_y)\frac{\partial w}{\partial y} + \left(\frac{\partial M_{xx}}{\partial x} + \frac{\partial M_{xy}}{\partial y}\right)n_x + \left(\frac{\partial M_{yy}}{\partial y} + \frac{\partial M_{xy}}{\partial x}\right)n_y\right]\delta w \\ & - (M_{xx} n_x + M_{xy} n_y)\frac{\partial \delta w}{\partial x} - (M_{xy} n_x + M_{yy} n_y)\frac{\partial \delta w}{\partial y} \\ & - (\mathcal{D}_x\, n_x + \mathcal{D}_y\, n_y)\delta \phi - (\mathcal{B}_x\, n_x + \mathcal{B}_y\, n_y)\delta \gamma \Bigg\} d\Gamma \end{aligned} \tag{22}$$

The external work due to the external boundary loads (where mechanical, electrical and magnetic loads are neglected) can be written as

$$\begin{aligned} \delta V = \int_{\Gamma} \Bigg\{ & (\hat{N}_{xx} n_x + \hat{N}_{xy} n_y)\delta u + (\hat{N}_{xy} n_x + \hat{N}_{yy} n_y)\delta v + \\ & - (\hat{M}_{xx} n_x + \hat{M}_{xy} n_y)\frac{\partial \delta w}{\partial x} - (\hat{M}_{xy} n_x + \hat{M}_{yy} n_y)\frac{\partial \delta w}{\partial y} + (\hat{Q}_x n_x + \hat{Q}_y n_y)\delta w \Bigg\} d\Gamma \end{aligned} \tag{23}$$

Variation of the Kinetic energy can be written as

$$\begin{aligned} \delta K = \int_{\mathcal{A}} \Bigg\{ & \left(-I_0 \ddot{u} + I_1 \frac{\partial \ddot{w}}{\partial x}\right)\delta u + \left(-I_0 \ddot{v} + I_1 \frac{\partial \ddot{w}}{\partial y}\right)\delta v \\ & + \left(-I_0 \ddot{w} - I_1 \frac{\partial \ddot{u}}{\partial x} - I_1 \frac{\partial \ddot{v}}{\partial y} + I_2 \frac{\partial \ddot{w}}{\partial x} + I_2 \frac{\partial \ddot{w}}{\partial y}\right)\delta w \Bigg\} d\mathcal{A} + \\ & + \int_{\Gamma} \left\{ I_1 \ddot{u} n_x + I_1 \ddot{v} n_y - I_2 \frac{\partial \ddot{w}}{\partial x} n_x - I_2 \frac{\partial \ddot{w}}{\partial y} n_y \right\} \delta w \, d\Gamma \end{aligned} \tag{24}$$

Introducing $\mathcal{N}(w)$ and $\mathcal{P}(w)$ as defined below

$$\begin{aligned}\mathcal{N}(w) &= \frac{\partial}{\partial x}\left(N_{xx}\frac{\partial w}{\partial x}\right)\delta w + \frac{\partial}{\partial y}\left(N_{yy}\frac{\partial w}{\partial y}\right)\delta w + \frac{\partial}{\partial y}\left(N_{xy}\frac{\partial w}{\partial x}\right)\delta w + \frac{\partial}{\partial x}\left(N_{xy}\frac{\partial w}{\partial y}\right)\delta w \\ \mathcal{P}(w) &= \left(N_{xx}\frac{\partial w}{\partial x} + N_{xy}\frac{\partial w}{\partial y}\right)n_x + \left(N_{xy}\frac{\partial w}{\partial x} + N_{yy}\frac{\partial w}{\partial y}\right)n_y\end{aligned} \tag{25}$$

the motion equations can be written as follow

$$\begin{gathered}\frac{\partial N_{xx}}{\partial x} + \frac{\partial N_{xy}}{\partial y} = I_0\ddot{u} - I_1\frac{\partial \ddot{w}}{\partial x} \\ \frac{\partial N_{yy}}{\partial y} + \frac{\partial N_{xy}}{\partial x} = I_0\ddot{v} - I_1\frac{\partial \ddot{w}}{\partial y} \\ \frac{\partial^2 M_{xx}}{\partial x^2} + 2\frac{\partial^2 M_{xy}}{\partial x \partial y} + \frac{\partial^2 M_{yy}}{\partial y^2} + \mathcal{N}(w) = I_0\ddot{w} + I_1\left(\frac{\partial \ddot{u}}{\partial x} + \frac{\partial \ddot{v}}{\partial y}\right) - I_2\left(\frac{\partial^2 \ddot{w}}{\partial x^2} + \frac{\partial^2 \ddot{w}}{\partial y^2}\right) \\ \frac{\partial \mathcal{D}_x}{\partial x} + \frac{\partial \mathcal{D}_y}{\partial y} + \mathcal{D}_z = 0 \\ \frac{\partial \mathcal{B}_x}{\partial x} + \frac{\partial \mathcal{B}_y}{\partial y} + \mathcal{B}_z = 0\end{gathered} \tag{26}$$

and relative boundary conditions become

$$\begin{aligned}\delta u = 0 &\quad \text{or} \quad (N_{xx} - \hat{N}_{xx})n_x + (N_{xy} - \hat{N}_{xy})n_y = 0 \\ \delta v = 0 &\quad \text{or} \quad (N_{yy} - \hat{N}_{yy})n_y + (N_{xy} - \hat{N}_{xy})n_x = 0 \\ \delta w = 0 &\quad \text{or} \quad \left(\frac{\partial M_{xx}}{\partial x} + \frac{\partial M_{xy}}{\partial y} - I_1\ddot{u} + I_2\frac{\partial \ddot{w}}{\partial x}\right)n_x + \\ &\quad + \left(\frac{\partial M_{yy}}{\partial y} + \frac{\partial M_{xy}}{\partial x} - I_1\ddot{v} + I_2\frac{\partial \ddot{w}}{\partial y}\right)n_y + \mathcal{P}(w) - (\hat{Q}_x + \hat{Q}_y) = 0 \\ \frac{\partial \delta w}{\partial x} = 0 &\quad \text{or} \quad (M_{xx} - \hat{M}_{xx})n_x + (M_{xy} - \hat{M}_{xy})n_y = 0 \\ \frac{\partial \delta w}{\partial y} = 0 &\quad \text{or} \quad (M_{yy} - \hat{M}_{yy})n_y + (M_{xy} - \hat{M}_{xy})n_x = 0 \\ \delta\phi = 0 &\quad \text{or} \quad \mathcal{D}_x\, n_x + \mathcal{D}_y\, n_y = 0 \\ \delta\gamma = 0 &\quad \text{or} \quad \mathcal{B}_x\, n_x + \mathcal{B}_y\, n_y = 0\end{aligned} \tag{27}$$

5. Navier Solution

Analytical solution is obtained using Navier's expansion. This type of solution allow to solve simply supported plate case. Navier expansion for the displacements take the form

$$\begin{Bmatrix} u \\ v \\ w \end{Bmatrix} = \sum_{m=1}^{M}\sum_{n=1}^{N}\begin{bmatrix} \cos\alpha x \sin\beta y & 0 & 0 \\ 0 & \sin\alpha x \cos\beta y & 0 \\ 0 & 0 & \sin\alpha x \sin\beta y \end{bmatrix}\begin{Bmatrix} U_{mn} \\ V_{mn} \\ W_{mn} \end{Bmatrix} \tag{28}$$

whereas, the electric and magnetic potentials are both approximated with a double sinusoidal trigonometric expansion.

$$\phi = \sum_{m=1}^{M}\sum_{n=1}^{N}\sin\alpha x \sin\beta y\, \Phi_{mn}, \quad \gamma = \sum_{m=1}^{M}\sum_{n=1}^{N}\sin\alpha x \sin\beta y\, \Gamma_{mn} \tag{29}$$

5.1. Buckling Analysis

Replacing the displacement field in the motion equations and performing the derivates the algebraic system is obtained

$$\begin{bmatrix} \hat{c}_{11} & \hat{c}_{12} & \hat{c}_{14} & \hat{c}_{15} & \hat{c}_{13} \\ \hat{c}_{12} & \hat{c}_{22} & \hat{c}_{24} & \hat{c}_{25} & \hat{c}_{23} \\ \hat{c}_{14} & \hat{c}_{24} & \hat{c}_{44} & \hat{c}_{45} & \hat{c}_{34} \\ \hat{c}_{15} & \hat{c}_{25} & \hat{c}_{45} & \hat{c}_{55} & \hat{c}_{35} \\ \hat{c}_{13} & \hat{c}_{23} & \hat{c}_{34} & \hat{c}_{35} & \hat{c}_{33}+\tilde{s}_{33} \end{bmatrix} \begin{Bmatrix} U_{mn} \\ V_{mn} \\ \Phi_{mn} \\ \Gamma_{mn} \\ W_{mn} \end{Bmatrix} = \begin{Bmatrix} 0 \\ 0 \\ 0 \\ 0 \\ 0 \end{Bmatrix} \tag{30}$$

The coefficients $\hat{c}_{ij}$ and $\tilde{s}_{33}$ are defined in Appendix A at Equation (A4). By introducing the quantities $N_0 = -\hat{N}_{xx}$, $\kappa = \hat{N}_{yy}/\hat{N}_{xx}$ and a_{mn} as

$$\begin{aligned} a_{mn} = \hat{c}_{33} &+ \alpha^2\left(\hat{N}^T_{xx} + \hat{N}^E_{xx} + \hat{N}^H_{xx}\right) + \beta^2\left(\hat{N}^T_{yy} + \hat{N}^E_{yy} + \hat{N}^H_{yy}\right) \\ &- \left\{\hat{c}_{13} \quad \hat{c}_{23} \quad \hat{c}_{34} \quad \hat{c}_{35}\right\} \begin{bmatrix} \hat{c}_{11} & \hat{c}_{12} & \hat{c}_{14} & \hat{c}_{15} \\ \hat{c}_{12} & \hat{c}_{22} & \hat{c}_{24} & \hat{c}_{25} \\ \hat{c}_{14} & \hat{c}_{24} & \hat{c}_{44} & \hat{c}_{45} \\ \hat{c}_{15} & \hat{c}_{25} & \hat{c}_{45} & \hat{c}_{55} \end{bmatrix}^{-1} \begin{Bmatrix} \hat{c}_{13} \\ \hat{c}_{23} \\ \hat{c}_{34} \\ \hat{c}_{35} \end{Bmatrix} \end{aligned} \tag{31}$$

we can write the solution of the eigenvalue problem as

$$N_0 = \frac{a_{mn}}{\alpha^2 + \kappa\beta^2} \tag{32}$$

The load that buckles the plate depends on m and n and in particular the critical load is the lowest of the buckling loads. The terms $\hat{N}^E_{xx}$, $\hat{N}^E_{yy}$, $\hat{N}^H_{xx}$, $\hat{N}^H_{yy}$ are defined below

$$\hat{N}^E_{xx} = \hat{N}^E_{yy} = \int_{-h/2}^{h/2} \tilde{e}_{31}(z)\frac{2V_0}{h}\,dz \quad , \quad \hat{N}^H_{xx} = \hat{N}^H_{yy} = \int_{-h/2}^{h/2} \tilde{q}_{31}(z)\frac{2\Omega_0}{h}\,dz \tag{33}$$

Note that the electric and magnetic in-plane loads have the same intensity since in the following applications the material is isotropic in-plane and anisotropic out-of-plane.

5.2. Thermal Free Vibration

In this paragraph it will be treated the problem of free vibrations of the FG plate simply supported. For this problem it is necessary to rewrite the solving system in a homogeneous form, and the rotational inertia I_1 are neglected. The system becomes then

$$\left(\begin{bmatrix} \hat{c}_{11} & \hat{c}_{12} & \hat{c}_{13} & \hat{c}_{14} & \hat{c}_{15} \\ \hat{c}_{12} & \hat{c}_{22} & \hat{c}_{23} & \hat{c}_{24} & \hat{c}_{25} \\ \hat{c}_{13} & \hat{c}_{23} & \hat{c}_{33} & \hat{c}_{34} & \hat{c}_{35} \\ \hat{c}_{14} & \hat{c}_{24} & \hat{c}_{34} & \hat{c}_{44} & \hat{c}_{45} \\ \hat{c}_{15} & \hat{c}_{25} & \hat{c}_{35} & \hat{c}_{45} & \hat{c}_{55} \end{bmatrix} - \omega^2 \begin{bmatrix} \hat{m}_{11} & 0 & 0 & 0 & 0 \\ 0 & \hat{m}_{22} & 0 & 0 & 0 \\ 0 & 0 & \hat{m}_{33} & 0 & 0 \\ 0 & 0 & 0 & 0 & 0 \\ 0 & 0 & 0 & 0 & 0 \end{bmatrix}\right) \begin{Bmatrix} U_{mn} \\ V_{mn} \\ W_{mn} \\ \Phi_{mn} \\ \Gamma_{mn} \end{Bmatrix} = \begin{Bmatrix} 0 \\ 0 \\ 0 \\ 0 \\ 0 \end{Bmatrix} \tag{34}$$

with $\hat{m}_{11} = \hat{m}_{22} = I_0$ e $\hat{m}_{33} = I_0 + I_2(\alpha^2 + \beta^2)$. This system can then be rewritten in a more compact matrix form as follows

$$\left(\begin{bmatrix} \boldsymbol{K}_{uu} & \boldsymbol{K}_{u\phi} \\ \boldsymbol{K}_{\phi u} & \boldsymbol{K}_{\phi\phi} \end{bmatrix} - \omega^2 \begin{bmatrix} \boldsymbol{M}_{uu} & \boldsymbol{0} \\ \boldsymbol{0} & \boldsymbol{0} \end{bmatrix}\right) \begin{Bmatrix} \boldsymbol{U} \\ \boldsymbol{\phi} \end{Bmatrix} = \begin{Bmatrix} \boldsymbol{0} \\ \boldsymbol{0} \end{Bmatrix} \tag{35}$$

Rewriting the matrix system by applying static condensation we get

$$(\bar{\boldsymbol{K}} - \omega^2 \boldsymbol{M}_{uu})\,\boldsymbol{U} = \boldsymbol{0} \tag{36}$$

where $\bar{K}$ is

$$\bar{K} = (K_{uu} - K_{u\phi}(K_{\phi\phi})^{-1}K_{\phi u}) \tag{37}$$

and U represents the ways of vibrating and ω natural frequencies.

6. Numerical Results

In this paper for the numerical solution has been considered a FG nanoplate composed of $CoFe_2O_4$ and $BaTiO_3$, the properties of the materials are shown in Table 1. Since it has not been possible to find in literature the hygrometric coefficients of the materials, the applications presented foresee only the case of thermal environment.

Table 1. Piezo-electro-magnetic-thermal properties of $BaTiO_3$ and $CoFe_2O_4$.

		$BaTiO_3$	$CoFe_2O_4$
C_{11}	[GPa]	166	286
C_{22}		166	286
C_{33}		162	269.5
C_{13}		78	170.5
C_{23}		78	170.5
C_{12}		77	173
C_{44}		43	45.3
C_{55}		43	45.3
C_{66}		44.5	56.5
e_{31}	[C/m^2]	−4.4	0
e_{32}		−4.4	0
e_{33}		18.6	0
q_{31}	[N/A·m]	0	580.3
q_{32}		0	580.3
q_{33}		0	699.7
ξ_{11}	[10^{-9}C^2/N·m^2]	11.2	0.08
ξ_{22}		11.2	0.08
ξ_{33}		12.6	0.093
$\zeta_{11} = \zeta_{22} = \zeta_{33}$	[s/m]	0	0
χ_{11}	[10^{-6}N·s^2/C]	5	−590
χ_{22}		5	−590
χ_{33}		10	157
$p_{11} = p_{22}$	[10^{-7} C/m^{2}K]	0	0
p_{33}		−11.4	0
$\lambda_{11} = \lambda_{22}$	[10^{-5} Wb/m^{2}K]	0	0
λ_{33}		0	−36.2
$\alpha_1 = \alpha_2$	[10^{-6}K^{-1}]	15.8	10
ρ	[kg/m^3]	5300	5800

The variation of the material properties along the thickness is regulated by the following relationship

$$P(z) = (P_t - P_b)\left(\frac{z}{h} + \frac{1}{2}\right)^{p_n} + P_b \tag{38}$$

where P_t and P_b represent the properties of the material placed on the top and bottom of the plate, respectively.

Note that if $p_n = 0$ the plate will be composed entirely of the material of the top side while if $p_n \to \infty$ the plate will be composed entirely of the material of the bottom side.

6.1. Buckling

In the following applications the values of the critical load will be presented in dimensionless form through the following relationship

$$\bar{N}_{cr} = \frac{N_{cr}a^2}{C_{11,m}h^3} \tag{39}$$

where $C_{11,m}$ is the average stiffness value of the two materials. As a first result the comparison with Li [74], Park and Han [75] is reported. The plate considered is isotropic with $a/h = 1000$ and the properties of the material used are as reported in the cited works (which are obtained as average values of two isotropic constituents, not as a functionally graded composite).

Figure 2 shows the critical buckling load by varying the electric and magnetic potentials. The present results agree well with the ones presented in the mentioned papers [74,75]. It is emphasized that the slight difference in the results is due to the fact that in the cited studies the potential is approximated using three contributions: one parabolic, one linear and one constant unlike the present study in which the potentials are approximated with a cosine function and a linear part. In addition, in the cited studies, the in-plane components E_x, E_y, H_x and H_y of electric and magnetic fields are null.

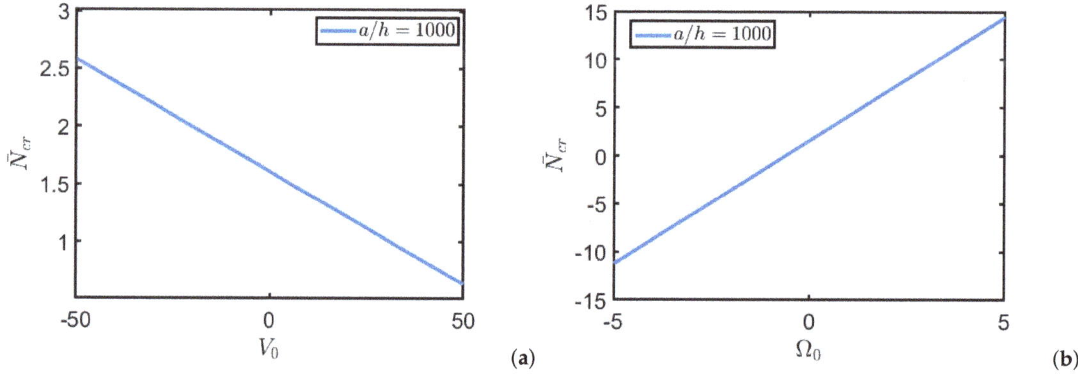

Figure 2. Critical load $\bar{N}_{cr}$ of a square Functionally Graded (FG) nanoplate for different values of electric potential V_0 (**a**) and magnetic potential Ω_0 (**b**).

In the applications below $a/h = 1000$ and $n_p = 1$ are always considered. Table 2 shows the dimensionless critical load of a square plate FG for different values of externally applied potentials and non-local parameter. It can be seen that as the non-local parameter increases, the value of the critical load increases. It can also be seen that by increasing the magnetic potential the critical load increases while the electric potential has the opposite effect. This last phenomenon is also clearly visible in Figure 3, where the value of the critical load is reported as the external potentials applied vary and for different values of the non-local parameter. Finally it is remarked that for same values of the electric and magnetic potentials the critical load takes a negative value, thus, buckling occurs for traction loads instead of compression. Figure 4 shows the dimensional critical load when the aspect ratio varies and for different values of the non-local parameter. As expected the critical load increases as the plane becomes of rectangular shape and as the nonlocal parameter increases.

Table 2. Dimensionless critical load $\bar{N}_{cr}$ of a square FG plate composed of $BaTiO_3/CoFe_2O_4$ for different electric and magnetic potentials and nonlocal parameter $(\ell/a)^2$. ($\kappa = 1, (m,n) = (1,1)$).

$(\ell/a)^2$	Ω_0 [A]	V_0 [V] −5	−2.5	0	2.5	5
0.00	1	2.1733	2.0124	1.8516	1.6907	1.5299
	0	1.4456	1.2848	1.1239	0.9631	0.8022
	−1	0.7180	0.5572	0.3963	0.2355	0.0746
0.05	1	3.2825	3.1217	2.9608	2.8000	2.6391
	0	2.5549	2.3940	2.2332	2.0723	1.9115
	−1	1.8273	1.6664	1.5056	1.3447	1.1838
0.10	1	4.3918	4.2309	4.0701	3.9092	3.7484
	0	3.6642	3.5033	3.3424	3.1816	3.0207
	−1	2.9365	2.7757	2.6148	2.4540	2.2931

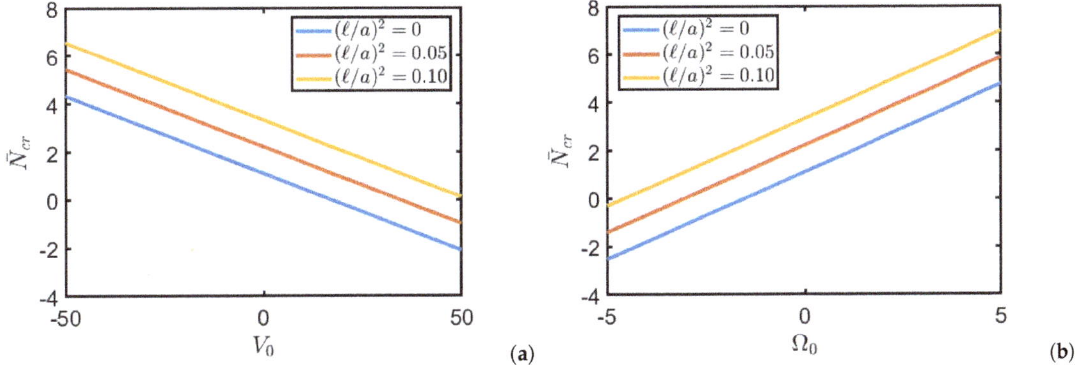

Figure 3. Critical load $\bar{N}_{cr}$ of a square FG nanoplate for different values of electric potential V_0 (**a**) and magnetic potential Ω_0 (**b**), for different values of non local parameter $(\ell/a)^2$.

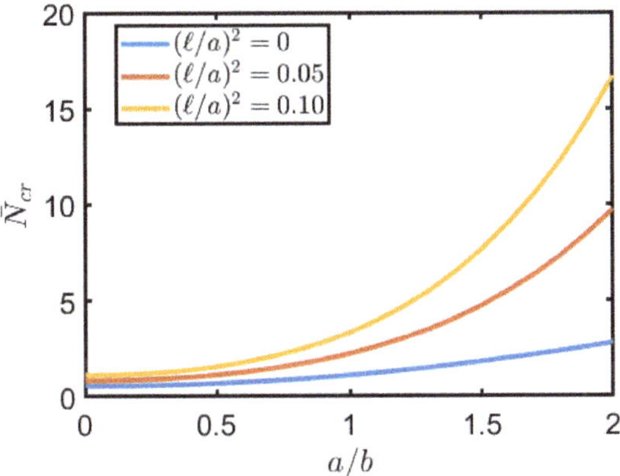

Figure 4. Critical load $\bar{N}_{cr}$ of square FG nanoplate for different values of ratio a/b and different values of non local parameter $(\ell/a)^2$.

6.2. Thermal Free Vibration

As first application for the free vibration of piezo-electro-magnetic-thermal plates a comparison with [76] is reported (Table 3). The plate is composed of $BaTiO_3/CoFe_2O_4$ and is of rectangular shape with $a = 2$ and $b = 1$, ratio $h/a = 0.1$ and properties vary linearly along the thickness ($n_p = 1$). The results are written in dimensionless form through the following formula

$$\bar{\omega} = \omega\left(\frac{a^2}{h}\right)\sqrt{\frac{\rho}{C_{11}}} \tag{40}$$

where ρ and C_{11} respectively represent the density and the $(1,1)$ position element of the material stiffness matrix on the underside of the plate. It should be noted that in the study just mentioned the Mindlin's moderately thick plates theory (FSDT) is used, and so the results deviate slightly and this difference increases as the vibration mode increases as the effects of shear become more relevant.

Table 3. Dimensionless natural frequencies $\bar{\omega}$ of a simply supported rectangular FG plate composed of $BaTiO_3/CoFe_2O_4$.

	Ref. [76]	**Present**
1	9.525	10.0244
2	28.762	32.5716
3	50.966	66.2842
4	131.186	104.0065
5	139.106	129.6477

Table 4 shows a comparison with article [57] for a thick magneto-electro-elastic square plate. The thickness of the plate is constant and the laminae all have the same thickness. The properties of the material are those reported in Table 1, except the density which is assumed constant for the two materials and equal to 1600 kg/m^3. The values calculated in this study differ slightly from those in the literature because in [57] a third order plate theory is used while in this study it is used thin plate theory.

Table 4. Natural frequencies (rad/s) of a simply supported square plate ($a = 1$ m; $h = 0.3$ m). B = $BaTiO_3$; F = $CoFe_2O_4$.

Mode	**Ref. [57]** $(\ell/a)^2 = 0$			
	B	F	B/F/B	F/B/F
1	12,863.98	15,185.24	13,024.78	15,043.32
2	25,106.78	28,177.03	25,401.26	27,880.80
	Present $(\ell/a)^2 = 0$			
1	15,044.28	17,253.16	15,281.37	17,159.99
2	34,945.88	39,415.58	28,264.36	39,871.43
	Present $(\ell/a)^2 = 0.05$			
1	21,065.74	24,351.13	20,932.93	24,069.44
2	63,337.77	73,727.92	62,619.25	72,511.91
	Present $(\ell/a)^2 = 0.10$			
1	25,723.93	29,805.02	25,499.40	29,410.18
2	82,655.46	96,547.86	81,654.82	94,727.82

Table 5 analyzes the influence of the non-local parameter on the natural frequencies of a square plate with $a = b = 1$ m and ratio $a/h = 100$ and $n_p = 1$. The results show an increase in natural frequencies, due to the stiffening of the plate, as the non-local parameter increases.

Figure 5 shows the influence of temperature and externally applied potentials on the natural frequency of a nanoplate composed by $BaTiO_3/CoFe_2O_4$. The plate considered is a simply supported square plate and the thickness ratio is $a/h = 100$. In particular, the critical temperature of each structure can be identified when the frequency becomes zero.

Table 5. Natural frequencies $\bar{\omega}$ of a simply supported square FG nanoplate composed by $BaTiO_3/CoFe_2O_4$.

	$(\ell/a)^2$		
	0	**0.05**	**0.10**
1	4.0913	5.7671	7.0554
2	10.2270	19.0434	24.9141
3	20.4499	49.8179	67.4200
4	34.7550	106.4926	146.5385
5	53.1353	197.6005	274.3517

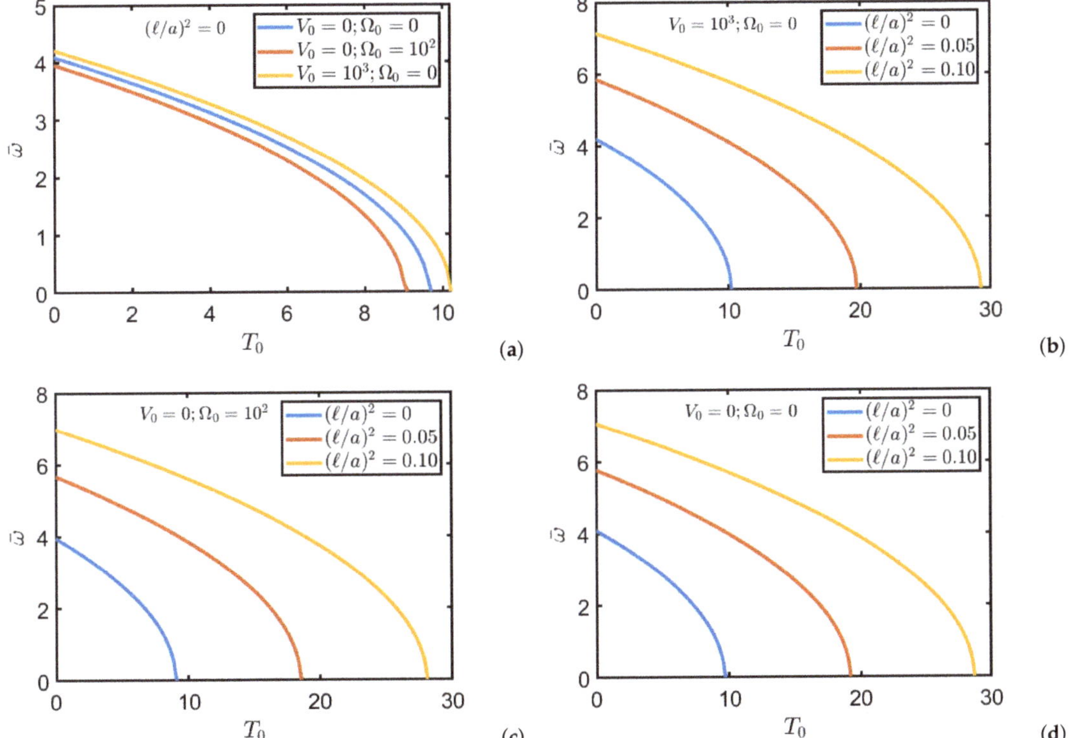

Figure 5. Natural frequencies $\bar{\omega}$ of a simply supported square FG nanoplate composed by $BaTiO_3/CoFe_2O_4$ to vary of temperature T_0 and for different values of magnetic and electric potentials and non-local parameter: (**a**) local configuration ($\ell = 0$); (**b**) nonlocal configuration with $V_0 \neq 0$; (**c**) nonlocal configuration with $\Omega_0 \neq 0$; (**d**) nonlocal configuration with $V_0 = \Omega_0 = 0$.

7. Conclusions

In this paper the dynamic and buckling problems of METE nanoplates have been analyzed. In particular, the interest focused on the coupling of magnet-electro-thermo-elastic effects and the influence that the external potentials applied to the plate have on the critical load and natural vibration frequencies. Through Hamilton's principle motion equations for FG METE thin plates are derived and analytical solution using Navier method is obtained. The materials that have been used in the simulations are $BaTiO_3$ and $CoFe_2O_4$ and the properties of the materials used are reported in the article. The results show that increasing the non-local parameter increases the critical load and natural vibration frequencies. For external potentials instead it was seen that the critical load increases with the increase of the negative electrical potential and the positive magnetic potential. Finally, from the graphs of the natural frequency of vibration it can be seen that the frequencies tend to increase by subjecting the plate to a positive magnetic potential and to decrease by subjecting it to a positive electrical potential. For what concerns the temperature instead we see how an increase of the latter leads to a reduction of the natural frequencies.

Author Contributions: Conceptualization, N.F.; methodology, N.F.; software, N.F. and G.T.M.; validation, N.F. and G.T.M.; formal analysis, N.F. and G.T.M.; investigation, N.F. and G.T.M.; resources, N.F., F.F. and R.L.; data curation, G.T.M.; writing—original draft preparation, N.F. and G.T.M.; writing—review and editing, N.F., G.T.M., F.F. and R.L.; visualization, F.F. and R.L.; supervision, N.F., F.F. and R.L.; project administration, N.F., F.F. and R.L.; funding acquisition, F.F. and R.L. All authors have read and agreed to the published version of the manuscript.

Funding: This research received no external funding.

Institutional Review Board Statement: Not applicable.

Informed Consent Statement: Not applicable.

Data Availability Statement: Data sharing not applicable.

Conflicts of Interest: The authors declare no conflict of interest.

Appendix A

Integrating last two integrals of Equation (22) along the thickness the following quantities can be defined

$$\begin{aligned}
\mathbf{A}_E^f &= \int_{-\frac{h}{2}}^{\frac{h}{2}} \tilde{\mathbf{e}}\mathbf{f}_E \, dz, \quad \mathbf{A}_E = \int_{-\frac{h}{2}}^{\frac{h}{2}} \tilde{\mathbf{e}} \, dz, \quad \mathbf{B}_E^f = \int_{-\frac{h}{2}}^{\frac{h}{2}} \tilde{\mathbf{e}}\mathbf{f}_E \, z \, dz \quad , \quad \mathbf{B}_E = \int_{-\frac{h}{2}}^{\frac{h}{2}} \tilde{\mathbf{e}} \, z \, dz \\
\mathbf{A}_H^f &= \int_{-\frac{h}{2}}^{\frac{h}{2}} \tilde{\mathbf{q}}\mathbf{f}_H \, dz, \quad \mathbf{A}_H = \int_{-\frac{h}{2}}^{\frac{h}{2}} \tilde{\mathbf{q}} \, dz, \quad \mathbf{B}_H^f = \int_{-\frac{h}{2}}^{\frac{h}{2}} \tilde{\mathbf{q}}\mathbf{f}_H \, z \, dz, \quad \mathbf{B}_H = \int_{-\frac{h}{2}}^{\frac{h}{2}} \tilde{\mathbf{q}} \, z \, dz
\end{aligned} \tag{A1}$$

$$\begin{aligned}
\mathbf{A}_\xi^f &= \int_{-\frac{h}{2}}^{\frac{h}{2}} \mathbf{f}_E^\top \tilde{\boldsymbol{\xi}} dz, \quad \mathbf{B}_\xi^f = \int_{-\frac{h}{2}}^{\frac{h}{2}} \mathbf{f}_E^\top \tilde{\boldsymbol{\xi}} \mathbf{f}_E dz, \quad \mathbf{A}_\zeta^f = \int_{-\frac{h}{2}}^{\frac{h}{2}} \mathbf{f}_E^\top \tilde{\boldsymbol{\zeta}} dz, \quad \mathbf{B}_\zeta^f = \int_{-\frac{h}{2}}^{\frac{h}{2}} \mathbf{f}_E^\top \tilde{\boldsymbol{\zeta}} \mathbf{f}_H dz \\
\mathbf{A}_p &= \int_{-\frac{h}{2}}^{\frac{h}{2}} \mathbf{f}_E^\top \mathbf{p} dz, \quad \mathbf{B}_p = \int_{-\frac{h}{2}}^{\frac{h}{2}} \mathbf{f}_E^\top \mathbf{p} z \, dz, \quad \mathbf{A}_h = \int_{-\frac{h}{2}}^{\frac{h}{2}} \mathbf{f}_E^\top \mathbf{h} dz, \quad \mathbf{B}_h = \int_{-\frac{h}{2}}^{\frac{h}{2}} \mathbf{f}_E^\top \mathbf{h} z \, dz
\end{aligned} \tag{A2}$$

$$\begin{aligned}
\mathbf{A}_\zeta^f &= \int_{-\frac{h}{2}}^{\frac{h}{2}} \mathbf{f}_H^\top \tilde{\boldsymbol{\zeta}} dz, \quad \mathbf{B}_\zeta^f = \int_{-\frac{h}{2}}^{\frac{h}{2}} \mathbf{f}_H^\top \tilde{\boldsymbol{\zeta}} \mathbf{f}_E dz, \quad \mathbf{A}_\chi^f = \int_{-\frac{h}{2}}^{\frac{h}{2}} \mathbf{f}_H^\top \tilde{\boldsymbol{\chi}} dz, \quad \mathbf{B}_\chi^f = \int_{-\frac{h}{2}}^{\frac{h}{2}} \mathbf{f}_H^\top \tilde{\boldsymbol{\chi}} \mathbf{f}_H dz \\
\mathbf{A}_\lambda &= \int_{-\frac{h}{2}}^{\frac{h}{2}} \mathbf{f}_H^\top \boldsymbol{\lambda} dz, \quad \mathbf{B}_\lambda = \int_{-\frac{h}{2}}^{\frac{h}{2}} \mathbf{f}_H^\top \boldsymbol{\lambda} z \, dz, \quad \mathbf{A}_\eta = \int_{-\frac{h}{2}}^{\frac{h}{2}} \mathbf{f}_H^\top \boldsymbol{\eta} dz, \quad \mathbf{B}_\eta = \int_{-\frac{h}{2}}^{\frac{h}{2}} \mathbf{f}_H^\top \boldsymbol{\eta} z \, dz
\end{aligned} \tag{A3}$$

Considering what reported in Equations (A1)–(A3) it can be write $\hat{c}_{ij}$ coefficients as follows

$$
\begin{aligned}
\hat{c}_{11} &= \alpha^2 A_{11} + \beta^2 A_{66} + \ell^2 \left[\alpha^4 A_{11} + \alpha^2\beta^2(A_{11} + A_{66}) + \beta^4 A_{66}\right] \\
\hat{c}_{12} &= \alpha\beta(A_{12} + A_{66}) + \ell^2 \left[\alpha^3\beta(A_{12} + A_{66}) + \alpha\beta^3(A_{12} + A_{66})\right] \\
\hat{c}_{13} &= -\alpha^3 B_{11} - \alpha\beta^2(B_{12} + 2B_{66}) \\
&\quad - \ell^2 \left[\alpha^5 B_{11} + \alpha^3\beta^2(B_{12} + B_{11} + 2B_{66}) + \alpha\beta^4(B_{12} + 2B_{66})\right] \\
\hat{c}_{22} &= \beta^2 A_{22} + \alpha^2 A_{66} + \ell^2 \left[\alpha^2\beta^2(A_{22} + A_{66}) + \alpha^4 A_{66} + \beta^4 A_{22}\right] \\
\hat{c}_{23} &= -\alpha^2\beta(B_{12} + 2B_{66}) - \beta^3 B_{22} \\
&\quad - \ell^2 \left[\alpha^4\beta(B_{12} + 2B_{66}) + \alpha^2\beta^3(B_{22} + B_{12} + 2B_{66}) + \beta^5 B_{22}\right] \\
\hat{c}_{33} &= \alpha^4 D_{11} + 2\alpha^2\beta^2(D_{12} + 2D_{66}) + \beta^4 D_{22} + \ell^2[\alpha^6 D_{11} + \alpha^4\beta^2(D_{11} + 2D_{12} + 4D_{66}) \\
&\quad + \alpha^2\beta^4(D_{22} + 2D_{12} + 4D_{66}) + \beta^6 D_{22}] \\
\hat{c}_{14} &= -\alpha A^f_{E,13} + \ell^2\left[-\alpha^3 A^f_{E,13} - \alpha\beta^2 A^f_{E,13}\right] \\
\hat{c}_{15} &= -\alpha A^f_{H,13} + \ell^2\left[-\alpha^3 A^f_{H,13} - \alpha\beta^2 A^f_{H,13}\right] \\
\hat{c}_{24} &= -\beta A^f_{E,23} + \ell^2\left[-\beta^3 A^f_{E,23} - \alpha^2\beta A^f_{E,23}\right] \\
\hat{c}_{25} &= -\beta A^f_{H,23} + \ell^2\left[-\beta^3 A^f_{H,23} - \alpha^2\beta A^f_{H,23}\right] \\
\hat{c}_{34} &= \alpha^2 B^f_{E,13} + \beta^2 B^f_{E,23} + \ell^2\left[\alpha^4 B^f_{E,13} + \alpha^2\beta^2\left(B^f_{E,13} + B^f_{E,23}\right) + \beta^4 B^f_{E,23}\right] \\
\hat{c}_{35} &= \alpha^2 B^f_{H,13} + \beta^2 B^f_{H,23} + \ell^2\left[\alpha^4 B^f_{H,13} + \alpha^2\beta^2\left(B^f_{H,13} + B^f_{H,23}\right) + \beta^4 B^f_{H,23}\right] \\
\hat{c}_{44} &= -B^f_{\zeta,33} - \alpha^2 B^f_{\zeta,11} - \beta^2 B^f_{\zeta,22} \\
&\quad - \ell^2\left[\alpha^4 B^f_{\zeta,11} + \alpha^2\beta^2\left(B^f_{\zeta,11} + B^f_{\zeta,22}\right) + \beta^4 B^f_{\zeta,22} + \alpha^2 B^f_{\zeta,33} + \beta^2 B^f_{\zeta,33}\right] \\
\hat{c}_{45} &= -B^{fE}_{\zeta,33} - \alpha^2 B^{fE}_{\zeta,11} - \beta^2 B^{fE}_{\zeta,22} \\
&\quad - \ell^2\left[\alpha^4 B^{fE}_{\zeta,11} + \alpha^2\beta^2\left(B^{fE}_{\zeta,11} + B^{fE}_{\zeta22}\right) + \beta^4 B^{fE}_{\zeta,22} + \alpha^2 B^{fE}_{\zeta,33} + \beta^2 B^{fE}_{\zeta,33}\right] \\
\hat{c}_{54} &= -B^{fH}_{\zeta,33} - \alpha^2 B^{fH}_{\zeta,11} - \beta^2 B^{fH}_{\zeta,22} \\
&\quad - \ell^2\left[\alpha^4 B^{fH}_{\zeta,11} + \alpha^2\beta^2\left(B^{fH}_{\zeta,11} + B^{fH}_{\zeta,22}\right) + \beta^4 B^{fH}_{\zeta,22} + \alpha^2 B^{fH}_{\zeta,33} + \beta^2 B^{fH}_{\zeta,33}\right] \\
\hat{c}_{55} &= -B^f_{\chi,33} - \alpha^2 B^f_{\chi,11} - \beta^2 B^f_{\chi22} \\
&\quad - \ell^2\left[\alpha^4 B^f_{\chi,11} + \alpha^2\beta^2\left(B^f_{\chi,11} + B^f_{\chi,22}\right) + \beta^4 B^f_{\chi,22} + \alpha^2 B^f_{\chi,33} + \beta^2 B^f_{\chi,33}\right] \\
\tilde{s}_{33} &= \alpha\left(\hat{N}_{xx} + \hat{N}^T_{xx} + \hat{N}^E_{xx} + \hat{N}^H_{xx}\right) + \beta\left(\hat{N}_{yy} + \hat{N}^T_{yy} + \hat{N}^E_{yy} + \hat{N}^H_{yy}\right)
\end{aligned} \tag{A4}
$$

In case the electrical and magnetic potentials have the same through-the-thickness expansion, as in the present study, $\mathbf{B}^{fE}_{\zeta} = \mathbf{B}^{fH}_{\zeta} = \mathbf{B}^{f}_{\zeta}$, hence, $\hat{c}_{45} = \hat{c}_{54}$.

References

1. Saji, V.S.; Choe, H.C.; Yeung, K.W. Nanotechnology in biomedical applications: A review. *Int. J. Nano-Biomater.* **2010**, *3*, 119–139. [CrossRef]
2. Berman, D.; Krim, J. Surface science, MEMS and NEMS: Progress and opportunities for surface science research performed on, or by, microdevices. *Prog. Surf. Sci.* **2013**, *88*, 171–211. [CrossRef]
3. Bhushan, B. Nanotribology and nanomechanics of MEMS/NEMS and BioMEMS/BioNEMS materials and devices. *Microelectron. Eng.* **2007**, *84*, 387–412. [CrossRef]
4. Ekinci, K.L.; Roukes, M.L. Nanoelectromechanical systems. *Rev. Sci. Instruments* **2005**, *76*, 061101. [CrossRef]
5. Barretta, R.; Fabbrocino, F.; Luciano, R.; de Sciarra, F.M.; Ruta, G. Buckling loads of nano-beams in stress-driven nonlocal elasticity. *Mech. Adv. Mater. Struct.* **2020**, *27*, 869–875. [CrossRef]

6. Bonanni, A.; del Valle, M. Use of nanomaterials for impedimetric DNA sensors: A review. *Anal. Chim. Acta* **2010**, *678*, 7–17. [CrossRef]
7. Wu, W. Inorganic nanomaterials for printed electronics: A review. *Nanoscale* **2017**, *9*, 7342–7372. [CrossRef]
8. Gohardani, O.; Elola, M.C.; Elizetxea, C. Potential and prospective implementation of carbon nanotubes on next generation aircraft and space vehicles: A review of current and expected applications in aerospace sciences. *Prog. Aerosp. Sci.* **2014**, *70*, 42–68. [CrossRef]
9. Singh, T. A review of nanomaterials in civil engineering works. *Inter. J. Struct. Civ. Eng. Res.* **2014**, *3*, 31–35.
10. Amabili, M. *Nonlinear Mechanics of Shells and Plates in Composite, Soft and Biological Materials*; Cambridge University Press: Cambridge, UK, 2018.
11. Lakes, R. Experimental microelasticity of two porous solids. *Int. J. Solids Struct.* **1986**, *22*, 55–63. [CrossRef]
12. Stölken, J.; Evans, A. A microbend test method for measuring the plasticity length scale. *Acta Mater.* **1998**, *46*, 5109–5115. [CrossRef]
13. Mancusi, G.; Fabbrocino, F.; Feo, L.; Fraternali, F. Size effect and dynamic properties of 2D lattice materials. *Compos. Part B Eng.* **2017**, *112*, 235–242. [CrossRef]
14. Fabbrocino, F.; Carpentieri, G. Three-dimensional modeling of the wave dynamics of tensegrity lattices. *Compos. Struct.* **2017**, *173*, 9–16. [CrossRef]
15. Eringen, A.; Edelen, D. On nonlocal elasticity. *Int. J. Eng. Sci.* **1972**, *10*, 233–248. [CrossRef]
16. Trovalusci, P. Molecular Approaches for Multifield Continua: Origins and Current Developments. In *Multiscale Modeling of Complex Materials*; Sadowski, T., Trovalusci, P., Eds.; Springer: Vienna, Austria, 2014; pp. 211–278.
17. Eringen, A.C. On differential equations of nonlocal elasticity and solutions of screw dislocation and surface waves. *J. Appl. Phys.* **1983**, *54*, 4703–4710. [CrossRef]
18. Aifantis, E. Update on a class of gradient theories. *Mech. Mater.* **2003**, *35*, 259–280. [CrossRef]
19. Meenen, J.; Altenbach, H.; Eremeyev, V.; Naumenko, K. A Variationally Consistent Derivation of Microcontinuum Theories. *Adv. Struct. Mater.* **2011**, *15*.
20. Mindlin, R.; Eshel, N. On first strain-gradient theories in linear elasticity. *Int. J. Solids Struct.* **1968**, *4*, 109–124. [CrossRef]
21. Karami, B.; Janghorban, M.; Rabczuk, T. Static analysis of functionally graded anisotropic nanoplates using nonlocal strain gradient theory. *Compos. Struct.* **2019**, *227*, 111249. [CrossRef]
22. Eremeyev, V.; Altenbach, H. On the Direct Approach in the Theory of Second Gradient Plates. In *Shell and Membrane Theories in Mechanics and Biology*; Altenbach, H., Mikhasev, G.I., Eds.; Springer: Cham, Switzerland, 2015; Volume 45, pp. 147–154.
23. Bacciocchi, M.; Fantuzzi, N.; Ferreira, A. Conforming and nonconforming laminated finite element Kirchhoff nanoplates in bending using strain gradient theory. *Comput. Struct.* **2020**, *239*, 106322. [CrossRef]
24. Barretta, R.; Feo, L.; Luciano, R.; Marotti de Sciarra, F.; Penna, R. Functionally graded Timoshenko nanobeams: A novel nonlocal gradient formulation. *Compos. Part B Eng.* **2016**, *100*, 208–219. [CrossRef]
25. Sahmani, S.; Aghdam, M.M.; Rabczuk, T. Nonlinear bending of functionally graded porous micro/nano-beams reinforced with graphene platelets based upon nonlocal strain gradient theory. *Compos. Struct.* **2018**, *186*, 68–78. [CrossRef]
26. Jamalpoor, A.; Hosseini, M. Biaxial buckling analysis of double-orthotropic microplate-systems including in-plane magnetic field based on strain gradient theory. *Compos. Part B Eng.* **2015**, *75*, 53–64. [CrossRef]
27. Apuzzo, A.; Barretta, R.; Faghidian, S.; Luciano, R.; Marotti de Sciarra, F. Free vibrations of elastic beams by modified nonlocal strain gradient theory. *Int. J. Eng. Sci.* **2018**, *133*, 99–108. [CrossRef]
28. Yang, F.; Chong, A.; Lam, D.; Tong, P. Couple stress based strain gradient theory for elasticity. *Int. J. Solids Struct.* **2002**, *39*, 2731–2743. [CrossRef]
29. Mühlhaus, H.; Oka, F. Dispersion and wave propagation in discrete and continuous models for granular materials. *Int. J. Solids Struct.* **1996**, *33*, 2841–2858. [CrossRef]
30. Leonetti, L.; Greco, F.; Trovalusci, P.; Luciano, R.; Masiani, R. A multiscale damage analysis of periodic composites using a couple-stress/Cauchy multidomain model: Application to masonry structures. *Compos. Part B Eng.* **2018**, *141*, 50–59. [CrossRef]
31. Farajpour, A.; Howard, C.Q.; Robertson, W.S. On size-dependent mechanics of nanoplates. *Int. J. Eng. Sci.* **2020**, *156*, 103368. [CrossRef]
32. Barretta, R.; Faghidian, S.A.; Marotti de Sciarra, F. Stress-driven nonlocal integral elasticity for axisymmetric nano-plates. *Int. J. Eng. Sci.* **2019**, *136*, 38–52. [CrossRef]
33. Trovalusci, P.; Bellis, M.D.; Ostoja-Starzewski, M. A Statistically-Based Homogenization Approach for Particle Random Composites as Micropolar Continua. *Adv. Struct. Mater.* **2016**, *42*.
34. Reccia, E.; De Bellis, M.L.; Trovalusci, P.; Masiani, R. Sensitivity to material contrast in homogenization of random particle composites as micropolar continua. *Compos. Part B Eng.* **2018**, *136*, 39–45. [CrossRef]
35. Fantuzzi, N.; Leonetti, L.; Trovalusci, P.; Tornabene, F. Some Novel Numerical Applications of Cosserat Continua. *Int. J. Comput. Methods* **2018**, *15*, 1850054. [CrossRef]
36. Rega, G.; Saetta, E.; Settimi, V. Modeling and nonlinear dynamics of thermomechanically coupled composite plates. *Int. J. Mech. Sci.* **2020**, *187*, 106106. [CrossRef]
37. Kim, J.Y. Micromechanical analysis of effective properties of magneto-electro-thermo-elastic multilayer composites. *Int. J. Eng. Sci.* **2011**, *49*, 1001–1018. [CrossRef]

38. Ansari, R.; Gholami, R. Size-Dependent Buckling and Postbuckling Analyses of First-Order Shear Deformable Magneto-Electro-Thermo Elastic Nanoplates Based on the Nonlocal Elasticity Theory. *Int. J. Struct. Stab. Dyn.* **2017**, *17*, 1750014. [CrossRef]
39. Ansari, R.; Gholami, R. Nonlocal free vibration in the pre- and post-buckled states of magneto-electro-thermo elastic rectangular nanoplates with various edge conditions. *Smart Mater. Struct.* **2016**, *25*, 095033. [CrossRef]
40. Lei, Z.; Liew, K.; Yu, J. Buckling analysis of functionally graded carbon nanotube-reinforced composite plates using the element-free kp-Ritz method. *Compos. Struct.* **2013**, *98*, 160–168. [CrossRef]
41. Mota, A.F.; Loja, M.A.R.; Barbosa, J.I.; Rodrigues, J.A. Porous Functionally Graded Plates: An Assessment of the Influence of Shear Correction Factor on Static Behavior. *Math. Comput. Appl.* **2020**, *25*, 25. [CrossRef]
42. Zhong, Z.; Yu, T. Vibration of a simply supported functionally graded piezoelectric rectangular plate. *Smart Mater. Struct.* **2006**, *15*, 1404–1412. [CrossRef]
43. Tomczyk, B.; Szczerba, P. Combined asymptotic-tolerance modelling of dynamic problems for functionally graded shells. *Compos. Struct.* **2018**, *183*, 176–184. In honor of Prof. Y. Narita. [CrossRef]
44. Tomczyk, B.; Szczerba, P. A new asymptotic-tolerance model of dynamic and stability problems for longitudinally graded cylindrical shells. *Compos. Struct.* **2018**, *202*, 473–481. Special issue dedicated to Ian Marshall. [CrossRef]
45. Kondaiah, P.; Shankar, K.; Ganesan, N. Pyroelectric and pyromagnetic effects on multiphase magneto–electro–elastic cylindrical shells for axisymmetric temperature. *Smart Mater. Struct.* **2012**, *22*, 025007. [CrossRef]
46. Malikan, M.; Eremeyev, V.A. On the Dynamics of a Visco–Piezo–Flexoelectric Nanobeam. *Symmetry* **2020**, *12*, 643. [CrossRef]
47. Malikan, M.; Eremeyev, V.A. On Nonlinear Bending Study of a Piezo-Flexomagnetic Nanobeam Based on an Analytical-Numerical Solution. *Nanomaterials* **2020**, *10*, 1762. [CrossRef] [PubMed]
48. Bacciocchi, M.; Tarantino, A. Natural Frequency Analysis of Functionally Graded Orthotropic Cross-Ply Plates Based on the Finite Element Method. *Math. Comput. Appl.* **2019**, *24*, 52. [CrossRef]
49. Uzun, B.; Civalek, O. Nonlocal FEM Formulation for Vibration Analysis of Nanowires on Elastic Matrix with Different Materials. *Math. Comput. Appl.* **2019**, *24*, 38. [CrossRef]
50. Ebrahimi, F.; Reza Barati, M. Magnetic field effects on buckling behavior of smart size-dependent graded nanoscale beams. *Eur. Phys. J. Plus* **2016**, *131*, 238. [CrossRef]
51. Xu, X.J.; Deng, Z.C.; Zhang, K.; Meng, J.M. Surface effects on the bending, buckling and free vibration analysis of magneto-electro-elastic beams. *Acta Mech.* **2016**, *227*, 1557–1573. [CrossRef]
52. Heyliger, P.; Ramirez, G. Free Vibration of Laminated Circular Piezoelectric Plates and Discs. *J. Sound Vib.* **2000**, *229*, 935–956. [CrossRef]
53. Mohammadimehr, M.; Rostami, R. Bending, buckling, and forced vibration analyses of nonlocal nanocomposite microplate using TSDT considering MEE properties dependent to various volume fractions of $CoFe_2O_4$-$BaTiO_3$. *J. Theor. Appl. Mech.* **2017**, *55*, 853. [CrossRef]
54. Farahmand, H.; Naseralavi, S.S.; Iranmanesh, A.; Mohammadi, M. Navier Solution for Buckling Analysis of Size-Dependent Functionally Graded Micro-Plates. *Lat. Am. J. Solids Struct.* **2016**, *13*, 3161–3173. [CrossRef]
55. Altenbach, H.; Eremeyev, V.A. On the shell theory on the nanoscale with surface stresses. *Int. J. Eng. Sci.* **2011**, *49*, 1294–1301. [CrossRef]
56. Ghobadi, A.; Golestanian, H.; Beni, Y.T.; Kamil Żur, K. On the size-dependent nonlinear thermo-electro-mechanical free vibration analysis of functionally graded flexoelectric nano-plate. *Commun. Nonlinear Sci. Numer. Simul.* **2020**, 105585. [CrossRef]
57. Simões Moita, J.M.; Mota Soares, C.M.; Mota Soares, C.A. Analyses of magneto-electro-elastic plates using a higher order finite element model. *Compos. Struct.* **2009**, *91*, 421–426. [CrossRef]
58. Jędrysiak, J. Free vibrations of thin periodic plates interacting with an elastic periodic foundation. *Int. J. Mech. Sci.* **2003**, *45*, 1411–1428. [CrossRef]
59. Zhou, X.; Yu, D.; Shao, X.; Wang, S.; Zhang, S. Simplified-super-element-method for analyzing free flexural vibration characteristics of periodically stiffened-thin-plate filled with viscoelastic damping material. *Thin-Walled Struct.* **2015**, *94*, 234–252. [CrossRef]
60. Wirowski, A.; Michalak, B.; Gajdzicki, M. Dynamic Modelling of Annular Plates of Functionally Graded Structure Resting on Elastic Heterogeneous Foundation with Two Modules. *J. Mech.* **2015**, *31*, 493–504. [CrossRef]
61. Michalak, B. 2D tolerance and asymptotic models in elastodynamics of a thin-walled structure with dense system of ribs. *Arch. Civ. Mech. Eng.* **2015**, *15*, 449–455. [CrossRef]
62. Tomczyk, B.; Szczerba, P. Tolerance and asymptotic modelling of dynamic problems for thin microstructured transversally graded shells. *Compos. Struct.* **2017**, *162*, 365–373. [CrossRef]
63. Xiang, H.J.; Shi, Z.F. Analysis of flexural vibration band gaps in periodic beams using differential quadrature method. *Comput. Struct.* **2009**, *87*, 1559–1566. [CrossRef]
64. Xu, Y.; Zhou, X.; Wang, W.; Wang, L.; Peng, F.; Li, B. On natural frequencies of non-uniform beams modulated by finite periodic cells. *Phys. Lett. A* **2016**, *380*, 3278–3283. [CrossRef]
65. Jędrysiak, J. Tolerance Modelling of Vibrations and Stability for Periodic Slender Visco-Elastic Beams on a Foundation with Damping. Revisiting. *Materials* **2020**, *13*, 3939. [CrossRef] [PubMed]
66. Jędrysiak, J. Tolerance modelling of free vibration frequencies of thin functionally graded plates with one-directional microstructure. *Compos. Struct.* **2017**, *161*, 453–468. [CrossRef]

67. Marczak, J.; Jędrysiak, J. Some remarks on modelling of vibrations of periodic sandwich structures with inert core. *Compos. Struct.* **2018**, *202*, 752–758. Special issue dedicated to Ian Marshall. [CrossRef]
68. Wu, Z.J.; Li, F.M.; Wang, Y.Z. Vibration band gap properties of periodic Mindlin plate structure using the spectral element method. *Meccanica* **2014**, *49*, 725–737. [CrossRef]
69. Cheng, Z.; Xu, Y.; Zhang, L. Analysis of flexural wave bandgaps in periodic plate structures using differential quadrature element method. *Int. J. Mech. Sci.* **2015**, *100*, 112–125. [CrossRef]
70. Zhou, X.; Yu, D.; Shao, X.; Wang, S.; Tian, Y. Band gap characteristics of periodically stiffened-thin-plate based on center-finite-difference-method. *Thin-Walled Struct.* **2014**, *82*, 115–123. [CrossRef]
71. Yu, D.; Wen, J.; Shen, H.; Xiao, Y.; Wen, X. Propagation of flexural wave in periodic beam on elastic foundations. *Phys. Lett. A* **2012**, *376*, 626–630. [CrossRef]
72. Chen, T. Investigations on flexural wave propagation of a periodic beam using multi-reflection method. *Arch. Appl. Mech.* **2013**, *83*, 315–329. [CrossRef]
73. Wang, Q. On buckling of column structures with a pair of piezoelectric layers. *Eng. Struct.* **2002**, *24*, 199–205. [CrossRef]
74. Li, Y. Buckling analysis of magnetoelectroelastic plate resting on Pasternak elastic foundation. *Mech. Res. Commun.* **2014**, *56*, 104–114. [CrossRef]
75. Park, W.T.; Han, S.C. Buckling analysis of nano-scale magneto-electro-elastic plates using the nonlocal elasticity theory. *Adv. Mech. Eng.* **2018**, *10*, 1687814018793335. [CrossRef]
76. Ramirez, F.; Heyliger, P.R.; Pan, E. Discrete Layer Solution to Free Vibrations of Functionally Graded Magneto-Electro-Elastic Plates. *Mech. Adv. Mater. Struct.* **2006**, *13*, 249–266. [CrossRef]

Article

Effect of Metallic or Non-Metallic Element Addition on Surface Topography and Mechanical Properties of CrN Coatings

Tatyana Kuznetsova [1], Vasilina Lapitskaya [1], Anastasiya Khabarava [1], Sergei Chizhik [1], Bogdan Warcholinski [2], Adam Gilewicz [2], Aleksander Kuprin [3], Sergei Aizikovich [4,*] and Boris Mitrin [4]

1 Nanoprocesses and Technology Laboratory, A.V. Luikov Heat and Mass Transfer Institute of the National Academy of Science of Belarus, 15 P. Brovki str., 220072 Minsk, Belarus; kuzn06@mail.ru (T.K.); vasilinka.92@mail.ru (V.L.); av.khabarova@mail.ru (A.K.); chizhik_sa@tut.by (S.C.)

2 Faculty of Mechanical Engineering, Koszalin University of Technology, 2 Sniadeckich, 75-453 Koszalin, Poland; bogdan.warcholinski@tu.koszalin.pl (B.W.); adam.gilewicz@tu.koszalin.pl (A.G.)

3 National Science Center Kharkov Institute of Physics and Technology, 1 Academic str, 61108 Kharkiv, Ukraine; kuprin@kipt.kharkov.ua

4 Research and Education Center "Materials", Don State Technical University, 1 Gagarin sq., 344003 Rostov-on-Don, Russia; bmitrin@dstu.edu.ru

* Correspondence: s.aizikovich@sci.donstu.ru; Tel.: +7-863-238-15-58

Received: 20 November 2020; Accepted: 25 November 2020; Published: 27 November 2020

Abstract: Alteration of the phase composition of a coating and/or its surface topography can be achieved by changing the deposition technology and/or introducing additional elements into the coating. Investigation of the effect of the composition of CrN-based coatings (including AlCrN and CrON) on the microparticle height and volume, as well as the construction of correlations between the friction coefficient at the microscale and the geometry of microparticles, are the goals of this study. We use atomic force microscopy (AFM), which is the most effective method of investigation with nanometer resolution. By revealing the morphology, AFM allows one to determine the diameter of the particles, their heights and volumes and to identify different phases in the studied area by contrasted properties. The evaluation of the distribution of mechanical properties (modulus of elasticity E and microhardness H) on the surfaces of multiphase coatings with microparticles is carried out by using the nanoindentation method. It is found that the roughness decreases with an increase in the Al concentration in AlCrN. For the CrON coatings, the opposite effect is observed. Similar conclusions are valid for the size of the microparticles and their height for both types of coating.

Keywords: AlCrN; CrON; topography; microparticles; roughness; AFM; elastic modulus; friction coefficient

1. Introduction

The application of multifunctional coatings is an effective way to control the surface properties of various products. A trend which has emerged in recent decades is the use of transition metal nitrides in the form of multicomponent systems, where each element of the additive enhances a specific function and simultaneously allows the achievement of excellent mechanical properties, corrosion and thermal resistance, wear and oxidation resistance [1–4]. The addition of components into the two-element transition metal–nitride system leads to the formation of various phases, grain refinement and various crystal lattices [5–8]. The addition of the third element to nitride systems expands their functional properties by changing the mechanical behavior of materials under the applied load depending on the element added [9–13]. This occurs due to the structural re-arrangement of multicomponent coatings that transform into nanocomposites, where both different crystal lattice polytypes and amorphous phases can be present [14–17]. Additionally, the surface microrelief of such coatings changes, which affects friction [18–22].

Chromium nitride is one of the most widely used coatings in industry. It provides high oxidation and corrosion resistance and good adhesion to steel substrates [1,23]. However, its hardness [24] and abrasion resistance are often insufficient. Alteration of the properties of CrN is achieved by alloying with metals and non-metals via the formation of ternary systems [25–27]. Al increases the thermal resistance, C and V reduce wear, the addition of B increases the hardness due to the grain refinement, and Si increases the oxidation resistance [1,5,28]. Due to the higher oxygen reactivity compared to nitrogen, even a small amount of oxygen in the presence of a growing metal nitride layer causes the formation of metal–oxygen ion bonds that appear in the matrix with a covalent metal–nitrogen bond and thus significantly changes the surface properties of CrN [29–33].

Thus, Al and O are the most effective and most known additives to CrN coatings [34,35]. The addition of Al to CrN increases its wear resistance at high temperatures due to formation of an oxide layer on worn surfaces [36]. The properties of AlCrN coating and the type of AlN phase lattice depend on the aluminum concentration in it. When Al content is below 75%, a cubic c-AlN phase is formed in AlCrN. An increase in the aluminum concentration to 80% promotes the formation of an h-AlN hexagonal phase [37]. Under mechanical loads and temperature, the transformation of the cubic c-AlN phase into the hexagonal h-AlN phase occurs spontaneously.

Technologists can effectively manipulate the surface properties of coatings by changing the phase composition of the coating by varying the technological parameters of the coating deposition (such as cathode current, bias, gas pressure) and introducing additional elements [4,38,39], which, to a great extent, depends on the microstructure of the surface—the size of microparticles (or micro-droplet phase), grains and the phase sizes on the surface [14,40,41].

One of the microstructural features of vacuum nitride coatings is that microparticles vary in size from 0.5 μm to several micrometers (according to other sources—a microdroplet phase) [42–44]. Particularly, microparticles are typical for the applied deposition method that is cathodic arc evaporation. Microparticles (or microdroplet phase) are important for tribological coatings, where they participate in the formation of "a modified tribolayer". This layer is formed under the load from microparticles and from the upper plastically deformed surface layer of the coating. It is actively involved in the friction processes, especially under the conditions of so-called "green" processing without the use of cutting fluids [45,46].

The amount of microparticles is determined by the deposition mode and the cathode composition. In coatings manufactured by the cathodic arc evaporation method, the microparticles (microdroplets) can cover the surface with a continuous pattern [28]. Molten metal droplets ejected from cathode spots can move into the substrate and accumulate there as microparticles as well as agglomerates formed in plasma from atoms or ions. The microparticles are more plastic than the nitride phase that forms on the surface, due to the presence of the pure cathode metal in them [47]. The size and height of microparticles are always of interest in tribological coatings, since they allow the prediction of the characteristics of the contact between the coating and the counterbody and calculation of the real contact area.

For the effective implementation of new technological solutions within the framework of existing methods, it is necessary to rely on fundamental knowledge and the results of the microstructure and surface properties research with nanometer resolution [1,41,48]. Atomic force microscopy (AFM) allows for the versatile characterization of the coating surface geometry. AFM provides a multiscale (from scanning areas of 100 × 100 μm^2 to 100 × 100 nm^2 and less) surface visualization, allowing one to study both the microdroplet phase and submicron and nanometer grains of a smooth surface, by analyzing the contrast in properties to identify various phases among them [49,50]. By revealing the morphology, AFM makes it possible to determine not only the particles' diameter but also their height and volume. More precisely, AFM allows one to measure all these characteristics of microparticles located "above the coating surface" but these microparticles are particularly important with respect to tribological contact. The height of the "above the coating surface" particles affects the value of the roughness, as well as the depth of the holes from chipping of microparticles. The effect of microparticle height on the roughness is higher because the microparticles are more common than the holes. Distribution of mechanical properties (elastic modulus E and microhardness H) on the surface of multiphase coatings with microparticles can be studied by nanoindentation (NI) [41,51].

The effect of the aluminum and oxygen concentration on the properties of AlCrN and CrON coatings was reported in [52–55] and [56,57], respectively. The structural, mechanical and tribological properties of the above coatings have been also studied by our group; for AlCrN, see [58–60], and for CrON, see [30,60–62]. It was determined that the microparticles affect the coatings' properties [42–44]. The connection between the topography and mechanical properties of the surfaces of AlCrN and CrON coatings and their composition was essentially not studied. The effect of addition of the metallic or non-metallic element to the CrN coatings on the microparticle height and volume has not been studied previously; neither has the correlation between the friction coefficients and the geometry of the microparticles.

The aim of this research is to compare the effect of aluminum and oxygen concentration on the height, diameter and volume of the microparticles and to determine the correlation between the friction coefficient and the geometry of the microparticles in AlCrN and CrON coatings obtained by cathodic arc evaporation.

2. Materials and Methods

2.1. Coating Deposition

All coatings were synthesized by cathodic arc evaporation in systems summarized in Table 1. All cathodes were characterized by purity of 99.99%. The substrates, HS6-5-2 steel (ArcelorMittal, Katowice, Poland), with a diameter of 32 mm and a thickness of 3 mm, were finished to a roughness parameter Ra of around 0.02 μm. Ra was controlled by a contact profilometer, Hommel Werke T8000 (Hommelwerke GmbH, Villingen-Schwenningen, Germany). They were separately cleaned in an alkaline and ultrasonic bath, rinsed with deionized water and dried with warm air. The substrates were placed on a rotary holder parallel to the surface of the evaporated cathode, in the working chamber at a distance from the arc sources, depending on the deposition system used.

The vacuum chamber was evacuated to the base pressure and then heated to the required temperature. After reaching the desired temperature, the process of ionic etching of the substrate began, aimed at removing the oxygen adsorbed on the substrate surface, as well as the oxide layer. After the etching time had elapsed, a thin chromium layer was deposited on the substrate to improve the adhesion of the coating to the substrate. The deposition process was carried out in accordance with the parameters listed in Table 1.

The details of the three-step process of substrate preparation and coating formation in the working chamber, i.e., substrate ion cleaning, deposition the layer increasing adhesion to the substrate (adhesive layer) and the proper layer, are summarized in Table 1.

In the case of CrON coatings, a gas mixture ($N_2 + O_2$) with different relative oxygen concentrations, $O_{2(x)} = O_2/(N_2 + O_2)\%$, was used; see Table 1. Taking this into account, deposited coatings were labeled as follows: CrO(x)N. This means that, for example, CrO(20)N coating was obtained at a relative concentration of oxygen of 20%. In addition, the notation CrO(0)N coating has been simplified to CrN.

Table 1. The parameters of the multistep coating preparation.

Parameter\Coating	AlCrN	CrON
Deposition system	TINA 900M [10] (Vakuumtechnik Dresden GmbH, Dresden, Germany)	BULAT 3T [30] (Kharkov Institute of Physics and Technology, Kharkiv, Ukraine)
Cathode	Cr, AlCr (50:50), (70:30) and (80:20)	Cr
Cathode diameter [mm]	100	60
Base pressure [Pa]	1×10^{-3}	2×10^{-3}
Cathode-substrate distance [mm]	180	300
Rotation [rev/min]	2	30
	Ion etching	
Bias [V]	−600	−1300
Argon pressure [Pa]	0.5	0.5
Cr arc current [A]	80	90
Etching time [min]	10	3
	Adhesion layer	
Type of the layer	Cr	Cr
Cathode current [A]	80	90
Argon pressure [Pa]	0.5	0.5
Deposition temperature [°C]	350	400
Bias [V]	−100	−100
Thickness [μm]	0.1	0.1
	Proper layer	
Cathode current [A]	80	90
Total pressure [Pa]	3	1.8
Nitrogen pressure [Pa]	3	-
Relative oxygen concentration $O_{2(x)} = O_2/(N_2 + O_2)$	-	0, 5, 20, 50%
Deposition temperature [°C]	350	400
Bias [V]	−100	−150
Thickness [μm]	3	3
	Investigated coatings	
Amount	4	4
Composition	CrN, $Al_{50}Cr_{50}N$, $Al_{70}Cr_{30}N$, $Al_{80}Cr_{20}N$,	CrN, CrO(5)N, CrO(20)N, CrO(50)N

2.2. *Coating Characterization*

The structure of AlCrN and CrON coatings was analyzed by means of X-ray diffraction (XRD) in a conventional symmetrical Bragg–Brentano configuration (θ/2θ) with Co-Kα radiation and a DRON4 device (Burevestnik, Saint Petersburg, Russia).

The surface morphology and microstructure were evaluated by electron microscopy (JEOL JSM-5500LV, JEOL Ltd., Tokyo, Japan). The chemical composition of the microparticles in the coatings was analyzed by energy dispersive X-ray spectroscope (EDX) Link ISIS 300 (Link Analytical/Oxford Instruments, High Wycombe, UK). Detection accuracy of analyzed elements was within 3.0–5.0%.

The surface investigations of the coatings were carried out using Dimension FastScan atomic-force microscope (Bruker, Santa Barbara, CA, USA) in the PeakForce Tapping QNM (Quantitative Nanoscale Mechanical Mapping) mode. The standard silicon cantilevers of MPP-12120-10 (Bruker, Karlsruhe, Germany), NSC11 and CSG10 (Micromasch, Tallinn, Estonia) types and standard diamond probes of DRS_10 (TipsNano, Moscow, Russia) type were used. Particle Analysis was used to determine diameter, *d*, area, *S*, and height, *h*, of the microparticles and Bearing Analysis was used to determine the volume, *V*, of the microparticles in NanoScope Analysis software of Dimension FastScan AFM (Bruker, Santa Barbara, CA, USA). The Particle Analysis command defines some features of interest based on the height of pixel data. Particles may be analyzed alone or in quantities. Particles in this context are considered as conjoined pixels above or below a given threshold height. Bearing Analysis provides a method of plotting and analyzing the distribution of the surface height over a sample. The volume is defined above the bearing depth plane. To determine the particle content (%), the ratio of the total area of the particles in the field to the area of the field is calculated. AFM was used to analyze the surface topography with the determination of the roughness (*Ra* = arithmetic average of the absolute values of the surface height deviations measured from the mean plane; *Rq* = the standard deviation of the Z values within the investigated area). The forces between the AFM tip and the sample can be precisely controlled to prevent the movement of even loosely fixed microparticles [63,64]. The images of three areas were taken and analyzed on each sample. Every area contained fields of 60 × 60, 30 × 30 and 10 × 10 μm^2. In the fields of 60 × 60 and 30 × 30 μm^2, the influence of the microparticles on roughness was significant, and in the fields of 10 × 10 μm^2, the roughness increased. Here, under the terms "height" and "diameter", we refer to "the height or diameter of the particles above the surface of the coating". The values *S* and *V* calculated in this work denote "the part of *S* and *V* of the microparticles above the coating surface". Sometimes, the microparticles can be dispersed in the coatings but their presence on the surface significantly affects the friction properties and can be easily measured by AFM.

Microhardness (*H*) and elastic modulus (*E*) were measured using Hysitron 750 Ubi nanoindentation device (Bruker, Minneapolis, MN, USA) equipped with the Berkovich indenter with a curvature radius of 200 nm. The tip radius was calibrated by indentation into a fused quartz calibration sample. The values of *H* and *E* were calculated by the Oliver–Pharr formula from the experimental curves of continuous recording of the applied load–indentation depth. Two modes of nanoindentation (NI) curves were used. The first one involved load/unload curves; the second one involved progressive partial load/unload curves with partial unloading [65]. Entire load/unload curves were used for the calculation of the *E* and *H* mean values according to 9 curves at the load of 5 mN on a nominally flat surface. The indentation depth in this case did not exceed 110 nm, with a coating thickness of around 3 μm. Progressive partial load/unload curves with partial unloading were used to investigate the dependences of *E* and *H* on the penetration depth into coatings at the load of 0.1–10 mN. The maps of *E* and *H* for the upper coating layer with an average depth of indentation of around 20 nm were obtained by using the load/unload curves. The *E* and *H* maps with the size of 20 × 20 μm^2 were obtained from 400 indentations at a load of 1 mN on the surface with the microparticles.

The content of elements Al and O, peak intensity of c-CrN and h-AlN phases in the coatings, was a criterion for correlation analysis. The relationship between two parameters (geometrical characteristics of the microparticles or mechanical properties and the friction coefficient) with respect to the third constant feature—an increase in the content of non-metallic (oxygen) and metallic (aluminum) additives or a decrease in the content (peak intensity) of c-CrN and h-AlN phases—was determined. The correlation coefficient of the parameters *x* and *y* was determined by the following formula:

$$r_{xy} = \frac{\sum (x_i - \overline{x}) \cdot (y_i - \overline{y})}{\sqrt{\sum (x_i - \overline{x})^2 \cdot \sum (y_i - \overline{y})^2}} \quad (1)$$

where *y* corresponded to *Cfr* at a certain element content, and *x* alternately took the values of *E*, *H*, *d*, *h*, *S*, *V*, *Ra* and *Rq*.

3. Results and Discussion

Structural properties of the deposited CrN, AlCrN and CrON coatings measured by XRD are shown in Figure 1. The main phase in the CrN coatings obtained by two instruments is the CrN phase with a cubic crystal lattice (c-CrN). The main peaks in the CrN coating obtained by TINA are (200), (111), (220), (311). In the CrN coating obtained using the BULAT system, only one high-intensity diffraction line (111) is presented. It is more intense compared to line (222). It indicates a strong (111) texture (crystallography orientation) in this CrN coating. The main phase in the $Al_{50}Cr_{50}N$ coating is c-CrN with (200) diffraction line. In the $Al_{50}Cr_{50}N$ coating, the diffraction lines of the hexagonal AlN phase (h-AlN) appear from planes (101) and (103). In the $Al_{70}Cr_{30}N$ coating, the intensity of the h-AlN phase lines increases. In the $Al_{80}Cr_{20}N$ coating, the peak in the h-AlN phase becomes a preferred line. The intensity of the (111) c- CrN line in the coating CrO(5)N is the highest one but its intensity is lower than in the CrN coating obtained using the BULAT system. The diffraction line originating from planes (311) became the preferred line in the CrO(20)N coating. Only the Cr_2O_5 phase line is established in the CrO(50)N coating (Figure 1). The diffraction peaks from the Cr sublayer are absent.

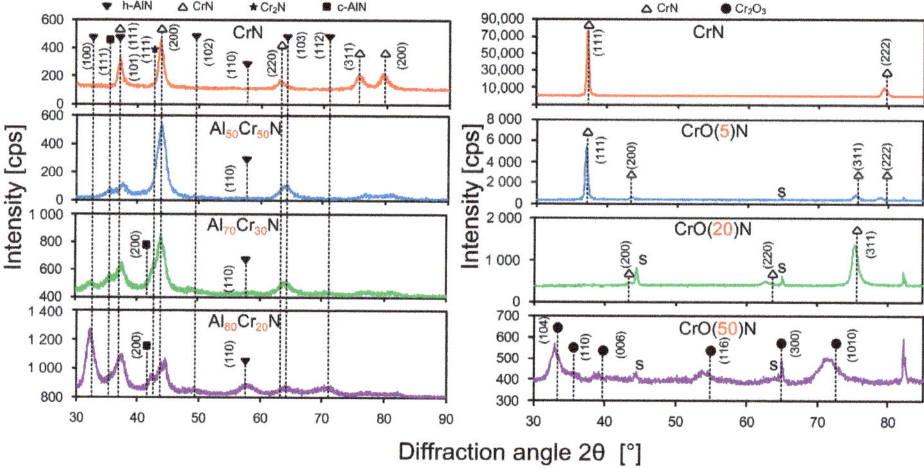

Figure 1. X-ray diffraction (XRD) diffraction patterns of AlCrN and CrON coatings.

The EDX analysis of the microparticles of coatings showed differences in the distribution of Al and Cr in AlCrN coatings and oxygen in CrON (Figure S1). This analysis is rather qualitative, aiming to show the differences in the element content between the microparticles and the surface. This difference in Al, Cr and oxygen content caused differences in the local distribution of the mechanical properties.

General assessment of the number of particles and their sizes on the CrN, AlCrN and CrON coating surfaces reveals the highest content and height for AlCrN, followed by CrN obtained at the TINA hardware, then CrON, and the smallest amount of microparticles with the lowest height for CrN was obtained using the BULAT unit (Figure 2). A significant difference was found in the amount (8.3% and 1.4%) and size (1.9 and 1.3 μm) of the microparticles in the CrN coatings deposited by either hardware. The XRD analysis demonstrated that the preferred orientation of the cubic CrN phase (c-CrN) in the coating obtained by TINA is (200). There are more other peaks of (111), (311), (220), whose intensities are significant in comparison with (200). The main peak of the c-CrN phase in the CrN coating obtained using BULAT is (111) and there is only one other peak of (222), which is insignificant in comparison with (111). In addition, the width of the peaks in the former coating relative to their width is significantly higher than in the latter. This qualitatively indicates a higher crystallinity degree in the second coating. The difference in the textures of the c-CrN phase can be explained by the different rates of atom and ion flow to the surfaces of the CrN coatings. For TINA, this flow was significantly higher due to the higher nitrogen pressure (more than two times) and shorter distance between the surface and the cathode. Etching was carried out in hardware in order to clean it. The time of cleaning for the first coating was almost three times longer, which could prepare a larger number of crystallization centers on the surface. Together, this resulted in more microparticles in the CrN coating produced by TINA. Based on the common statement about the effect of roughness on the friction coefficient (*Cfr*) and on the basis of this preliminary estimate, one could assume that the lowest values of *Cfr* should be on CrN obtained using the BULAT device, then slightly higher for CrON, succeeded by CrN obtained using the TINA unit, and finally, one should expect the highest *Cfr* on AlCrN.

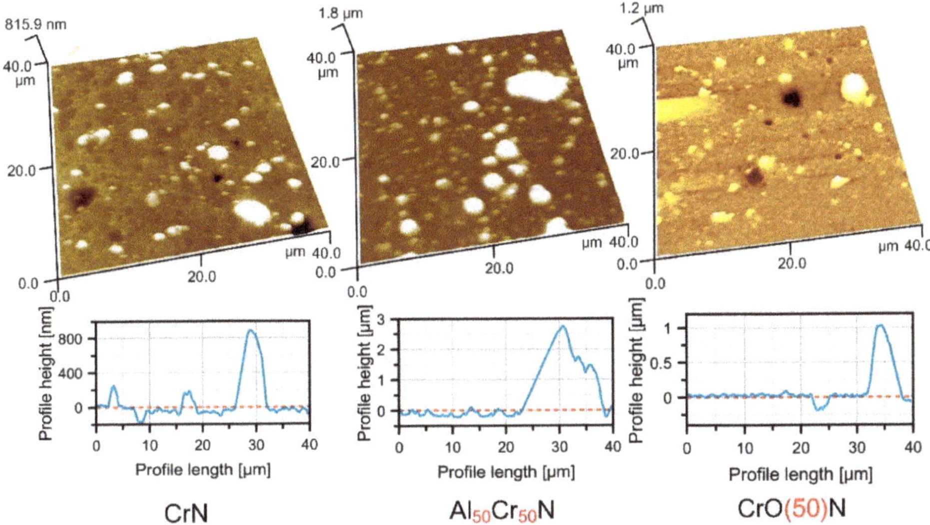

Figure 2. Three-dimensional atomic force microscopy (AFM) images and surface profiles of CrN, $Al_{50}Cr_{50}N$ and CrO(50)N coatings with microparticles.

The experimental results disagree with this assumption, since the phase composition, mechanical properties and surface microgeometry have to be taken into account simultaneously. Figure 3 shows the dependences of *Cfr* from the Al content in AlCrN and from the oxygen content in CrON coatings, constructed on the basis of data obtained in [28,51] under the same conditions of sliding friction without lubrication.

The AFM capabilities make it possible to reveal the details of a significant difference in the morphology of the coating surfaces (Figures 4 and 5). Cathodic arc evaporation forms a "smooth" coating surface consisting of connected cells (Figure 4a, blue arrows). The "ribs" of the cells are the protruding edges of the crystallites. The CrN coating obtained by the TINA unit demonstrates cell diameters of 0.5–2 μm. For the CrN coating obtained by the BULAT system, the cell diameter is approximately two times lower and is in the range of 0.2–1 μm (Figure 4a).

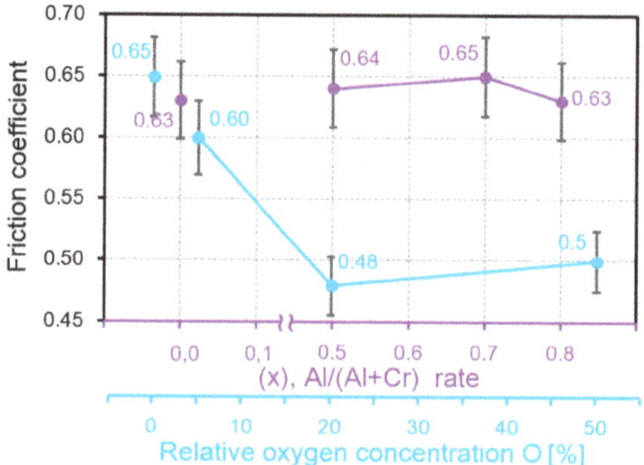

Figure 3. The dependences of the friction coefficient from the addition content in the CrN coatings.

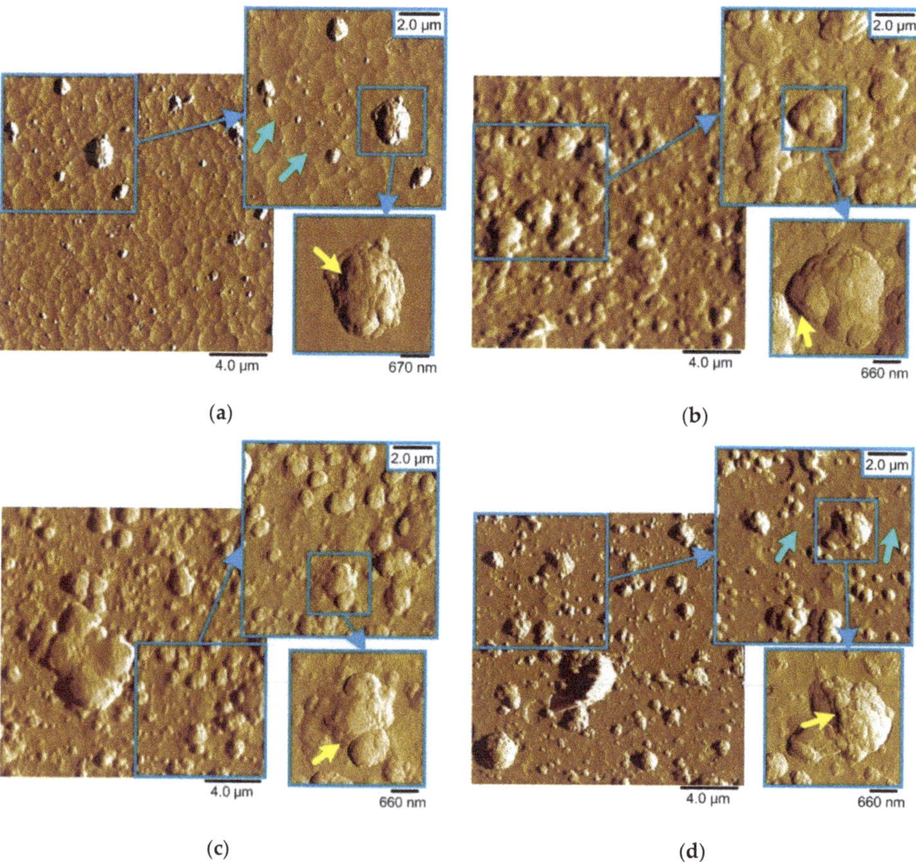

Figure 4. Two-dimensional AFM images of the AlCrN coatings, area 20 × 20 µm^2: (**a**) CrN; (**b**) $Al_{50}Cr_{50}N$; (**c**) $Al_{70}Cr_{30}N$; (**d**) $Al_{80}Cr_{20}N$.

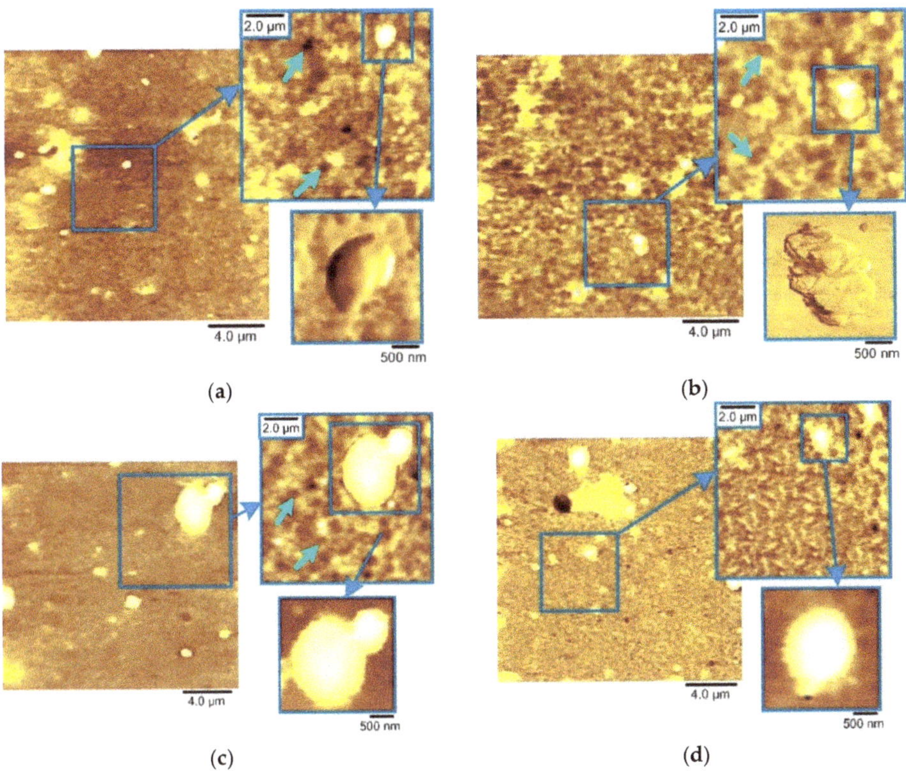

Figure 5. Two-dimensional AFM images of the CrON coatings, area 20 × 20 µm^2: (**a**) CrN; (**b**) CrO(5)N; (**c**) CrO(20)N; (**d**) CrO(50)N.

The surface images shown in Figure 4 are presented in the Error mode. This channel produces a map of the peak force measured during the scanning process. Because the PeakForce QNM mode uses the peak force as the feedback signal, this channel is essentially the Peak Force Setpoint with the error. It is recorded simultaneously with the topography. In this mode, one can better observe the boundaries between the layers in the microparticle images. The height scale corresponds to the real topography of the microparticles. The use of the Topography mode in the illustration of AlCrN microparticles, due to their height, gives the image "flare", i.e., very tall particles look white and uniform and there is no way to show their internal microstructure.

The microparticle shapes in AlCrN and CrON coatings are different. In the AlCrN coatings, there are layered particles of irregular shape. The layers in the microparticles are marked in Figure 4 by yellow arrows and in Figure S2 by violet and yellow arrows. Their morphology shows that they were formed by the merging of gradually layering flat islets. The long axes of these flat islets are oriented differently with respect to the coating surface plane (Figure 4). With an increase in the Al content, the amount of microparticles increases. In CrON, both irregular, differently oriented and layered particles and regular spherical particles are observed. The correct spherical shape of such particles indicates that it is unlikely that they were formed by multiple hits of clusters of already grouped atoms outside the coating. It is possible that the regular spherical particles were formed as a result of the crystallization of single microdroplets in the working gas environment (Figure 5).

The percentage of particles in the AlCrN coating is ten times higher than that in the CrON coating (Figure 6). Assessing the geometric parameters of the microparticles, we used two sizes of AFM images, namely 40 × 40 µm^2 and 20 × 20 µm^2. The 40 × 40 µm^2 field allows for a better assessment of the total amount of the microparticles, and better accuracy in determining the diameter and height of the microparticles is achieved using 20 × 20 μm^2 fields. The dependences of the microparticle content on the oxygen amount in the CrON coating in the fields of 40 × 40 μm^2 and 20 × 20 μm^2 are almost linear (Figure 6). For the AlCrN coatings, the dependence of the content of the microparticles from the content of aluminum in the fields of 20 × 20 μm^2 is close to linear, and in the fields of 40 × 40 μm^2, it is non-monotonic, with a maximum at 70% Al. Such a difference in the form of the dependences of the particle content from the amount of Al in the AlCrN coating at different scales of fields shows the heterogeneity of the distribution of the microparticles and the importance of using two sizes of fields in their study.

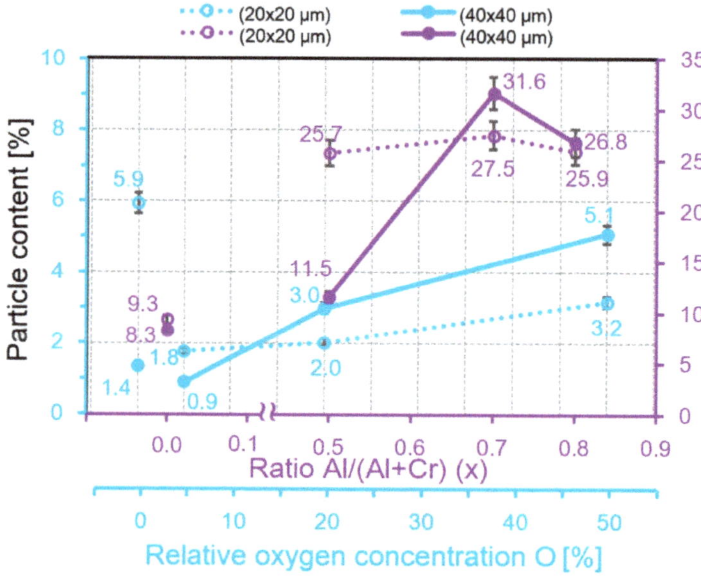

Figure 6. The dependences of the particle content from the addition content in the CrN coatings.

It is difficult to perceive any difference for the $CrO_{20}N$ and $CrO_{50}N$ coatings by their distribution by diameter (Figure 7), but this can be identified by the average values. The microparticle distribution by height in the AlCrN and CrON coatings is shown in Figure 8. The higher sensitivity of the microparticle height to both the content of additives and the scale of the investigated microstructure is clearly manifested. The differences in the most frequently encountered intervals of the microparticle heights are much more significant than those in the diameters of microparticles (Figure 8). A difference in the particle heights in different fields of 40 × 40 μm^2 and 20 × 20 μm^2 in the distribution of heights is not detected for $Al_{70}Cr_{30}N$ but is clearly seen from the average values.

Roughness quite accurately reflects the different tendency in the microparticle height change of the AlCrN and CrON coatings (Figure 9), which is in good agreement with previous work [35]. For the AlCrN coatings, *Ra* decreases from 263 nm for $Al_{50}Cr_{50}N$ to 183 nm for $Al_{80}Cr_{20}N$ in the field of 40 × 40 μm^2. In the field of 20 × 20 μm^2, the height difference is smoother, from 272 nm for $Al_{50}Cr_{50}N$ to 241 nm for $Al_{80}Cr_{20}N$. For the CrON coatings, the reverse process occurs: *Ra* increases when the oxygen content increases from 10 nm for CrO(5)N to 28 nm for CrO(50)N in the field of 40 × 40 μm^2, and from 4 nm for CrO(5)N to 48 nm for CrO(50)N in the field of 20 × 20 μm^2.

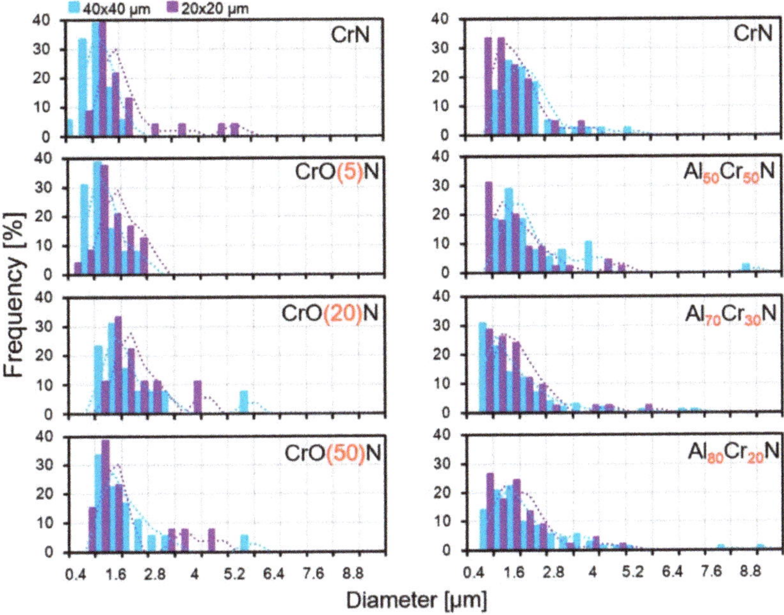

Figure 7. Histograms of the microparticle diameters of the AlCrN and CrON coatings in the fields of $40 \times 40\ \mu m^2$ and $20 \times 20\ \mu m^2$.

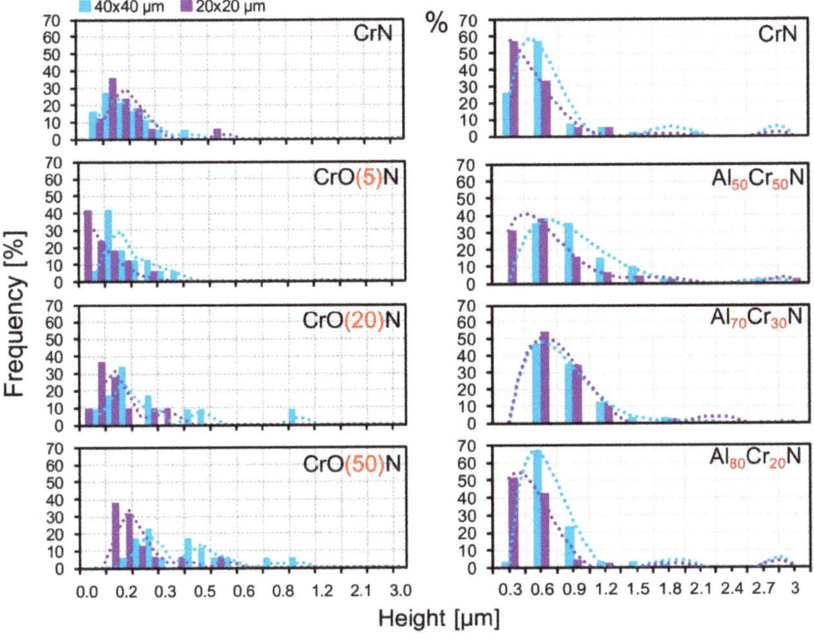

Figure 8. Histograms of the microparticle heights of the AlCrN and CrON coatings in the fields of $40 \times 40\ \mu m^2$ and $20 \times 20\ \mu m^2$.

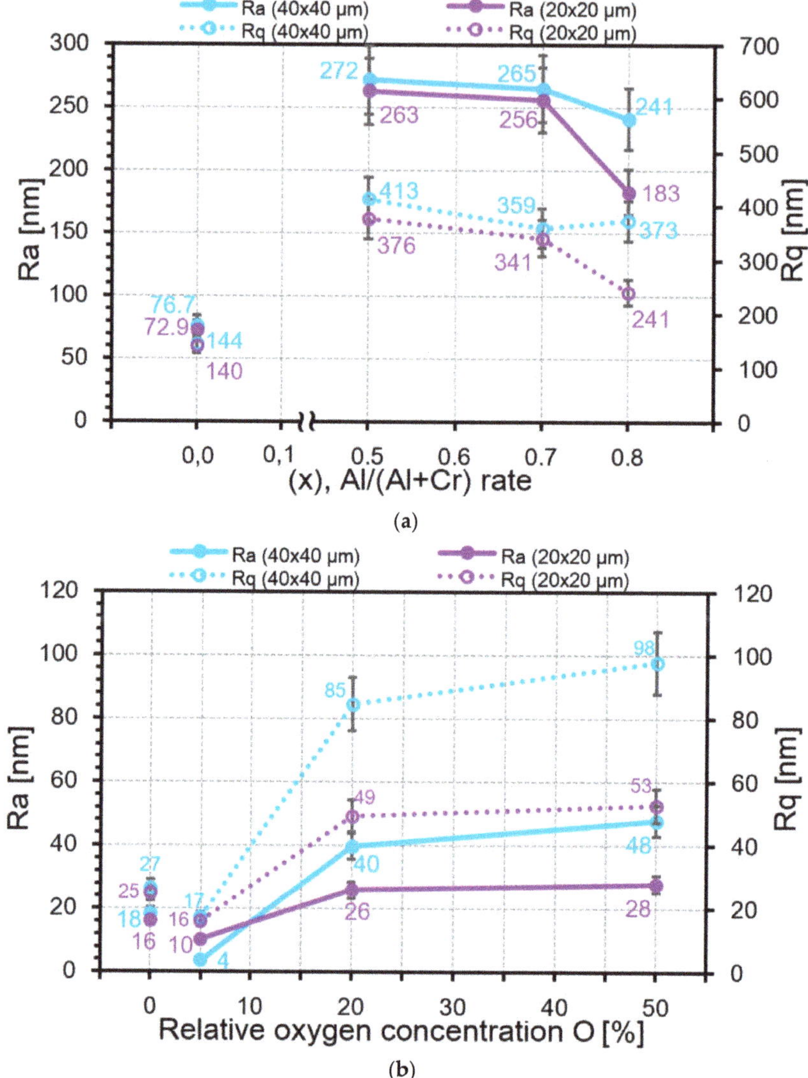

Figure 9. The dependences of *Ra* and *Rq* from the element concentration: (**a**) aluminum in the AlCrN coatings; (**b**) oxygen in the CrON coatings.

Comparison of the height and diameter dependences from the Al amount in the AlCrN coatings clearly shows high sensitivity of the particle height to the Al content in contrast to the particle diameter. The particle diameter remains almost unchanged in the fields of 40 × 40 μm^2 and does not change at all in the fields of 20 × 20 μm^2 (Figure 10). In the AlCrN coatings, with an increase in the Al amount, the particle heights decrease from 845 nm for $Al_{50}Cr_{50}N$ to 547 nm for $Al_{80}Cr_{20}N$ in the fields of 40 × 40 μm^2 and from 571 nm for $Al_{50}Cr_{50}N$ to 333 nm for $Al_{80}Cr_{20}N$ in the fields of 20 × 20 μm^2 (Figure 10). Since the diameter of particles in the AlCrN coatings practically does not change with the increase in the Al amount, a decrease in the volume of the microparticles in the field of 20 × 20 μm^2

from 17.7 μm^3 for $Al_{50}Cr_{50}N$ to 10.0 μm^3 for $Al_{80}Cr_{20}N$ is associated with a decrease in their heights (Figure 11).

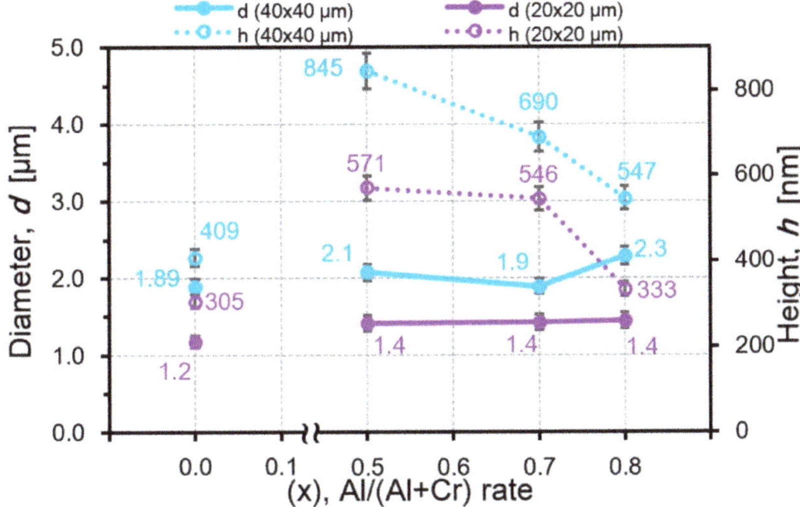

Figure 10. The dependences of the particle diameter and height from the aluminum content in the AlCrN coatings.

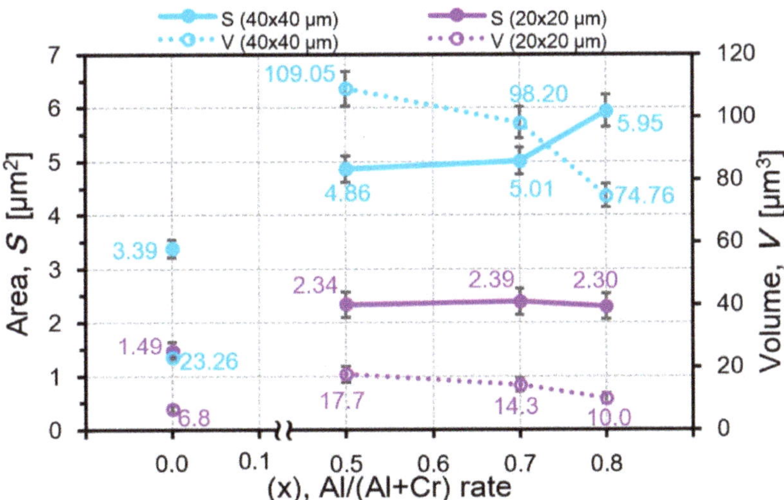

Figure 11. The dependences of the particle area and volume from the aluminum content in the AlCrN coatings.

However, the decrease in their total volume is possible not only with a constant diameter and decrease in the height of the particles. Even in the case when the particle diameter increases from 1.9 for $Al_{70}Cr_{30}N$ to 2.3 μm for $Al_{80}Cr_{20}N$, the decrease in the particle height from 690 to 547 nm leads to the decrease in the particle volume in the field of 40 × 40 μm^2 and the volume decreases from 98.2 μm^3 for $Al_{70}Cr_{30}N$ up to 74.7 μm^3 for $Al_{80}Cr_{20}N$ (Figure 11). In the CrON coatings, the particle diameter increases roughly twice (in CrO(20)N and CrO(50)N) with an increase in the oxygen amount but the particle height increases much more, around 2.5 times with respect to CrO(5)N (Figure 12). This leads to the growth of S and V of the microparticles (Figure 13). The largest increase in the volume of particles for the CrO(50)N coating was 0.75 μm^3.

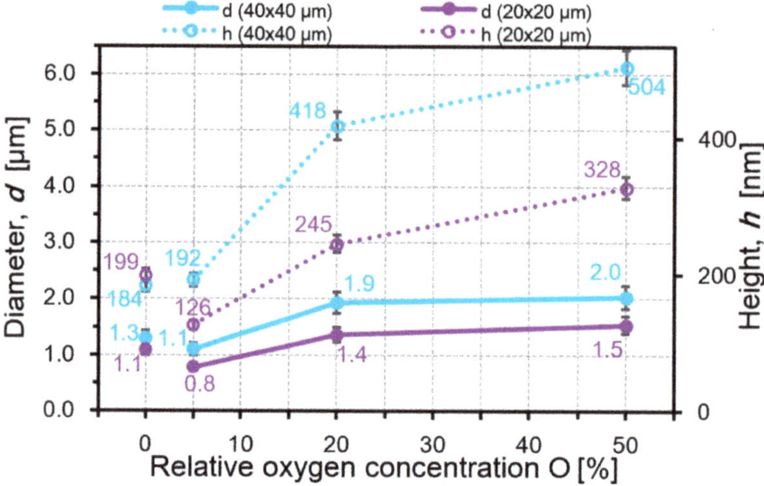

Figure 12. The dependences of the particle height and diameter from the oxygen content in the CrON coatings.

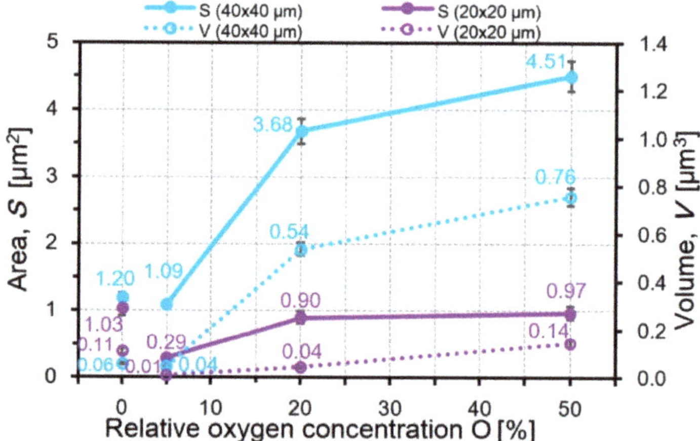

Figure 13. The dependences of the particle area and volume from the oxygen content in the CrON coatings.

When estimating the effect of additives on the amount of microparticles and the microparticle geometry, the AlCrN and CrON coatings should be compared within the groups, given that, due to the different deposition parameters (bias voltage, arc current, distance to the sample, pressure in the chamber) in the TINA and BULAT equipment, even the CrN coatings have significantly different amounts of microparticles.

Thus, the CrN coating obtained via TINA contains 9.3% of particles, Ra of 77 nm, Rq of 144 nm, average diameter of 1.9 μm, average height of 409 nm, average area of 3.39 μm^2 and average volume of 1.49 μm^3. The CrN coating obtained via BULAT contains 5.9% of particles, *Ra* of 18 nm, *Rq* of 18 nm, average diameter of 1.3 μm, average height of 184 nm, average area of 1.2 μm^2 and volume of 0.1 μm^3.

The microparticles on the surfaces of the cathodic arc deposited coatings significantly depend on the modes of the technological process [66]. Collisions between atoms or ions become more frequent under the influence of high pressure, and during the formation of the AlCrN coatings, it is almost twice as high as compared to CrN, which leads to the formation of agglomerates of a larger number of atoms before deposition on the substrate [28].

In addition to roughness and microparticle amount (subsequently transforming into a modified material), one should take into account the mechanical properties of the "smooth" surface of the coatings when evaluating the tribological properties of coatings. The elastic modulus *E* at the depth of up to 100 nm for the CrON coatings is 292–335 GPa, which is higher compared to 157–204 GPa for AlCrN (Figure 14). For the AlCrN coatings, with the increase in the Al content till 80%, *E* decreases, which is well traced on the "load–depth" curves, which, with the Al content increase, gradually shifts to the right (Figure S3a). According to the curves, the CrON coatings are divided into two groups, CrN and CrO(50)N are referred to as coatings with "low *E*" and CrO(5)N and CrO(20)N as coatings with "high *E*". According to the shape of the "load–depth" curves (the part of plastic deformation), we can conclude that the CrON coatings are more plastic than AlCrN (Figure S3b). These curves were obtained on a "smooth" surface at a penetration depth of up to 110 nm.

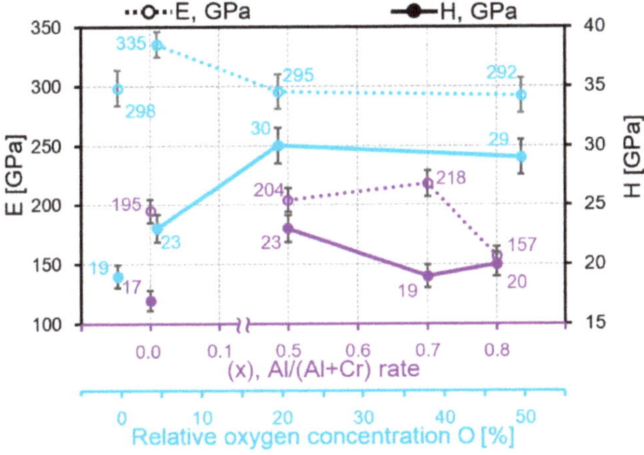

Figure 14. The dependences of *E* and *H* from the aluminum content in the AlCrN coatings and on the oxygen content in the CrON coatings.

The dependences of *E* and *H* on the indentation depth in the CrN and $Al_{50}Cr_{50}N$ coatings are shown in Figure 15. For the NI progressive partial load/unload curves with partial unloading of the CrN and $Al_{50}Cr_{50}N$ coatings used for the calculation of *E* and *H*, these dependences are shown in Figure S4. The value of H in the upper layer of the coatings is significantly lower than at the depth of 50–100 nm, where the values are stable. For the $Al_{50}Cr_{50}N$ coating, this difference is more than four times (Figure 15).

To assess the distribution of mechanical properties over the surfaces of the coatings, maps of *E* and *H* were constructed, with an average indentation depth of around 20 nm (Figure 16). The maps fit well the morphology of the coatings (Figure 4a,b). The softer areas on the maps, marked in brown, refer to the microparticles. They clearly demonstrate an increase in the number of softer particles in the AlCrN coatings and their distribution over the surface (Figure 16). The average *E* values of the CrN coating obtained with TINA system were 180 ± 69 GPa and for *H* were 14.4 ± 6.6 GPa. The average values of *E* for the $Al_{50}Cr_{50}N$ coating were 142 ± 109 GPa and for *H* were 8.2 ± 8.0 GPa.

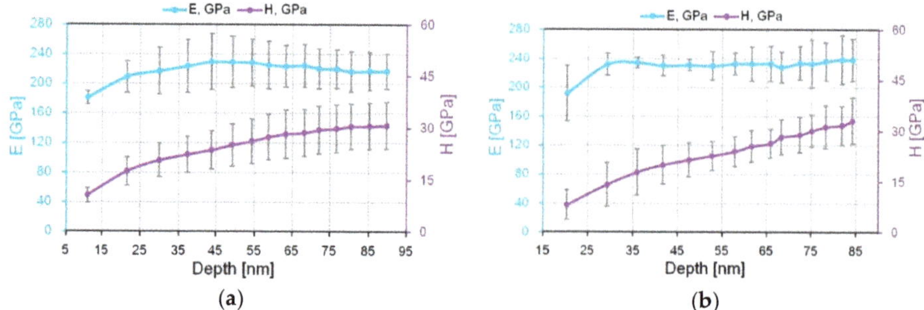

Figure 15. The dependences of *E* and *H* from the indentation depth in CrN (**a**) and $Al_{50}Cr_{50}N$ coatings (**b**).

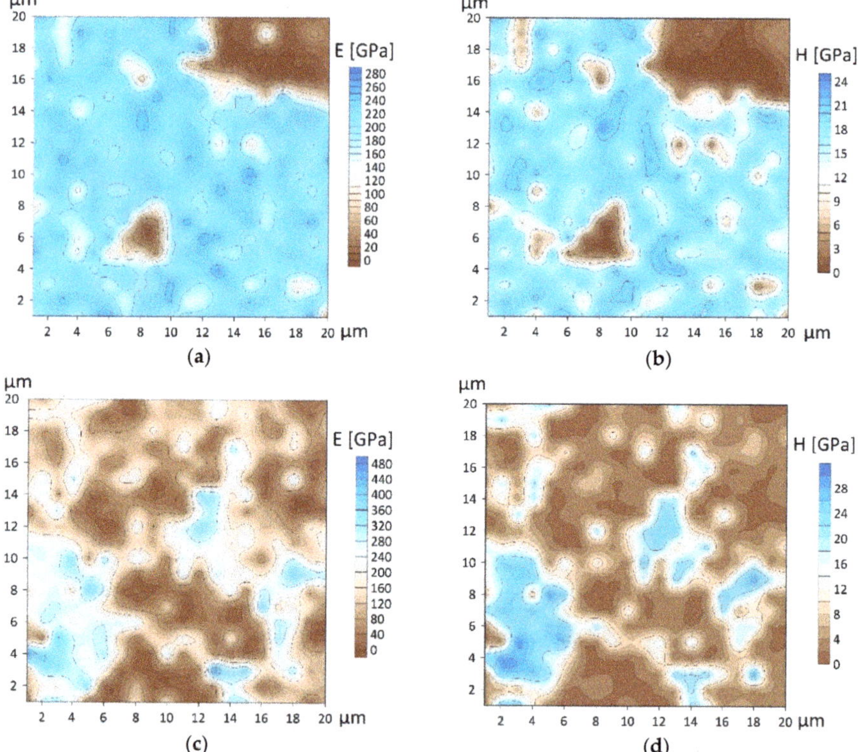

Figure 16. Elastic modulus (**a**,**c**) and microhardness (**b**,**d**) maps of CrN (**a**,**b**) and $Al_{50}Cr_{50}N$ (**c**,**d**) coatings, area of 20 × 20 μm^2.

The larger microparticle volumes on the AlCrN coating surface lead to the formation of a modified soft layer thicker which was than that of the CrON coatings [33] (Figure S3). The presence of such a layer can reduce friction [33]. However, it is impossible to draw a full analogy with lubrication without taking into account the properties of the hard underlying layers of the wear-resistant coating. For example, in the case of the elastohydrodynamic lubricant, a reduction in the friction coefficient with the increasing coating thickness was mentioned in some theoretical studies. Thus, according to [67], in the case of sliding contact, the increasing soft coating thickness leads to a reduction in both normal and tangential stresses; correspondingly, the friction force will decrease. In this case, a thinner modified layer of microparticles with a harder CrON coating showed lower values of the friction coefficient and the forces than in the case of a thicker layer of the AlCrN coating. The microstructure of the "smooth" coating surface with smaller cells in CrON than in AlCrN coatings also played a role, which helped to distribute evenly and retain the modified layer.

The influence degree of the microparticle geometry and h-AlN, c-CrN phase intensity on the friction coefficient was shown by the correlation coefficients (Figures S4 and S5, Tables S1–S4).

For all geometrical characteristics of particles (*d*, *h*, *S*, *V*, *Ra*), high correlations ($r > 0.7$) with *Cfr* were found. For *h* and *V*, the correlation coefficient of the CrON coatings was especially high—0.84 and −0.90. This result indicates a significant effect of *h* and *V* of the microparticles on *Cfr* and that, along with *E*, the geometric characteristics of the coating microparticles are significant features of the friction process and can be used in analytical models to solve the problem of *Cfr* determination. Most of the correlations for the AlCrN coatings are positive, while for the CrON coatings, they are negative. This means that in order to reduce *Cfr* for coatings in the AlCrN group, *h*, *S*, *V*, *Ra* and *E* should be reduced, while for coatings in the CrON group, *E* should be reduced but *d*, *h*, *S*, *V*, *Ra* should be increased (Figures S6 and S7).

The use of the peak intensity of c-CrN and h-AlN phases in the coatings as a criterion for the correlation analysis gave stronger correlations between the geometrical characteristics of the microparticles and the friction coefficient ($r > 0.9$). The highest is obtained for *h* (−1.0) and *V* (−0.99) (Tables S1–S4).

4. Conclusions

Two sets of CrN-based coatings, one with the addition of aluminum and the other with the addition of oxygen, were deposited by cathodic arc evaporation. The effect of metal and non-metal addition on the geometry of surface defects, i.e., diameter, height and volume of the microparticles, all above the coating surface, was assessed. The main findings are as follows:

(1) Strong correlations with a correlation coefficient of 0.82–1.00 between the friction coefficient obtained under conditions of sliding friction without a lubricant and the geometric characteristics of the microparticles on the coating surface (content, roughness, diameter, height, area and volume) for AlCrN coatings with 50%, 70%, and 80% aluminum and CrON coatings containing 5%, 20% and 50% oxygen have been established.
(2) It was found that the friction coefficient does not change significantly with the increase in aluminum content but significantly decreases with the increase in the oxygen content.
(3) The roughness parameters decrease with the increase of the Al concentration in AlCrN. For the CrON coatings, the opposite effect is observed. Similar relationships are observed for the size of the microparticles and their height for both types of coating.

This allows us to consider the geometric characteristics of the microparticles as significant features in the problem of friction coefficient determination.

Supplementary Materials: The following are available online at http://www.mdpi.com/2079-4991/11/3/274/s1, Figure S1: SEM images and spectra of microparticles and surface in $Al_{70}Cr_{30}N$ and CrO(5)N coatings, Figure S2: AFM image with the "layered" microstructure of microparticle from Figure 4a, Figure S3: NI complete load/unload curves in AlCrN coatings (a) and in CrON coatings (b), Figure S4: NI progressive partial load/unload curves with partial unloading of CrN (a) and $Al_{50}Cr_{50}N$ (b) coatings, Figure S5: Upper modified layer formed under tribological load on the CrN (a) and CrO(50)N (b) coatings surface, area of 20 × 20 μm², Figure S6: Correlation coefficients between *Cfr* and *E* and particles characteristics for AlCrN coatings, Figure S7: Correlation coefficients between *Cfr* and *E* and particles characteristics for CrON coatings, Table S1: Correlation between intensity of h-AlN (100) (101), c-CrN (200) and particles characteristics (areas 40×40 μm^2 and 20×20 μm^2), *E*, *H*, *Cfr* for AlCrN coatings, Table S2: Correlation between intensity of h-AlN (100) (101), c-CrN (200) and particle content, *E*, *H*, *Cfr* for AlCrN coatings, Table S3: Correlation between intensity of c-CrN (111), Cr_2O_3 (104) and particles characteristics (areas 40×40 μm^2 and 20×20 μm^2), *E*, *H*, *Cfr* for CrON coatings, Table S4: Correlation between intensity of c- CrN (111), Cr_2O_3 (104) and particle content, *E*, *H*, *Cfr* for CrON coatings.

Author Contributions: Conceptualization, T.K., S.C. and B.W.; methodology, V.L., T.K. and B.W.; software, A.K. (Anastasiya Khabarava), V.L. and B.M.; validation, B.M. and A.K. (Aleksander Kuprin); formal analysis, V.L. and A.K. (Anastasiya Khabarava); investigation, V.L., A.K. (Anastasiya Khabarava), T.K., A.G., B.W. and A.K. (Aleksander Kuprin); resources, A.G., B.W. and A.K. (Aleksander Kuprin); data curation, B.M. and S.A.; writing—original draft preparation, T.K.; writing—review and editing, B.W. and S.A.; visualization, V.L. and A.K. (Anastasiya Khabarava); supervision, S.C.; project administration, T.K., B.M. and B.W.; funding acquisition, S.A. All authors have read and agreed to the published version of the manuscript.

Funding: This research was supported by the grant of Belarusian Republican Foundation for Fundamental Research BRFFR–SCST–Poland No. T18PLDG-002 and No. F18R-239. Publication was partly financed by the National Centre for Research and Development, Poland, BIOSTRATEG3/344303/14/NCBR/2018. B.M. and S.M. were supported by the Government of Russia (grant No. 14.Z50.31.0046).

Conflicts of Interest: The authors declare no conflict of interest.

References

1. Jäger, N.; Meindlhumer, M.; Spor, S.; Hruby, H.; Julin, J.; Stark, A.; Nahif, F.; Keckes, J.; Mitterer, C.; Daniel, R. Microstructural evolution and thermal stability of AlCr(Si)N hard coatings revealed by in-situ high-temperature high-energy grazing incidence transmission X-ray diffraction. *Acta Mater.* **2020**, *186*, 545–554. [CrossRef]
2. Tritremmel, C.; Daniel, R.; Lechthaler, M.; Rudigier, H.; Polcik, P.; Mitterer, C. Microstructure and mechanical properties of nanocrystalline Al-Cr-B-N thin films. *Surf. Coat. Technol.* **2012**, *213*, 1–7. [CrossRef]
3. Tillmann, W.; Grisales, D.; Stangier, D.; Butzke, T. Tribomechanical behaviour of TiAlN and CrAlN coatings deposited onto AISI H11 with different pre-treatments. *Coatings* **2019**, *9*, 519. [CrossRef]
4. Bobzin, K.; Brögelmann, T.; Kruppe, N.C.; Carlet, M. Wear behavior and thermal stability of HPPMS (Al,Ti,Cr,Si)ON, (Al,Ti,Cr,Si)N and (Ti,Al,Cr,Si)N coatings for cutting tools. *Surf. Coat. Technol.* **2020**, *385*, 125370. [CrossRef]
5. Franz, R.; Mitterer, C. Vanadium containing self-adaptive low-friction hard coatings for high-temperature applications: A review. *Surf. Coat. Technol.* **2013**, *228*, 1–13. [CrossRef]
6. Voevodin, A.A.; Muratore, C.; Aouadi, S.M. Hard coatings with high temperature adaptive lubrication and contact thermal management: Review. *Surf. Coat. Technol.* **2014**, *257*, 247–265. [CrossRef]
7. Bobzin, K.; Brögelmann, T.; Kruppe, N.C.; Carlet, M. Nanocomposite (Ti,Al,Cr,Si)N HPPMS coatings for high performance cutting tools. *Surf. Coat. Technol.* **2019**, *378*, 124857. [CrossRef]
8. Grigoriev, S.; Vereschaka, A.; Milovich, F.; Tabakov, V.; Sitnikov, N.; Andreev, N.; Sviridova, T.; Bublikov, J. Investigation of multicomponent nanolayer coatings based on nitrides of Cr, Mo, Zr, Nb, and Al. *Surf. Coat. Technol.* **2020**, *401*, 126258. [CrossRef]
9. Gilewicz, A.; Jedrzejewski, R.; Myslinski, P.; Warcholinski, B. Influence of substrate bias voltage on structure, morphology and mechanical properties of AlCrN coatings synthesized using cathodic Arc evaporation. *Tribol. Ind.* **2019**, *41*, 484–497. [CrossRef]
10. Warcholinski, B.; Gilewicz, A. Multilayer coatings on tools for woodworking. *Wear* **2011**, *271*, 2812–2820. [CrossRef]
11. Brögelmann, T.; Bobzin, K.; Kruppe, N.C.; Arghavani, M. Understanding the deformation and cracking behavior of Cr-based coatings deposited by hybrid direct current and high power pulse magnetron sputtering: From nitrides to oxynitrides. *Thin Solid Films* **2019**, *688*, 137354. [CrossRef]
12. Volkhonsky, A.O.; Blinkov, I.V.; Belov, D.S. The effect of the metal phase on the compressive and tensile stresses reduction in the superhard nitride coatings. *Coatings* **2020**, *10*, 798. [CrossRef]
13. Ahmad, F.; Zhang, L.; Zheng, J.; Sidra, I.; Zhang, S. Characterization of AlCrN and AlCrON coatings deposited on plasma nitrided AISI H13 steels using ion-source-enhanced arc ion plating. *Coatings* **2020**, *10*, 306. [CrossRef]
14. Kuznetsova, T.A.; Andreev, M.A.; Markova, L.V.; Chekan, V.A. Wear Resistance of Composite Chrome Coatings With Additives of Utlradispersed Diamonds. *J. Frict. Wear* **2001**, *22*, 423–428.
15. Andreyev, M.; Anishchik, V.; Markova, L.; Kuznetsova, T. Ion-beam coatings based on Ni and Cr with ultradispersed diamonds—Structure and properties. *Vacuum* **2005**, *78*, 451–454. [CrossRef]
16. Vityaz', P.A.; Komarov, A.I.; Komarova, V.I.; Kuznetsova, T.A. Peculiarities of triboformation of wear-resistant layers on the surface of a MAO-coating1 modified by fullerenes. *J. Frict. Wear* **2011**, *32*, 231–241. [CrossRef]

17. Liu, Y.; Wang, T.-G.; Lin, W.; Zhu, Q.; Yan, B.; Hou, X. Microstructure and properties of the AlCrSi(O)N tool coatings by arc ion plating. *Coatings* **2020**, *10*, 841. [CrossRef]
18. Hogmark, S.; Jacobson, S.; Larsson, M. Design and evaluation of tribological coatings. *Wear* **2000**, *246*, 20–33. [CrossRef]
19. Holmberg, K.; Matthews, A.; Ronkainen, H. Coatings tribology—Contact mechanisms and surface design. *Tribol. Int.* **1998**, *31*, 107–120. [CrossRef]
20. Kuznetsova, T.A.; Lapitskaya, V.A.; Chizhik, S.A.; Warcholinski, B.; Gilewicz, A.; Kuprin, A.S. Influence of the third element additives on the surface morphology of the wear-resistant ZrN coatings. *IOP Conf. Ser. Mater. Sci. Eng.* **2018**, *443*, 8–13. [CrossRef]
21. Kumar, S.; Maity, S.R.; Patnaik, L. Friction and tribological behavior of bare nitrided, TiAlN and AlCrN coated MDC-K hot work tool steel. *Ceram. Int.* **2020**, *46*, 17280–17294. [CrossRef]
22. Macías, H.A.; Yate, L.; Coy, L.E.; Aperador, W.; Olaya, J.J. Influence of Si-addition on wear and oxidation resistance of TiWSixN thin films. *Ceram. Int.* **2019**, *45*, 17363–17375. [CrossRef]
23. Long, Y.; Zeng, J.; Yu, D.; Wu, S. Microstructure of TiAlN and CrAlN coatings and cutting performance of coated silicon nitride inserts in cast iron turning. *Ceram. Int.* **2014**, *40*, 9889–9894. [CrossRef]
24. Wan, X.S.; Zhao, S.S.; Yang, Y.; Gong, J.; Sun, C. Effects of nitrogen pressure and pulse bias voltage on the properties of Cr-N coatings deposited by arc ion plating. *Surf. Coat. Technol.* **2010**, *204*, 1800–1810. [CrossRef]
25. Reiter, A.E.; Derflinger, V.H.; Hanselmann, B.; Bachmann, T.; Sartory, B. Investigation of the properties of Al1-xCrxN coatings prepared by cathodic arc evaporation. *Surf. Coat. Technol.* **2005**, *200*, 2114–2122. [CrossRef]
26. Mishra, S.K.; Ghosh, S.; Aravindan, S. Investigations into friction and wear behavior of AlTiN and AlCrN coatings deposited on laser textured WC/Co using novel open tribometer tests. *Surf. Coat. Technol.* **2020**, *387*, 125513. [CrossRef]
27. Chen, W.; Hu, T.; Hong, Y.; Zhang, D.; Meng, X. Comparison of microstructures, mechanical and tribological properties of arc-deposited AlCrN, AlCrBN and CrBN coatings on Ti-6Al-4V alloy. *Surf. Coat. Technol.* **2020**, *404*. [CrossRef]
28. Gilewicz, A.; Dobruchowska, E.; Murzynski, D.; Kuznetsova, T.A.; Lapitskaya, V.A. Influence of the Chemical Composition of AlCrN Coatings on Their Mechanical, Tribological, and Corrosion Characteristics. *J. Frict. Wear* **2020**, *41*, 383–392. [CrossRef]
29. Brögelmann, T.; Bobzin, K.; Kruppe, N.C.; Carlet, M. Incorporation of oxygen at column boundaries in (Cr,Al)ON hard coatings. *Thin Solid Films* **2019**, *685*, 275–281. [CrossRef]
30. Warcholinski, B.; Gilewicz, A.; Lupicka, O.; Kuprin, A.S.; Tolmachova, G.N.; Ovcharenko, V.D.; Kolodiy, I.V.; Sawczak, M.; Kochmanska, A.E.; Kochmanski, P.; et al. Structure of CrON coatings formed in vacuum arc plasma fluxes. *Surf. Coat. Technol.* **2017**, *309*, 920–930. [CrossRef]
31. Lee, S.H.; Son, B.S.; Park, G.T.; Ryu, J.S.; Lee, H. Investigation of short-term, high-temperature oxidation of AlCrN coating on WC substrate. *Appl. Surf. Sci.* **2020**, *505*, 144587. [CrossRef]
32. Singh, A.; Ghosh, S.; Aravindan, S. Investigation of oxidation behaviour of AlCrN and AlTiN coatings deposited by arc enhanced HIPIMS technique. *Appl. Surf. Sci.* **2020**, *508*, 144812. [CrossRef]
33. Zhang, J.; Li, Z.; Wang, Y.; Zhou, S.; Wang, Y.; Zeng, Z.; Li, J. A new method to improve the tribological performance of metal nitride coating: A case study for CrN coating. *Vacuum* **2020**, *173*, 109158. [CrossRef]
34. Romero, J.; Gómez, M.A.; Esteve, J.; Montalà, F.; Carreras, L.; Grifol, M.; Lousa, A. CrAlN coatings deposited by cathodic arc evaporation at different substrate bias. *Thin Solid Films* **2006**, *515*, 113–117. [CrossRef]
35. Reiter, A.E.; Mitterer, C.; De Figueiredo, M.R.; Franz, R. Abrasive and adhesive wear behavior of Arc-evaporated Al 1-x Cr x N hard coatings. *Tribol. Lett.* **2010**, *37*, 605–611. [CrossRef]
36. Lin, J.; Mishra, B.; Moore, J.J.; Sproul, W.D. Microstructure, mechanical and tribological properties of Cr1-xAlxN films deposited by pulsed-closed field unbalanced magnetron sputtering (P-CFUBMS). *Surf. Coat. Technol.* **2006**, *201*, 4329–4334. [CrossRef]
37. Wang, L.; Zhang, S.; Chen, Z.; Li, J.; Li, M. Influence of deposition parameters on hard Cr-Al-N coatings deposited by multi-arc ion plating. *Appl. Surf. Sci.* **2012**, *258*, 3629–3636. [CrossRef]
38. Tang, J.F.; Lin, C.Y.; Yang, F.C.; Tsai, Y.J.; Chang, C.L. Effects of nitrogen-argon flow ratio on the microstructural and mechanical properties of AlCrN coatings prepared using high power impulse magnetron sputtering. *Surf. Coat. Technol.* **2020**, *386*, 125484. [CrossRef]

39. Zheng, J.; Zhou, H.; Gui, B.; Luo, Q.; Li, H.; Wang, Q. Influence of power pulse parameters on the microstructure and properties of the AlCrN Coatings by a modulated pulsed power magnetron sputtering. *Coatings* **2017**, *7*, 216. [CrossRef]
40. Li, T.; Li, M.; Zhou, Y. Phase segregation and its effect on the adhesion of Cr-Al-N coatings on K38G alloy prepared by magnetron sputtering method. *Surf. Coat. Technol.* **2007**, *201*, 7692–7698. [CrossRef]
41. Kuznetsova, T.; Lapitskaya, V.; Khabarava, A.; Chizhik, S.; Warcholinski, B.; Gilewicz, A. The influence of nitrogen on the morphology of ZrN coatings deposited by magnetron sputtering. *Appl. Surf. Sci.* **2020**, *522*, 146508. [CrossRef]
42. Panjan, P.; Kek-Merl, D.; Zupanič, F.; Čekada, M.; Panjan, M. SEM study of defects in PVD hard coatings using focused ion beam milling. *Surf. Coat. Technol.* **2008**, *202*, 2302–2305. [CrossRef]
43. Panjan, P.; Čekada, M.; Panjan, M.; Kek-Merl, D. Growth defects in PVD hard coatings. *Vacuum* **2009**, *84*, 209–214. [CrossRef]
44. Panjan, P.; Gselman, P.; Kek-Merl, D.; Čekada, M.; Panjan, M.; Dražić, G.; Bončina, T.; Zupanič, F. Growth defect density in PVD hard coatings prepared by different deposition techniques. *Surf. Coat. Technol.* **2013**, *237*, 349–356. [CrossRef]
45. Creasey, S.; Lewis, D.B.; Smith, I.J.; Münz, W.D. SEM image analysis of droplet formation during metal ion etching by a steered arc discharge. *Surf. Coat. Technol.* **1997**, *97*, 163–175. [CrossRef]
46. Münz, W.D.; Smith, I.J.; Lewis, D.B.; Creasey, S. Droplet formation on steel substrates during cathodic steered arc metal ion etching. *Vacuum* **1997**, *48*, 473–481. [CrossRef]
47. Kuznetsova, T.A.; Lapitskaya, V.A.; Chizhik, S.A.; Warcholinski, B.; Gilewicz, A.; Aizikovich, S.M. Uniformity Evaluation for the Mechanical Properties of an AlCrN Coating for Tribological Application Using Probe Methods. *J. Surf. Investig. X-ray Synchrotron Neutron Tech.* **2020**, *14*, 1032–1039. [CrossRef]
48. Chizhik, S.A.; Rymuza, Z.; Chikunov, V.V.; Kuznetsova, T.A.; Jarzabek, D. Micro-and nanoscale testing of tribomechanical properties of surfaces. In *Recent Advances in Mechatronics*; Springer: Berlin/Heidelberg, Germany, 2007; pp. 541–545. [CrossRef]
49. Zhdanok, S.A.; Sviridenok, A.I.; Ignatovskii, M.I.; Krauklis, A.V.; Kuznetsova, T.A.; Chizhik, S.A.; Borisevich, K.O. On the properties of a steel modified with carbon nanomaterials. *J. Eng. Phys. Thermophys.* **2010**, *83*, 1–5. [CrossRef]
50. Anishchik, V.M.; Uglov, V.V.; Kuleshov, A.K.; Filipp, A.R.; Rusalsky, D.P.; Astashynskaya, M.V.; Samtsov, M.P.; Kuznetsova, T.A.; Thiery, F.; Pauleau, Y. Electron field emission and surface morphology of a-C and a-C:H thin films. *Thin Solid Films* **2005**, *482*, 248–252. [CrossRef]
51. Warcholinski, B.; Gilewicz, A.; Kuprin, A.S.; Tolmachova, G.N.; Ovcharenko, V.D.; Kuznetsova, T.A.; Zubar, T.I.; Khudoley, A.L.; Chizhik, S.A. Mechanical properties of Cr-O-N coatings deposited by cathodic arc evaporation. *Vacuum* **2018**, *156*, 97–107. [CrossRef]
52. Hirai, M.; Ueno, Y.; Suzuki, T.; Jiang, W.; Grigoriu, C.; Yatsui, K. Characteristics of (Cr1-x, Alx)N films prepared by pulsed laser deposition. *Jpn. J. Appl. Phys. Part. 1 Regul. Pap. Short Notes Rev. Pap.* **2001**, *40*, 1056–1060. [CrossRef]
53. Tang, J.F.; Lin, C.Y.; Yang, F.C.; Chang, C.L. Influence of nitrogen content and bias voltage on residual stress and the tribological and mechanical properties of CrAlN films. *Coatings* **2020**, *10*, 546. [CrossRef]
54. Tlili, B.; Mustapha, N.; Nouveau, C.; Benlatreche, Y.; Guillemot, G.; Lambertin, M. Correlation between thermal properties and aluminum fractions in CrAlN layers deposited by PVD technique. *Vacuum* **2010**, *84*, 1067–1074. [CrossRef]
55. Ghrib, T.; Tlili, B.; Nouveau, C.; Benlatreche, Y.; Lambertin, M.; Yacoubi, N.; Ennasri, M. Experimental investigation of the mechanical micro structural and thermal properties of thin CrAIN layers deposited by PVD technique for various aluminum percentages. *Phys. Procedia* **2009**, *2*, 1327–1336. [CrossRef]
56. Yun, J.S.; Hong, Y.S.; Kim, K.H.; Kwon, S.H.; Wang, Q.M. Characteristics of ternary Cr-O-N coatings synthesized by using an arc ion plating technique. *J. Korean Phys. Soc.* **2010**, *57*, 103–110. [CrossRef]
57. Castaldi, L.; Kurapov, D.; Reiter, A.; Shklover, V.; Schwaller, P.; Patscheider, J. Effect of the oxygen content on the structure, morphology and oxidation resistance of Cr–O–N coatings. *Surf. Coat. Technol.* **2008**, *203*, 545–549. [CrossRef]
58. Warcholinski, B.; Gilewicz, A.; Myslinski, P.; Dobruchowska, E.; Murzynski, D. Structure and Properties of AlCrN Coatings Deposited Using Cathodic Arc Evaporation. *Coatings* **2020**, *10*, 793. [CrossRef]

59. Gilewicz, A.; Jedrzejewski, R.; Myslinski, P.; Warcholinski, B. Structure, Morphology, and Mechanical Properties of AlCrN Coatings Deposited by Cathodic Arc Evaporation. *J. Mater. Eng. Perform.* **2019**, *28*, 1522–1531. [CrossRef]
60. Warcholinski, B.; Gilewicz, A.; Kuprin, A.S.; Tolmachova, G.N.; Ovcharenko, V.D.; Kuznetsova, T.A.; Lapitskaya, V.A.; Chizhik, S.A. Comparison of Mechanical and Tribological Properties of Nitride and Oxynitride Coatings Based on Chrome and Zirconium Obtained by Cathodic Arc Evaporation. *J. Frict. Wear* **2019**, *40*, 163–170. [CrossRef]
61. Kuznetsova, T.; Lapitskaya, V.; Chizhik, S.; Kuprin, A.S.; Tolmachova, G.N.; Ovcharenko, V.D.; Gilewicz, A.; Lupicka, O.; Warcholinski, B. Friction and Wear of Cr-O-N Coatings Characterized by Atomic Force Microscopy. *Tribol. Ind.* **2019**, *41*, 274–285. [CrossRef]
62. Kuprin, A.S.; Kuznetsova, T.A.; Gilewicz, A.; Tolmachova, G.N.; Ovcharenko, V.D.; Abetkovskaia, S.O.; Zubar, T.I.; Khudoley, A.L.; Chizhik, S.A.; Lupicka, O.; et al. Tribological properties of vacuum arc Cr-O-N coatings in macro- and microscale. *Probl. At. Sci. Technol.* **2016**, *106*, 211–214.
63. Lapitskaya, V.A.; Kuznetsova, T.A.; Melnikova, G.B.; Chizhik, S.A.; Kotov, D.A. The Changes in Particle Distribution over the Polymer Surface under the Dielectric Barrier Discharge Plasma. *Int. J. Nanosci.* **2019**, *18*, 1–4. [CrossRef]
64. Khort, A.; Romanovski, V.; Lapitskaya, V.; Kuznetsova, T.; Yusupov, K.; Moskovskikh, D.; Haiduk, Y.; Podbolotov, K. Graphene@Metal Nanocomposites by Solution Combustion Synthesis. *Inorg. Chem.* **2020**, *59*, 6550–6565. [CrossRef] [PubMed]
65. Coy, E.; Yate, L.; Kabacińska, Z.; Jancelewicz, M.; Jurga, S.; Iatsunskyi, I. Topographic reconstruction and mechanical analysis of atomic layer deposited Al_2O_3/TiO_2 nanolaminates by nanoindentation. *Mater. Des.* **2016**, *111*, 584–591. [CrossRef]
66. Sanchette, F.; Ducros, C.; Schmitt, T.; Steyer, P.; Billard, A. Nanostructured hard coatings deposited by cathodic arc deposition: From concepts to applications. *Surf. Coat. Technol.* **2011**, *205*, 5444–5453. [CrossRef]
67. Kudish, I.I.; Pashkovski, E.; Volkov, S.S.; Vasiliev, A.S.; Aizikovich, S.M. Heavily loaded line EHL contacts with thin adsorbed soft layers. *Math. Mech. Solids* **2020**, *25*, 1011–1037. [CrossRef]

Article

Improved Hardness and Thermal Stability of Nanocrystalline Nickel Electrodeposited with the Addition of Cysteine

Tamás Kolonits [1,2,*], Zsolt Czigány [2], László Péter [3], Imre Bakonyi [3] and Jenő Gubicza [1]

1 Department of Materials Physics, ELTE Eötvös Loránd University, Pázmány Péter sétány 1/A, H-1117 Budapest, Hungary; jeno.gubicza@ttk.elte.hu
2 Centre for Energy Research, Institute for Technical Physics and Materials Science, Konkoly-Thege M. út 29-33, H-1121 Budapest, Hungary; czigany.zsolt@energia.hu
3 Wigner Research Centre for Physics, Konkoly-Thege út 29-33, H-1121 Budapest, Hungary; peter.laszlo@wigner.hu (L.P.); bakonyi.imre@wigner.hu (I.B.)
* Correspondence: tamas.kolonits@ttk.elte.hu

Received: 20 October 2020; Accepted: 10 November 2020; Published: 13 November 2020

Abstract: Experiments were conducted for the study of the effect of cysteine addition on the microstructure of nanocrystalline Ni films electrodeposited from a nickel sulfate-based bath. Furthermore, the thermal stability of the nanostructure of Ni layers processed with cysteine addition was also investigated. It was found that with increasing cysteine content in the bath, the grain size decreased, while the dislocation density and the twin fault probability increased. Simultaneously, the hardness increased due to cysteine addition through various effects. Saturation in the microstructure and hardness was achieved at cysteine contents of 0.3–0.4 g/L. Moreover, the texture changed from (220) to (200) with increasing the concentration of cysteine. The hardness of the Ni films processed with the addition of 0.4 g/L cysteine (~6800 MPa) was higher than the values obtained for other additives in the literature (<6000 MPa). This hardness was further enhanced to ~8400 MPa when the Ni film was heated up to 500 K. It was revealed that the hardness remained as high as 6000 MPa even after heating up to 750 K, while for other additives, the hardness decreased below 3000 MPa at the same temperature.

Keywords: electrodeposition; nickel; cysteine; microstructure; hardness; thermal stability

---

1. Introduction

Nanocrystalline materials such as electrodeposited nickel [1–4] are frequently investigated due to their unique properties as compared to their coarse-grained counterparts [5,6]. Both the correlation between the physical properties and the microstructure of electrodeposited nickel and the effect of deposition parameters (such as pH and temperature of the bath, applied current density, stirring rate, and composition of the bath) were intensively studied during the past decades. A short summary of these studies can be found in our previous article. [7].

Reduction of the grain size and increase of the defect density in nanocrystalline materials usually improve the hardness [8], but decrease the thermal stability [9], especially in the case of pure metals [9–13]. If alloying with other elements [14–19] or codeposition of secondary-phase particles (e.g., SiC) [20–23] are not applicable; the only way to hinder the grain growth during annealing is the application of additives during electrodeposition, which causes the inclusion of a small amount of impurities in the deposited layer. Even a small impurity content can improve the thermal stability [9,24–27], due to the segregation of impurity atoms to the grain boundaries and decreasing both the boundary mobility [28–31] and the extra enthalpy of the material caused by the presence of grain boundaries [12,32].

Cysteine is considered to be a promising additive to create nanocrystalline nickel with a very small grain size and with an improved thermal stability, due to the following reasons:

- As demonstrated earlier, incorporating sulfur can reduce the grain size, and increase the hardness [7,8,33]. Sulfur may also yield a moderate strengthening during annealing at low temperatures (between 400 K and 600 K) [9,24]. Cysteine contains sulfur and can easily decompose into sulfur-containing components [34]. Since the sulfur atom in the cysteine molecule is relatively weakly bound as compared to the most common additive saccharin, cysteine can be a good candidate as sulfur-donor bath additive.
- Additives with –SH functional groups (i.e., thiols) tend to adsorb on metal surfaces with the sulfur atom as the anchoring entity of the adsorbed molecule. Thiols can even form monolayers that block the direct interaction between the metal and the solution. This phenomenon has a rich literature [35–39], especially for noble metals like Au and Ag. Although this "organic shield" does not completely block the electron transfer between the metal and the reactive solute species, the adsorbed layer inhibits the growth of the already existing grains of the deposit and promotes the formation of new nucleation sites [7]. It has been well evidenced that the metal deposition can take place through the defect sites of adsorbed thiol layers [35–39]. Additionally, the interaction of the polar/ionic groups of the cysteine molecule (i.e., the amino and the carboxylate groups) can stabilize the solution side of the adsorbed layer by pinning the adsorbed molecules with zwitterion formation.
- Cysteine can easily form complexes with Ni [40–42].

The electrodeposition of different metals, such as Cu [43], CuZn [44] or As [41], in the presence of cysteine has already been studied. Investigations were conducted on the electrodeposition of gold with cysteine addition. Different morphologies like monocrystaline surfaces [45], nanoparticles [46,47] and dendrites [48] were observed. In the case of Ni deposited from cysteine containing baths, mainly the electrokinetic behaviour was investigated [42,49]. Although, Ebadi et al. concluded that cysteine could reduce the grain size and the surface roughness of Ni [49], a detailed analysis of the microstructure and its thermal stability is missing from the literature.

The aim of the present study was to investigate the effect of various cysteine concentrations in a nickel-sulfate based bath on the microstructure and hardness of Ni electrodeposits. At one selected cysteine concentration (0.4 g/L), the evolution of the defect structure (dislocation density and twin fault probability), grain size and hardness of Ni films were investigated at different annealing temperatures up to 1000 K. The microstructure was studied by the complementary usage of transmission electron microscopy (TEM), electron backscatter diffraction (EBSD) and X-ray line profile analysis (XLPA).

2. Materials and Methods

2.1. Electrodeposition of Nickel Samples

In order to investigate the effect of cysteine on the microstructure of Ni electrodeposit, first cysteine was added to an organic additive-free nickel electrolyte denoted as "NOA" and described in our former paper Ref. [8]. Then, it turned out that the original pH and current density conditions were unfavourable when cysteine was added to bath "NOA", since the as-deposited samples were fragmented and had a poor adhesion to the substrate. Therefore, several preliminary trials were performed to find more appropriate electrodeposition conditions. On the basis of these experiments, a direct current density of $j = -25$ mA/cm^2 and an increased concentration of boric acid were selected and the pH was adjusted to 6.1 using NaOH. Crack-free Ni films with a thickness as large as ~30 μm could be successfully deposited by adjusting the deposition time, assuming 96% current efficiency [50]. The composition of the new bath is given in Table 1.

Table 1. Composition of the electrolyte bath containing cysteine.

Component	Concentration (g/L)	Role
$NiSO_4 \cdot 6H_2O$	155	Ni source
H_3BO_3	30	pH-buffer
$NaSO_4 \cdot 10H_2O$	96	Supporting electrolyte
H_2NSO_3H	10	wetting agent
Cysteine $HSCH_2CH(NH_2)COOHS$	0; 0.1; 0.2; 0.3; 0.4	additive
pH	6.1	
Current density	-25 mA/cm^2	

Samples were deposited at room temperature onto an 8 μm thick rolled Cu sheet acting as the cathode at the bottom of a 8 × 20 mm^2 tubular cell [51]. Nickel resupply was provided by a soluble Ni anode. After deposition, the substrate was removed with electrochemical dissolution in the same way as described in Ref. [9]. The samples were denoted as CYS00, CYS01, CYS02, CYS03 and CYS04, corresponding to the cysteine contents of 0.0, 0.1, 0.2, 0.3 and 0.4 g/L, respectively. In this study, the notations "es" and "ss" were used to distinguish between the electrolyte and substrate sides of the as-deposited films.

2.2. Heat Treatment

As sample CYS04 (produced with a cysteine content of 0.4 g/L) exhibited the highest hardness (see section "Results" below), this material was used for the study of the effect of cysteine on the thermal stability of the as-deposited Ni nanostructure. Therefore, CYS04 samples were heated up to the temperatures of 400, 500, 600, 750 and 1000 K in an Ar gas atmosphere using a Perkin-Elmer 2 type differential scanning calorimeter (DSC) (Waltham, MA, USA). These temperatures were the same as the values used for the investigation of the thermal stability of Ni films processed with other bath additives in our former study [9], since we wanted to compare the influence of cysteine on the stability with the effects of other additives. The samples were heated up to the selected temperatures at a rate of 40 K/min, and then were cooled down to room temperature at a rate of 300 K/min (this is the highest applicable rate in the calorimeter used). Since there was no significant difference between the two sides "es" and "ss" of sample CYS04, after the heat treatment, only the electrolyte side ("es") was studied. It should be noted that, in the heat treatment experiments, the parts of the same film were used for annealing.

2.3. X-ray Diffraction

X-ray diffraction (XRD) characterization of the microstructure consisted of three kinds of measurements: (i) phase analysis, (ii) X-ray diffraction line profile analysis (XLPA) of the lattice defect structure and (iii) study of the crystallographic texture by pole figures.

The phase analysis was performed on the XRD patterns taken by a Smartlab X-ray diffractometer (Rigaku, Tokyo, Japan), using CuK$_\alpha$ radiation (wavelength $\lambda = 0.15418$ nm). These measurements were taken in Bragg–Brentano geometry. The texture of the samples was characterized by the same Smartlab diffractometer, applying parallel-beam optics and in-plane geometry (referred to as pole figure measurements). The XLPA measurements for the characterization of the defect structure were performed by a RA-MultiMax9 rotating anode diffractometer (Rigaku, Tokyo, Japan) (with CuK$_{\alpha_1}$ radiation, wavelength $\lambda = 0.15406$ nm). The diffraction patterns were evaluated by the extended convolutional multiple whole profile (eCMWP) fitting method [52,53]. A brief description of this method is given below. During this evaluation, the whole experimental diffraction pattern was fitted by the sum of a background and theoretical peak profile functions. The latter profiles were obtained as the convolution of the measured instrumental peak and the calculated XRD profiles caused by the finite crystallite size, dislocations and twin faults. The instrumental profiles were measured on a LaB_6 standard material. The peak profile function related to the crystallite size effect was calculated assuming a log-normal distribution of spherical-shaped crystallites [7], which was characterized by the median (m) and the log-normal variance (σ^2). The peak profile related to the strain effect depended on the density of dislocations (ρ) and their arrangement, while the function related to twin faults was influenced by the twin fault probability. The latter quantity is defined as the fraction of twin boundaries among the (111) lattice planes, denoted as β. From the fitting of the XRD patterns, the mean crystallite size, the dislocation density and the twin fault probability were determined. Details of the eCMWP procedure can be found in Refs. [52,53].

It is important to note that the size of the coherently scattering domains provided by the XLPA evaluation (consequently called "crystallite size" in this paper) is often smaller than the size of the regions bounded by high-angle grain boundaries (HAGBs) determined by transmission electron microscopy (TEM). The latter quantity is consequently called "grain size" in this paper. The difference between the grain and crystallite sizes is pronounced for Ni films with larger grain sizes and is manifested in an increasing $d/<x>$ ratio with grain size, as depicted in Figure 10 of [8].

2.4. Transmission Electron Microscopy

TEM samples were prepared with mechanical polishing and Ar^+ ion-beam milling with the use of a Technoorg Linda ionmill (Technoorg Linda, Budapest, Hungary). Thinning was started at high energy (10 keV) and was finished at an ion energy of 3 keV in order to minimize the structural changes induced by the ion beam [54]. A detailed description of sample preparation can be found in Ref. [7].

To investigate both sides of the as-deposited samples with different cysteine contents, cross-sectional (cv) samples were prepared. Since there was no difference between the two sides of the annealed samples, for this films, only the electrolyte side was investigated by plane view (pv) samples. The TEM measurements were performed by a Philips CM20 transmission electron microscope (Thermo Fisher, Waltham, MA, USA) operated at an acceleration voltage of 200 kV. Selected area electron diffraction (SAED) patterns were taken at a camera length of 1 m on areas with linear dimensions between 1 and 5 μm.

2.5. Scanning Electron Microscopy and Energy-Dispersive X-ray Spectroscopy

The chemical composition of the samples was investigated by energy-dispersive X-ray spectroscopy (EDS) using a Hitachi TM4000Plus scanning electron microscope (SEM) (Tokyo, Japan) equipped with an AztecOne EDS system (made by the Oxford Instruments – Bognor Regis, UK). The grain size for the sample heated up to 1000 K was studied by electron-backscatter diffraction (EBSD) using a FEI Quanta 3D SEM (Hillsboro, OR, USA). The size of the grains was determined from the grain maps obtained by EBSD using the linear intercept method.

2.6. Hardness Tests

Vickers hardness tests were performed by a Zwick/Roell ZHμ indenter (Ulm, Germany), using a load of 10 g. The investigated samples were not polished since this surface treatment could modify the as-deposited and the annealed microstructures. In order to avoid the influence of the substrate on the hardness values, the load was selected to a sufficiently small value so that it yielded indentation diagonal not higher than 66% of the thickness of the film (according to ISO 6507 standard). Thus, the indentation diagonal for all samples was kept to be less than 15 μm, since the film thickness was about 30 μm. Fifty indentations were made on each sample.

3. Results

3.1. The Effect of Cysteine Addition on the Chemical Composition and the Microstructure

Within the detection limit of the EDS analysis (~0.1 at.% for both metallic impurities and sulfur), no foreign element was found in the samples deposited in the absence of cysteine, i.e., these specimens were pure nickel. In contrast, all the Ni samples deposited from baths with cysteine contained about 0.6–0.7 at.% sulfur irrespectively of the cysteine concentration of the bath.

Figure 1 illustrates the effect of cysteine on the microstructure of Ni electrodeposits. These TEM images were taken on the cross-section of the films close to the electrolyte side. Although the TEM study was performed for all cysteine concentrations, Figure 1 shows only the cysteine-free sample and the films processed with 0.1 and 0.4 g/L cysteine, illustrating the trend of grain refinement with increasing additive concentration. The grain size values obtained from the TEM images (dTEM) for both the "es" and "ss" sides are listed in Table 2 and plotted in Figure 2. For the cysteine-free sample (CYS00), a columnar grain structure developed with increasing column diameter toward the "es" side. The column width (called grain size hereafter) was about 67 nm at the "ss" side, which increased to about 215 nm at the "es" side. With the addition of 0.1 g/L cysteine to the electrolyte, the grain size decreased to about 37 nm and the difference between the electrolyte and substrate sides of the sample disappeared. Columnar microstructure was not observed in the TEM images taken on the films deposited from the baths containing cysteine. With an increasing amount of cysteine in the bath, only a slight decrease in the grain size was observed and for the highest cysteine concentration applied (0.4 g/L), the grain size decreased to 25–27 nm.

The crystallite sizes obtained by XLPA for the different cysteine contents are listed in Table 2. For the cysteine-free case, the crystallite size values for both sides of the film are smaller than the grain sizes determined by TEM with a factor of about three. A similar effect has been observed formerly for materials in which the grains were divided into subgrains (see Ref. [8]). As XRD is sensitive to small misorientations, the crystallite size characterizes rather the size of subgrains and not the grains. For the films processed from the bath containing cysteine, the values of the crystallite size are similar to the grain size (see Table 2), which suggests that the grains with the size of about 20–30 nm are not divided into subgrains.

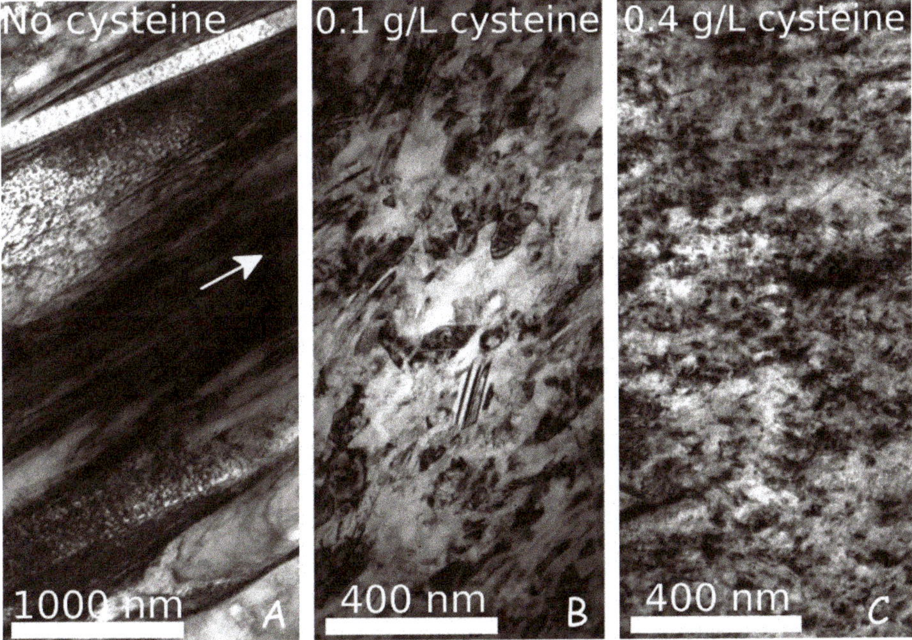

Figure 1. TEM images showing the microstructures in the cross section of the Ni films (close to the electrolyte side) processed without cysteine (**A**), 0.1 g/L cysteine (**B**) and 0.4 g/L cysteine (**C**). The white arrow in (**A**) indicates the growth direction of the film.

Table 2. Microstructural parameters and Vickers hardness (HV) of samples deposited from baths containing different amounts of cysteine. Both the electrolyte side (es) and the substrate side (ss) were investigated. The table contains the crystallite size (<x>), the dislocation density (ρ) and the twin fault probability (β) obtained by XLPA, as well as the mean grain size obtained from TEM (d_{TEM}).

Concentration (g/L)	Side	$<x>$ (nm)	ρ (10^{14}m^{-2})	β (%)	d_{TEM} (nm)	HV (MPa)
0	es	64 ± 6	34 ± 4	0 ± 0.1	215 ± 70	2800 ± 350
0	ss	23 ± 3	39 ± 5	0 ± 0.1	67 ± 12	3700 ± 180
0.1	es	34 ± 3	380 ± 40	1.6 ± 0.2	37 ± 3	5100 ± 270
0.1	ss	36 ± 4	360 ± 40	1.6 ± 0.2	38 ± 2	5000 ± 280
0.2	es	25 ± 3	300 ± 30	1.5 ± 0.2	19 ± 4	6700 ± 490
0.2	ss	28 ± 3	360 ± 40	1.9 ± 0.2	31 ± 3	5900 ± 340
0.3	es	28 ± 3	550 ± 60	2.5 ± 0.3	21 ± 3	5800 ± 170
0.3	ss	28 ± 3	550 ± 60	3.1 ± 0.3	20 ± 3	6100 ± 340
0.4	es	27 ± 3	620 ± 70	2.8 ± 0.3	27 ± 3	6800 ± 400
0.4	ss	28 ± 3	610 ± 60	3.3 ± 0.3	25 ± 3	6500 ± 160

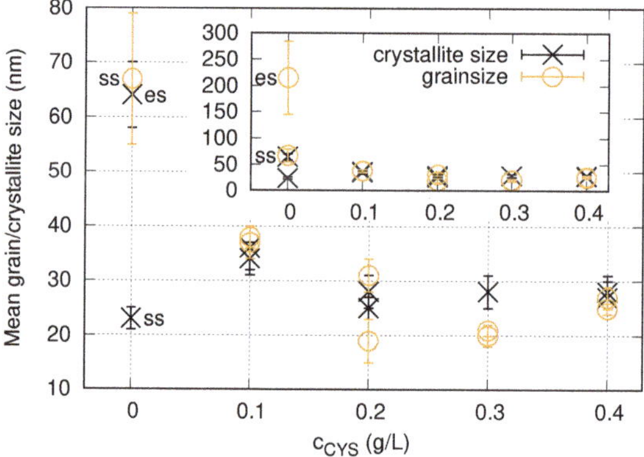

Figure 2. Grain size determined by TEM and crystallite size obtained by XLPA for the "es" and "ss" sides of the films versus the cysteine content of the bath. Subfigure was inserted to reveal the details of the figure without the outlier data point corresponding to the grain size of the electrolyte side of sample CYS00.

It is noted that the TEM investigations confirmed the log-normal shape of the size distribution function assumed in the XLPA evaluation. Namely, Pearson's chi-square test [55] revealed that the grain size distributions for the different samples follow log-normal distribution, with a probability higher than 95% in most of the cases. A comparison between the directly determined median of the data and the calculated median assuming a log-normal distribution was also made, and the difference between them was always smaller than 5%. A detailed description of this evaluation procedure can be found in Ref. [8].

The dislocation density and the twin fault probability determined by XLPA are summarized in Table 2. For both sides of the cysteine-free sample, the dislocation density was about $34 - 39 \times 10^{14}$ m^{-2}, while the twin fault probability was zero. The latter observation is not in contradiction with the appearance of twins in the cross-sectional TEM image (see the top of Figure 1A). These twin boundaries are commonly perpendicular to the film surface (i.e., parallel to the diffraction vector), therefore they are practically invisible to the XLPA method. In addition, their average spacing is higher than the detection limit of XLPA (about 200 nm). It should also be noted that the microstructure of this cysteine-free layer differs significantly from the values obtained for the additive-free film used in our former studies (denoted as "NOA" in Ref. [8]). Table 3 shows the difference between the conditions of electrodeposition and the microstructural parameters for these two additive-free samples. The dislocation density for sample CYS00 is much higher than that for the film NOA that might have contributed to the higher hardness of the former film.

Table 3. The difference between the conditions of electrodeposition of the additive-free sample from Ref. [8] (denoted as "NOA") and the additive-free sample used in this study (denoted as "CYS00") and a comparison between the microstructural parameters, texture and hardness.

Sample	Boric Acid (g/L)	pH	Curr. Dens. (mA/cm^2)	<x> (nm)	ρ (10^{14}m^{-2})	β (%)	d_{TEM} (nm)	HV (MPa)	Texture
NOA-es	15	3.25	−6.25	42 ± 5	12 ± 1	0 ± 0.1	127 ± 45	1877 ± 211	(220)
NOA-ss	15	3.25	−6.25	27 ± 3	14 ± 2	0 ± 0.1	91 ± 14	2837 ± 344	NA
CYS00-es	30	6.1	−25	64 ± 6	34 ± 4	0 ± 0.1	215 ± 70	2800 ± 350	(220)
CYS00-ss	30	6.1	−25	23 ± 3	39 ± 5	0 ± 0.1	67 ± 12	3700 ± 180	NA

The density of lattice defects (dislocations and twin faults) was determined by XLPA using the eCMWP fitting method. Figure 3 illustrates the eCMWP fitting for the substrate side of sample CYS01. The evolution of the dislocation density and the twin fault probability obtained by XLPA versus the cysteine content in the bath is plotted in Figure 4. The addition of 0.1 g/L cysteine to the bath resulted in one order of magnitude enhancement of the dislocation density from ~39 $\times 10^{14}$ m^{-2} to $360 - 380 \times 10^{14}$ m^{-2}. In addition, the twin fault probability increased from 0 to 1.6 ± 0.2%. Both the dislocation density and the twin fault probability increased monotonously with the addition of cysteine. Initially, the increase was intense and then the saturation of both parameters was observed for the higher cysteine concentrations. The saturation value was about 620×10^{14} m^{-2} for the dislocation density and about 3% for the twin fault probability.

Figure 5 shows the evolution of the crystallographic texture on the "es" side of the films with increasing the cysteine content of the bath. Without cysteine, a (220) out-of-plane fiber texture was formed. When 0.1–0.2 g/L cysteine was added to the bath, a (111) texture was developed. With increasing the concentration of cysteine to 0.3–0.4 g/L, the dominating texture turned into a (200) one with a weaker (111) component. On the "ss" side of the Ni films, a similar texture evolution was observed with the increase of cysteine content.

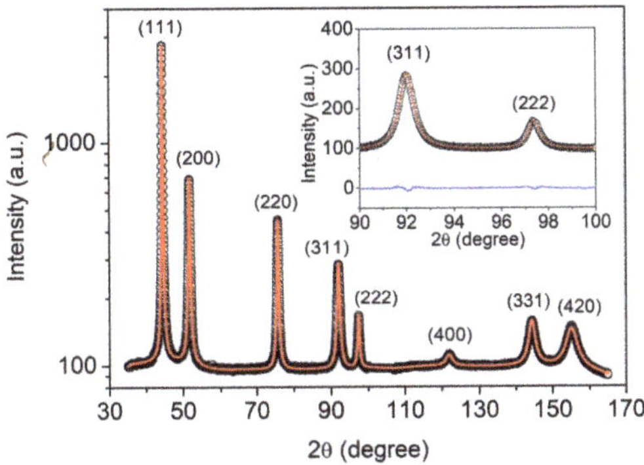

Figure 3. eCMWP fitting on the pattern taken on the "ss" side of the film CYS01. The black open circles and the red solid line represent the measured and the calculated XRD patterns, respectively. The intensity is displayed on a logarithmic scale. The inset shows a magnified part of the pattern with linear intensity scale. The difference between the measured and the calculated data is represented by the blue line.

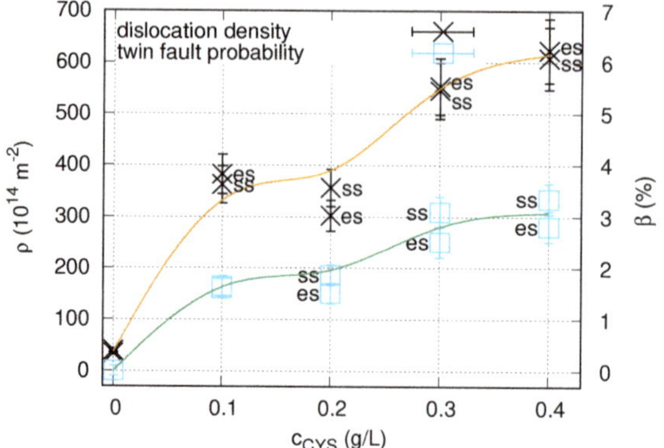

Figure 4. Dislocation density (ρ) and twin fault probability (β) versus the cysteine content of the bath for the "es" and "ss" sides of the films. The line serves as a guide for the eye only.

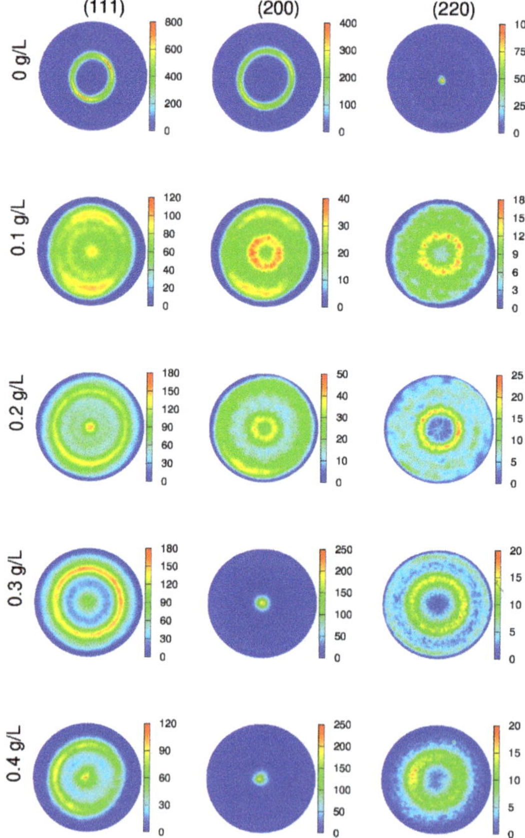

Figure 5. (111), (200) and (220) pole figures obtained by XRD for the "es" side of the Ni films deposited with different cysteine contents.

3.2. Hardness Evolution with Increasing Cysteine Content

The hardness values obtained on the "es" and "ss" sides of the Ni films processed from the baths containing different amounts of cysteine are listed in Table 2. For the cysteine-free sample, the hardness on the side "ss" was higher (~3700 MPa) than that measured on the "es" side (~2800 MPa), most probably due to the smaller grain size. For the films deposited from the baths containing cysteine, a difference between the hardness values obtained on the two sides was not observed. The change of the hardness as a function of the cysteine content is shown in Figure 6. The hardness initially increased monotonously with the addition of cysteine. Initially, the increase was intense. Saturation of the hardness was observed for the higher cysteine concentrations, with a saturation value of about 6800 MPa.

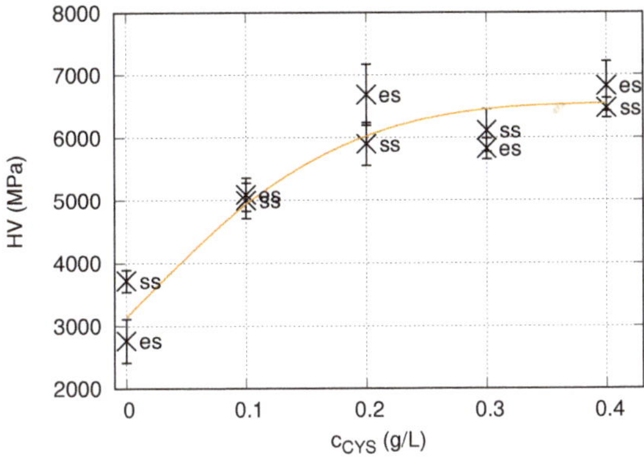

Figure 6. The Vickers hardness (HV) as a function of the cysteine content in the bath for the "es" and "ss" sides of the Ni films. The line serves as a guide for the eye only.

3.3. Thermal Stability of the Microstructure in the Ni Film Processed with the Cysteine Content of 0.4 g/L

Figure 7 shows TEM micrographs taken on the "es" side of CYS04 sample heated up to 400, 500, 600 and 750 K. The microstructure obtained on the film heated up to 1000 K is depicted by the EBSD image in Figure 8. The grain sizes determined from the TEM and EBSD images are listed in Table 4 and plotted as a function of temperature in Figure 9. It can be seen that the grain size remained unchanged after heating up to 400 K. This is also true for the dislocation density and the twin fault probability (see Figure 10). When the annealing temperature was increased to 500 K, a slight decrease in the defect densities was observed. In addition, both the grain and crystallite sizes determined by TEM and XLPA decreased to about half of their original values. A further increase of the heat treatment temperature resulted in a gradual enhancement of the grain and crystallite sizes, as well as a simultaneous reduction of the dislocation density and the twin fault probability. It is worth noting that for 600 and 750 K, the crystallite size determined by XLPA is slightly different from the grain size determined by TEM. This observation can be explained by the fact that in the present case, the TEM characterized the grain size parallel to the film surface while XLPA determined the crystallite size perpendicular to the layer due to the applied diffraction geometry. Thus, the difference between the crystallite and grain sizes may be caused by the formation of grains elongated along the thickness after annealing at 600 and 750 K. Between 750 and 1000 K, the sample was fully recrystallized, which was indicated by the grain growth to 43 μm and the decrease of the defect density under the detection limit of XLPA (10^{13} m^{-2} and 0.1% for the dislocation density and the twin fault probability, respectively).

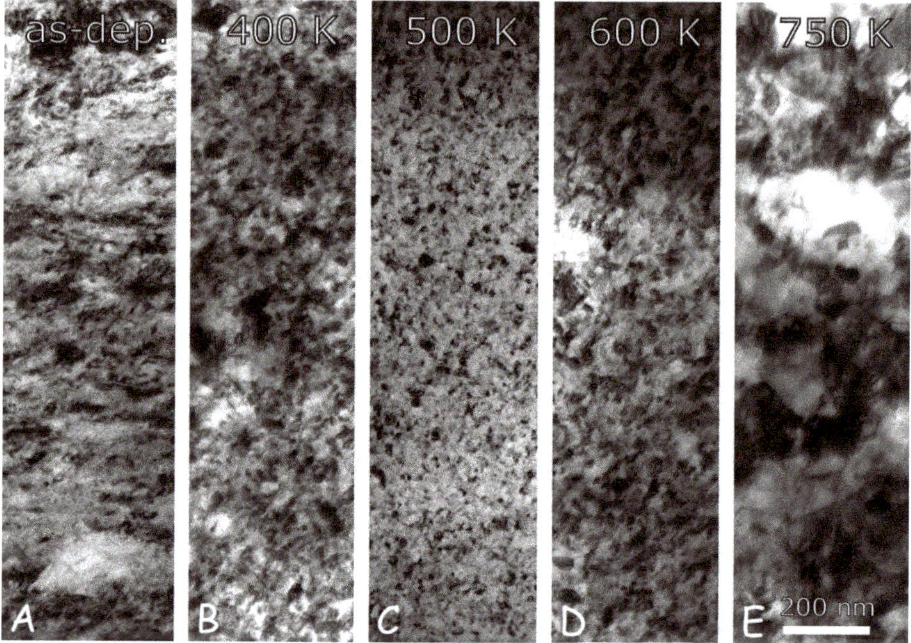

Figure 7. Plane view TEM images showing the microstructures on the "es" side of the CYS04 film in the as-deposited state (**A**) and after annealing at 400 (**B**), 500 (**C**), 600 (**D**) and 750 K (**E**).

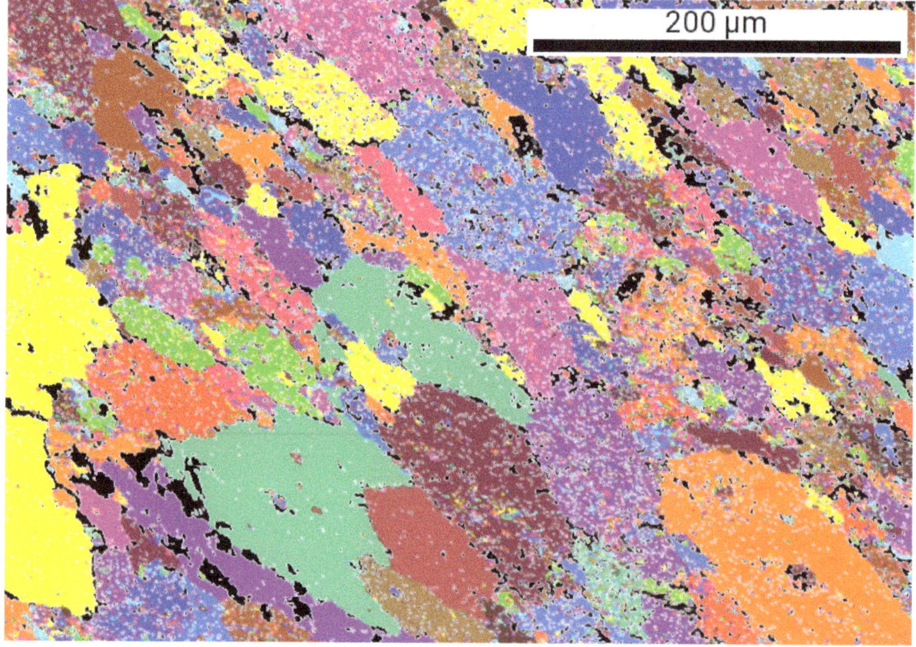

Figure 8. EBSD image on the microstructure obtained on the "es" side of the CYS04 film heated up to 1000 K.

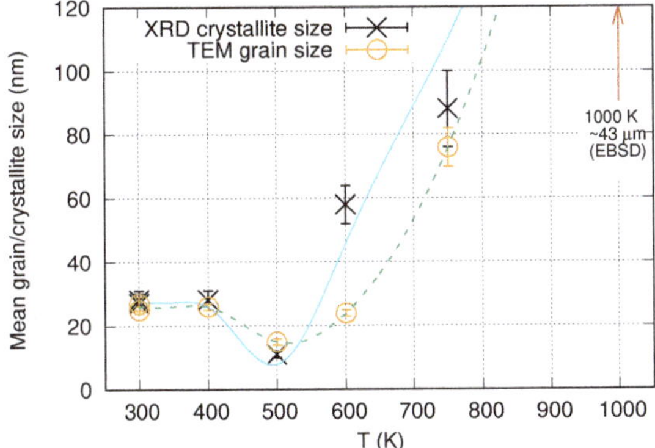

Figure 9. Grain size determined by TEM and crystallite size obtained by XLPA for the "es" side of the film CYS04 versus the annealing temperature. The line serves as a guide for the eye only.

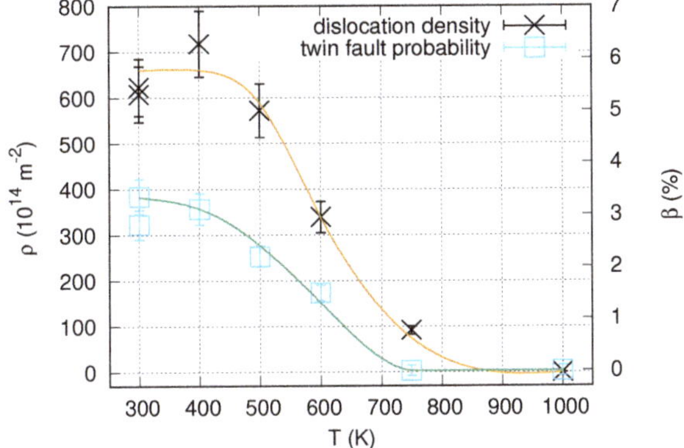

Figure 10. Dislocation density (ρ) and twin fault probability (β) for the "es" side of the film CYS04 versus the annealing temperature. The line serves as a guide for the eye only.

Table 4. The microstructural parameters and the hardness of the samples deposited from the bath containing 0.4 g/L cysteine and heat treated up to different temperatures. Only the electrolyte side (es) of the samples was investigated. The table contains the mean crystallite size (<x>), the dislocation density (ρ) and the twin fault probability (β) obtained XLPA, the mean grain size obtained by TEM (d_{TEM}) and the Vickers hardness (HV).

Temperature (K)	<x> (nm)	ρ (10^{14}m^{-2})	β (%)	d_{TEM} (nm)	HV (MPa)
300	27 ± 3	620 ± 70	2.8 ± 0.3	27 ± 3	6800 ± 400
400	28 ± 3	720 ± 70	3.1 ± 0.3	26 ± 2	6600 ± 340
500	11 ± 2	570 ± 60	2.2 ± 0.2	15 ± 2	8400 ± 820
600	58 ± 6	340 ± 40	1.5 ± 0.2	24 ± 2	7100 ± 350
750	88 ± 14	91 ± 10	0 ± 0.1	76 ± 6	6000 ± 260
1000	NA	NA	NA	43,000 ±6600 [1]	2300 ± 280

[1] Grain size for sample annealed up to 1000 K was determined by EBSD.

During the heat treatment of sample CYS04, both texture components (200) and (111) remained strong and component (111) took the dominance (see Figure 11). It should be noted that after heating up to 500 K, both texture components are less sharp than for either 400 or 600 K.

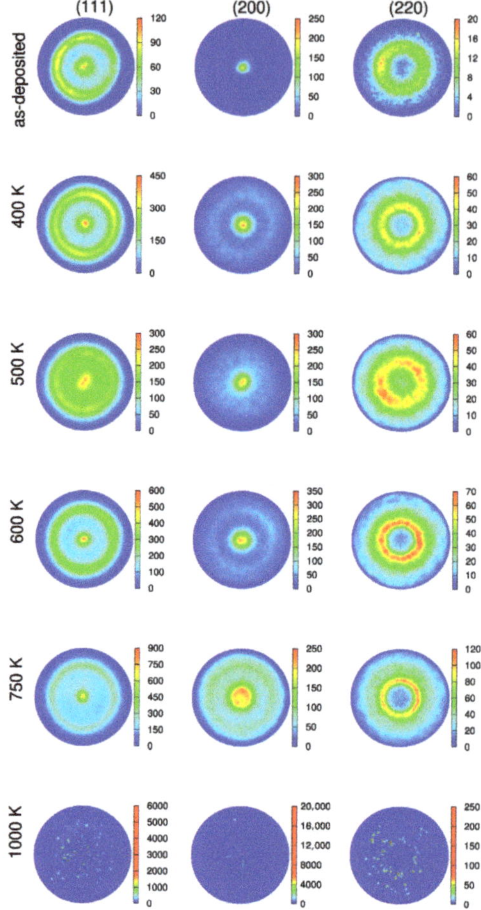

Figure 11. (111), (200) and (220) pole figures obtained by XRD for the "es" side of the CYS04 film heated up to different temperatures.

3.4. Change of the Hardness During Annealing of the Ni Deposited With Cysteine Additive

Figure 12 shows the evolution of the hardness as a function of the temperature of heat treatment. The hardness values are also listed in Table 4. Up to 400 K, a significant change in the hardness was not observed. At the same time, after heating up to 500 K, the hardness increased from ~6800 MPa to ~8400 MPa. At higher temperatures, the hardness gradually decreased, but even at 750 K, the hardness still had a high value of about 6000 MPa. For the highest temperature of 1000 K, the hardness decreased to ~2300 MPa (see Table 4). The sample annealed to 500 K exhibited the most interesting behaviour, therefore this experiment was repeated on another film. The same microstructural parameters and hardness values were obtained in both cases.

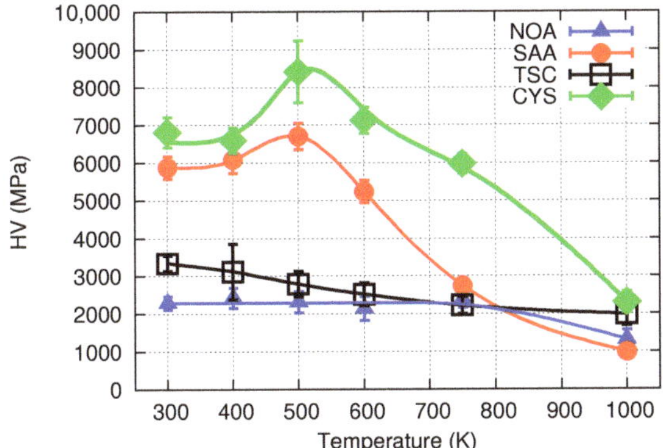

Figure 12. Vickers hardness (HV) as a function of the annealing temperature for the "es" side of the film CYS04 (denoted as CYS). In addition, the hardness evolution versus the temperature is also plotted for other additives, published formerly in Ref. [9]. NOA: no additive, SAA: saccharin, TSC: trisodium-citrate. The line serves as a guide for the eye only.

4. Discussion

4.1. Effect of Cysteine Addition on the Texture, Grain Size and Defect Density

The grain size of an electrodeposited metal is determined by the ratio of the rates of new grain nucleation and growth of the existing grains. During electrodeposition of nickel in the presence of cysteine, there are various ways to reduce nickel: (i) reduction of a free (hydrated) Ni^{2+} ion, (ii) reduction of a Ni^{2+} ion formerly part of a Ni-cysteine complex [40,49]. The first process is certainly dominant, since the concentration ratio of Ni^{2+}:cysteine is larger than 300. However, the surface is covered by the adsorbed cysteine molecules, independently of the Ni deposition process. A similar behaviour was demonstrated for Au-thiol adsorbate systems [45,47,56]. Here, we have to count with Ni-thiol surface groups, which is analogous to the Au-thiol adsorbates [57–61]. The surface coverage with the thiols means that the near-equilibrium growth sites are mostly blocked by the adsorbed molecules, and the maintenance of the growth rate with the constant current must lead to the nucleation of new crystallites that are not covered with an adsorbate layer at the moment of their formation, but also become thiol-covered soon thereafter.

During the deposition of a face-centered cubic (fcc) material such as Ni, the different crystal planes grow at different rates due to the surface energy anisotropy. The order of formation rate is (111) > (100) > (110) [62]. However, since nickel electrodeposition is always accompanied by hydrogen codeposition (because of the value of the equilibrium potential of the Ni(II)/Ni redox system), these formation rates can change [8,62,63]. In the absence of other strongly adsorbing additives, the different lattice planes have different hydrogen adsorption affinity, resulting in different hydrogen coverages [63–65]. Hydrogen codeposition favours the crystal planes with lower planar packing density. Thus, during electrodeposition of additive-free nickel, due to the hydrogen codeposition, the order of the formation rate of the crystalline planes is (110) > (100) > (111), which is in accordance with the experimentally observed strong (220) texture in specimen CYS00.

As mentioned above, during the electrodeposition of Au in the presence of cysteine, cysteine molecules could covalently attach to the metal surface in a preference order (110) > (100) > (111), causing an (111) type texture, since cysteine blocks the growth of the strongly covered crystallographic planes. Assuming the same effect for Ni, the strong (111) texture formed in the presence of a small amount of cysteine (0.1–0.2 g/L) can be understood.

When increasing the amount of cysteine in the bath, other planes (e.g., the less favoured (111) planes) are also covered by cysteine molecules, thereby blocking the growth of these planes, too.

The effectiveness of this blocking depends on the number of the attached molecules (which favours the order 110 > 100 > 111) and the packing density of lattice points on the planes (which increases in the reverse order, 111 > 100 > 110). Thus, the combination of these two effects can result in the dominant (200) texture observed for higher cysteine concentrations (0.3–0.4 g/L).

Moreover, quantum mechanical calculations proved that during co-adsorption of S and H on Ni deposits, the adsorption of S at a three-fold site on plane (111) has a tendency to block the adsorption of H at the nearby surface sites [66], thus suppressing the effect of hydrogen codeposition on the (111) planes. Then, this effect may cause the development of a (111) texture component in accordance with the observation for CYS03 and CYS04.

The coverage of the crystal planes by sulfur atoms and cysteine molecules can influence not only the texture, but also the grain size and the defect density. If sulfur atoms are adsorbed on the surface of a growing crystal during deposition, the strong Ni–S covalent bonds and their preferred bonding angles [60] can increase the probability of grain boundary formation. Similarly, the cysteine molecules attached to the surface can form an adsorbed monolayer acting as an "organic shield" on the crystal planes, thereby hindering the ion transfer, but not blocking the electron transfer (though increasing the overvoltage of the Ni^{2+} ion reduction process at the same time). Thus, a decrease of the grain size was observed due to cysteine addition to the bath. Twin faults are special grain boundaries with low energy. Therefore, when a new crystal is nucleated on a surface for which the growth was blocked by sulfur atoms and/or cysteine molecules, the new and old crystals are often separated by coherent twin faults. In addition, the non-coherent grain boundaries may contain dislocations. Thus, both the dislocation density and the twin fault probability increased as a result of cysteine addition (see Figure 4).

It is worth comparing the microstructure developed in the present Ni layers deposited from cysteine containing bath with Ni films processed with other additives, such as saccharin, nickel-chloride or trisodium citrate [8]. Our former study revealed that the smallest grain size and the highest defect density were obtained when saccharin was added to the electrolyte bath [8]. For the present Ni samples processed with 0.3–0.4 g/L cysteine, similarly small grain size (20–27 nm) and twin fault probability (about 3%) were achieved; however the dislocation density was much higher ($\sim$620 $\times$ 10^{14} m^{-2}) than that for the former Ni deposited from the saccharin-containing bath ($\sim$160 $\times$ 10^{14} m^{-2}). In nanocrystalline materials, the majority of dislocations can be found in the grain boundaries [67]. Since the grain size is the same for both Ni films processed with saccharin and cysteine while the dislocation density is much higher for the CYS04 sample, a unit area in the grain boundaries contains more dislocations in the latter specimen. Such dislocations may form in order to relax the stresses caused by the segregation of additives to the grain boundaries. Therefore, the higher dislocation density in the Ni film processed with cysteine suggests that a higher distortion was developed at the grain boundaries during electrodeposition compared to the layer produced from the saccharin-containing bath.

4.2. Influence of Concentration of Cysteine on the Hardness

Figure 6 shows that the hardness increased monotonously with increasing cysteine content of the bath and a saturation was achieved at 0.2 g/L. The enhancement of the hardness due to the addition of 0.1–0.2 g/L cysteine can be attributed to the reduced grain size, the increased defect density and the development of (111) texture. Indeed, former studies have shown that (111) texture yielded a higher hardness than that for other crystallographic directions in fcc metals [68,69]. Although the defect density increased when the cysteine content was raised from 0.2 to 0.3 g/L, further enhancement in the hardness was not observed. Most probably, the defect-induced hardening was compensated by the texture softening, since for the cysteine contents of the 0.3–0.4 g/L texture component (002) became dominant (see Figure 5). A significant difference between the hardness values for 0.3 and 0.4 g/L cysteine contents was not observed, since all microstructural parameters and the texture for these two samples agree within the experimental error.

The hardness of the CYS04 sample (~6800 MPa) is considerably higher than that for any other Ni film deposited formerly with additives [8]. In this previous study, it was found that saccharin addition caused the highest hardness among the Ni layers processed with different organic additives. The present investigation revealed that, for comparable grain sizes, cysteine may cause a hardness even higher than the saccharin additive. The difference between the grain boundary structures of the Ni films processed with cysteine and saccharin might have caused the higher hardness for the former sample. Indeed, the emission of dislocations from grain boundaries decorated with a high amount of additive atoms and/or molecules is difficult during hardness testing since dislocations help to relax the stresses caused by these additives, i.e., there is an attractive interaction between the dislocations and the additives in the boundaries. Therefore, the plastic deformation is difficult, resulting in an elevated hardness for the CYS04 sample.

4.3. *Effect of Cysteine on the Thermal Stability*

Figure 12 shows that the hardness of the Ni film processed with cysteine did not decrease up to 600 K or even increased significantly, by 24% at 500 K. The latter phenomenon has already been observed for Ni and Ni alloy electrodeposits and called anneal-hardening [24,70–78]. This effect was explained by the structural relaxation of grain boundaries (e.g., by annihilation of excess dislocations which are geometrically not necessary) and the segregation of impurities and solute atoms to grain boundaries in alloy samples. These changes in the grain boundaries result in a more difficult emission of dislocations from grain boundaries and a hindered grain boundary sliding, which are the main deformation mechanisms in nanocrystalline materials with larger and smaller grains, respectively. Due to the impeded plastic deformation, hardening occurs which depends on the grain size. The smaller the grain size, the higher the anneal-hardening effect. For Ni with a grain size of about 30 nm, the hardness increased by 15%; however, when the grain size decreased to about 3 nm (for Ni–Mo electrodeposits), the hardness enhancement reached 120% [70]. Moreover, the temperature of annealing, for which the maximum hardness was observed, increased from 500 to 800 K when the grain size decreased from 30 to 3 nm (the duration of heat treatment was 1 h for all samples) [70].

It should be noted that in the former studies, anneal-hardening is accompanied by a slight grain growth, but the hardening caused by the change of grain boundaries compensated the softening due to coarsening. At the same time, in the present case, heating of the sample CYS04 up to 500 K resulted in a decrease of the grain size from ~26 to ~15 nm (see Figure 9). This is an interesting observation, since the annealing of nanocrystalline materials usually results in grain coarsening. However, the following thermodynamic consideration may explain this unusual behaviour. Namely, the segregation of impurities and solute alloying elements (both of them can be called foreign atoms) can decrease the free energy of grain boundaries [79]. For a given foreign atom concentration and temperature, the minimum free energy of the material is achieved at a certain grain size (i.e., for a certain grain boundary area) [80]. If the grain size obtained during processing is larger than this optimal value, there is a thermodynamic driving force for the reduction of the grain size, i.e., for the increase of the amount of grain boundaries. It was found that the grain boundary area can increase by faceting of the existing boundaries without the nucleation of new grains [81]. The faceting of boundaries can be achieved by grain boundary migration, therefore this phenomenon may occur at relatively low temperatures of annealing. If the faceted grain boundaries come in contact with each other, grain size reduction may occur, as it was observed in the present case. Figure 13 shows a model of grain refinement caused by faceting of grain boundaries.

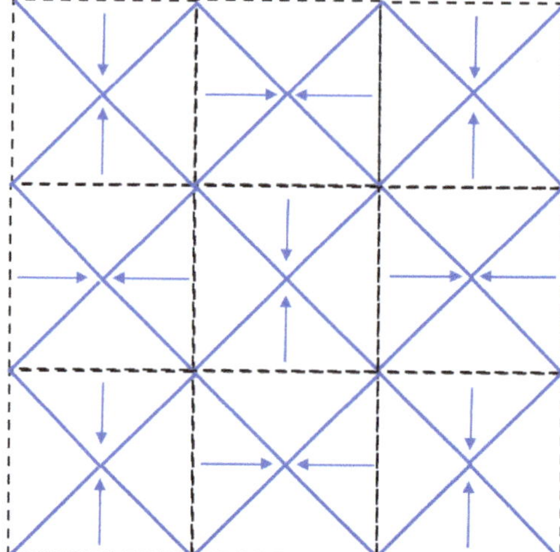

Figure 13. A schematic showing how the grain-boundary faceting can cause the reduction of the grain size. The black dashed lines illustrate the original grain boundaries while the blue solid lines represent the new grain boundaries after faceting. The arrows indicate the movement direction of the boundaries.

The grain refinement at 500 K may be expected to contribute to the hardening, since for nanocrystalline Ni, the Hall–Petch formula is valid down to the grain size of about 14 nm [82]. On the other hand, significant twin fault probability was found in this sample by XLPA and twin faults are similarly strong obstacles against dislocation glide as general grain boundaries. Therefore, when the twin fault spacing is smaller than the grain size, the former value is suggested to be used in the Hall–Petch formula. The twin fault spacing can be obtained from the twin fault probability as $100 \times d_{111}/\beta$, where d111 is the lattice spacing for planes (111). Since for sample CYS04 annealed at 500 K, the twin fault spacing (~9 nm) is lower than the grain size (~15 nm), the decrease of the grain size has no contribution to the increase of the hardness. It is noted that the twin fault spacing slightly increased when film CYS04 was heated up to 500 K.

For revealing the correlation between the microstructure and the hardness, the Hall–Petch plot for the films processed with cysteine is shown in Figure 14. The data obtained on the heat-treated samples are also plotted. In addition, the hardness values obtained on Ni films processed with other additives (e.g., saccharin and trisodium citrate) are also shown. These data were taken from Ref. [9]. It can be seen that the data points obtained for as-deposited films from a cysteine-containing bath are lying only at a slightly higher hardness level than the earlier data reported for as-deposited Ni layers processed with the use of other additives. At the same time, the samples processed with cysteine and heat-treated at 500 K or higher temperatures exhibit considerably higher hardness than that predicted by the Hall–Petch plot obtained for Ni films, either as-deposited or annealed, prepared by using other additives. This observation suggests that the annealing of the specimens deposited from cysteine containing bath increased the strength of the grain boundaries, i.e., the deformation mechanisms (e.g., dislocation emission from the boundaries) became more difficult due to the change of the grain boundary structure, as discussed in the first paragraph of this section. The increase of the dominance of (111) texture during annealing of the samples processed with cysteine might have also contributed to the high hardness of the heat-treated films compared to the Hall–Petch trend determined for other specimens.

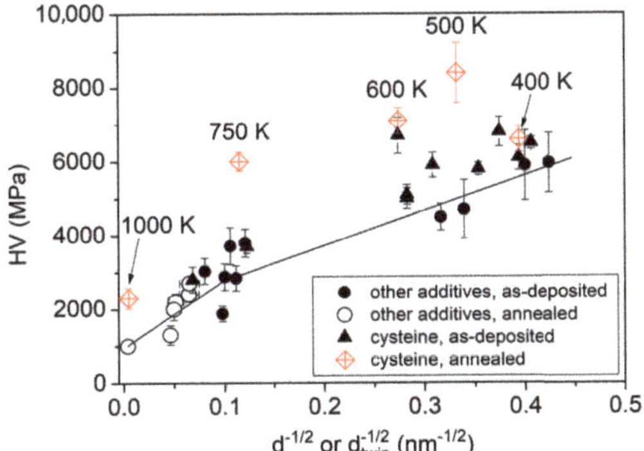

Figure 14. Hall–Petch plot for the films processed with cysteine for both the as-deposited and annealed states. In addition, the hardness values obtained on Ni films processed with other additives (e.g., saccharin and trisodium citrate) are also shown for comparison. from Ref. [9]. The two straight lines for grain sizes smaller and larger than 100 nm were obtained by fitting to the data determined for the Ni films processed with additives other than cysteine (published in [9]).

As demonstrated by Figure 12, the thermal stability of Ni films processed with cysteine is much better than that for other additives. Indeed, the hardness remained as high as ~6000 MPa even at 750 K, while for other additives, the hardness fell below ~3000 MPa at such high annealing temperatures. The higher hardness of the Ni film deposited from the cysteine-containing bath can be explained by the more stable grain size and defect density. Indeed, while for other additives the grain size increased above ~240 nm at 750 K [9], for the present sample, the grain size remained below 100 nm. Similarly, the remaining dislocation density at 750 K was about one order of magnitude higher for CYS04 sample than for the other Ni films deposited with other additives.

5. Conclusions

Nanocrystalline Ni electrodeposits with cysteine additive were successfully processed. The microstructure and the hardness as a function of the cysteine content were investigated. In addition, the thermal stability of the Ni film produced with the highest applied cysteine content was studied. The following conclusions were drawn from the results:

1. The optimal parameters for electrodeposition of nickel from a cysteine-containing bath differ from the ones used for deposition from a conventional nickel-sulfate based bath. A nearly neutral pH with the value of 6.1, a high current density of –25 mA/cm^2 and an increased concentration of boric acid (30 g/L) in the bath are recommended for obtaining nice deposits with cysteine.
2. With increasing the cysteine content up to 0.3 g/L, the density of lattice defects (dislocations and twin faults) increased, while the grain size decreased in the Ni films. In addition, the (220) texture in the additive-free Ni layer changed to (200) texture. When the cysteine content was enhanced from 0.3 to 0.4 g/L, further significant variation in the microstructure was not observed.
3. The Ni film obtained with the addition of 0.4 g/L cysteine exhibited a higher hardness (~6800 MPa) than the values reported for other additives in the literature. The thermal stability of the nanostructured Ni film fabricated with cysteine was exceptional; namely, the hardness of the Ni layer deposited from cysteine-containing bath remained as high as ~6000 MPa, even after heating up to 750 K. The relatively large hardness was retained upon annealing even at 1000 K, while the (111) texture became stronger.
4. When the Ni film processed with cysteine was heated up to 500 K, the grain size decreased from ~27 to ~15 nm and the hardness increased to ~8400 MPa, which is an exceptionally high value

among Ni electrodeposits. This very high hardness might be attributed rather to the structural changes in the grain boundaries and the strengthening of the (111) texture than to the reduction of the grain size.

Author Contributions: Formal analysis, T.K.; electrochemcial and TEM sample preparation, T.K.; investigation, T.K.; resources, L.P., Z.C. and J.G.; writing—original draft preparation, T.K. and J.G.; writing—review and editing, T.K., J.G., Z.C., L.P, and I.B.; visualization, T.K. and J.G.; supervision, J.G. and Z.C.; funding acquisition, J.G. All authors have read and agreed to the published version of the manuscript.

Funding: This work was completed in the ELTE Institutional Excellence Program (TKP2020-IKA-05) financed by the Hungarian Ministry of Human Capacities.

Acknowledgments: Ábel Szabó and Gábor Varga from ELTE Eötvös Loránd University are acknowledged for the assistance in the EBSD measurements.

Conflicts of Interest: The authors declare no conflict of interest.

References

1. Kumar, K.S.; Suresh, S.; Chisholm, M.F.; Horton, J.A.; Wang, P. Deformation of electrodeposited nanocrystalline nickel. *Acta Mater.* **2003**, *51*, 387–405.
2. Torre, F.D.; Swygenhoven, H.V.; Victoria, M. Nanocrystalline electrodeposited Ni: Microstructure and tensile properties. *Acta Mater.* **2002**, *50*, 3957–3970.
3. Budrovic, Z.; Swygenhoven, H.V.; Derlet, P.M.; Petegem, S.V.; Schmitt, B. Plastic Deformation with Reversible Peak Broadening in Nanocrystalline Nickel. *Science* **2004**, *304*, 273–276.
4. Schwaiger, R.; Moser, B.; Dao, M.; Chollacoop, N.; Suresh, S. Some critical experiments on the strain-rate sensitivity of nanocrystalline nickel. *Acta Mater.* **2003**, *51*, 5159–5172.
5. Meyers, M.A.; Mishra, A.; Benson, D.J. Mechanical properties of nanocrystalline materials. *Prog. Mater. Sci.* **2006**, *51*, 427–556.
6. Kumar, K.S.; Swygenhoven, H.V.; Suresh, S. Mechanical behavior of nanocrystalline metals and alloys. *Acta Mater.* **2003**, *51*, 5743–5774.
7. Kolonits, T.; Jenei, P.; Tóth, B.G.; Czigány, Z.; Gubicza, J.; Péter, L.; Bakonyi, I. Characterization of Defect Structure in Electrodeposited Nanocrystalline Ni Films. *J. Electrochem. Soc.* **2016**, *163*, D107–D114, doi:10.1149/2.0911603jes.
8. Kolonits, T.; Jenei, P.; Péter, L.; Bakonyi, I.; Czigány, Z.; Gubicza, J. Effect of bath additives on the microstructure, lattice defect density and hardness of electrodeposited nanocrystalline Ni films. *Surf. Coat. Technol.* **2018**, *349*, 611–621, doi:10.1016/j.surfcoat.2018.06.052.
9. Kolonits, T.; Czigány, Z.; Péter, L.; Bakonyi, I.; Gubicza, J. Influence of Bath Additives on the Thermal Stability of the Nanostructure and Hardness of Ni Films Processed by Electrodeposition. *Coatings* **2019**, *9*, 644, doi:10.3390/coatings9100644.
10. Gertsman, V.Y.; Birringer, R. On the room-temperature grain growth in nanocrystalline copper. *Scr. Metall. Mater.* **1994**, *30*, 577–581.
11. Klement, U.; Erb, U.; El-Sherik, A.M.; Aust, K.T. Thermal stability of nanocrystalline Ni. *Mater. Sci. Eng. A* **1995**, *203*, 177–186.
12. Dake, J.M.; Krill, C.E. Sudden loss of thermal stability in Fe-based nanocrystalline alloys. *Scr. Mater.* **2012**, *66*, 390–393.
13. Ames, M.; Markmann, J.; Karos, R.; Michels, A.; Tschope, A.; Birringer, R. Unraveling the nature of room temperature grain growth in nanocrystalline materials. *Acta Mater.* **2008**, *56*, 4255–4266.
14. Boylan, K.; Ostrander, D.; Erb, U.; Palumbo, G.; Aust, K.T. An in situ TEM study of the thermal stability of nanocrystalline NiP. *Scr. Metall. Mater.* **1991**, *25*, 2711–2716.
15. Chookajorn, T.; Murdoch, H.A.; Schuh, C.A. Design of stable nanocrystalline alloys. *Science* **2012**, *337*, 951–954.
16. Darling, K.A.; VanLeeuwen, B.K.; Semones, J.E.; Koch, C.C.; Scattergood, R.O.; Kecskes, L.J.; Mathaudhu, S.N. Stabilized nanocrystalline iron-based alloys: Guiding efforts in alloy selection. *Mater. Sci. Eng. A* **2011**, *528*, 4365–4371.
17. Choi, P.; da Silva, M.; Klement, U.; Al-Kassab, T.; Kirchheim, R. Thermal stability of electrodeposited nanocrystalline Co–1.1 at.P. *Acta Mater.* **2005**, *53*, 4473–4481.

18. Chookajorn, T.; Schuh, C.A. Nanoscale segregation behavior and high-temperature stability for nanocrystalline W–20 at.% Ti. *Acta Mater.* **2014**, *73*, 128–138.
19. Darling, K.A.; VanLeeuween, B.K.; Koch, C.C.; Scattergood, R.O. Thermal stability of nanocrystalline Fe–Zr alloys. *Mater. Sci. Eng. A* **2010**, *527*, 3572–3580.
20. Burzyńska, L.; Rudnik, E.; Koza, J.; Błaż, L.; Szymański, W. Electrodeposition and heat treatment of nickel/silicon carbide composites. *Surf. Coat. Technol.* **2008**, *202*, 2545–2556.
21. Broszeit, E. Mechanical, thermal and tribological properties of electro- and chemodeposited composite coatings. *Thin Solid Films* **1982**, *95*, 133–142.
22. Gyftou, P.; Stroumbouli, M.; Pavlatou, E.A.; Spyrellis, N. Electrodeposition of Ni/SiC Composites by Pulse Electrolysis. *Trans. IMF* **2002**, *80*, 88–91.
23. Wang, S.C.; Wei, W.C.J. Characterization of electroplated Ni/SiC and Ni/Al_2O_3 composite coatings bearing nanoparticles. *J. Mater. Res.* **2003**, *18*, 1566–1574.
24. Wang, Y.M.; Cheng, S.; Wei, Q.M.; Ma, E.; Nieh, T.G.; Hamza, A. Effects of annealing and impurities on tensile properties of electrodeposited nanocrystalline Ni. *Scr. Mater.* **2004**, *51*, 1023–1028.
25. Heuer, J.K.; Okamoto, P.R.; Lam, N.Q.; Stubbins, J.F. Relationship between segregation-induced intergranular fracture and melting in the nickel–sulfur system. *Appl. Phys. Lett.* **2000**, *76*, 3403–3405.
26. Hibbard, G.D.; McCrea, J.L.; Palumbo, G.; Aust, K.T.; Erb, U. An initial analysis of mechanisms leading to late stage abnormal grain growth in nanocrystalline Ni. *Scr. Mater.* **2002**, *47*, 83–87.
27. Xiao, C.; Mirshams, R.A.; Whang, S.H.; Yin, W.M. Tensile behavior and fracture in nickel and carbon doped nanocrystalline nickel. *Mater. Sci. Eng. A* **2001**, *301*, 35–43.
28. Lücke, K.; Detert, K. A quantitative theory of grain-boundary motion and recrystallization in metals in the presence of impurities. *Acta Metall.* **1957**, *5*, 628–637.
29. Michels, A.; Krill, C.E.; Ehrhardt, H.; Birringer, R.; Wu, D.T. Modeling the influence of grain-size-dependent solute drag on the kinetics of grain growth in nanocrystalline materials. *Acta Mater.* **1999**, *47*, 2143–2152.
30. Hillert, M. Inhibition of grain growth by second-phase particles. *Acta Metall.* **1998**, *36*, 3177–3181.
31. Gottstein, G.; Shvindlerman, L.S. *Grain Boundary Migration in Metals: Thermodynamics, Kinetics, Applications Second Edn*; CRC Press: Boca Raton, FL, USA, 2010.
32. Hondros, E.D.; Seah, M.P.; Hofmann, S.; Lejcek, P. Interfacial and surface microchemistry. In *Physical Metallurgy Fourth Edn*; Elsevier: North-Holland, The Netherlands 1996; pp. 1202–1289.
33. Bicelli, L.; Bozzini, B.; Mele, C.; D'Urzo, L. A Review of Nanostructural Aspects of Metal Electrodeposition. *Int. J. Electrochem. Sci.* **2008**, *3*, 356–408.
34. Kouncheva, M.; Raichevski, G.; Vitkova, S.; Prazak, M. The effect of sulphur and carbon inclusions on the corrosion resistance of electrodeposited Ni-Fe alloy coatings. *Surf. Coat. Technol.* **1987**, *31*, 137–142, doi:10.1016/0257-8972(87)90066-1.
35. Hagenström, H.; Schneeweiss, M.; Kolb, D. Modification of a Au (111) electrode with ethanethiol. 2. Copper electrodeposition. *Langmuir* **1999**, *15*, 7802–7809, doi:10.1021/la9904307.
36. Petri, M.; Kolb, D.M.; Memmert, U.; Meyer, H. Adsorption of mercaptopropionic acid onto Au(1 1 1): Part II. Effect on copper electrodeposition. *Electrochim. Acta* **2003**, *49*, 183–189, doi:10.1016/j.electacta.2003.07.011.
37. Langerock, S.; Ménard, H.; Rowntree, P.; Heerman, L. Electrocrystallization of Rhodium clusters on thiolate-covered polycrystalline gold. *Langmuir* **2005**, *21*, 5124–5133, doi:10.1021/la050078z.
38. Pattanaik, G.; Shao, W.; Swami, N.; Zangari, G. Electrolytic Gold Deposition on Dodecanethiol-Modified Gold Films. *Langmuir* **2009**, *25*, 5031–5038, doi:10.1021/la803907p.
39. Nisanci, F.B.; Demir, U. Size-Controlled Electrochemical Growth of PbS Nanostructures into Electrochemically Patterned Self-Assembled Monolayers. *Langmuir* **2012**, *28*, 8571–8578, doi:10.1021/la301377r.
40. Ross, S.A.; Burrows, C.J. Nickel Complexes of Cysteine- and Cystine-Containing Peptides: Spontaneous Formation of Disulfide-Bridged Dimers at Neutral pH. *Inorg. Chem.* **1998**, *37*, 5358–5363, doi:10.1021/ic971075f.
41. Cox, J.A.; Gray, T.J. Controlled-potential electrolysis of bulk solutions at a modified electrode: Application to oxidations of cysteine, cystine, methionine, and thiocyanate. *Anal. Chem.* **1990**, *62*, 2742–2744.
42. Chen, D.; Giroud, F.; Minteer, S.D. Nickel Cysteine Complexes as Anodic Electrocatalysts for Fuel Cells. *J. Electrochem. Soc.* **2014**, *161*, F933–F939, doi:10.1149/2.0811409jes.
43. Matos, J.; Pereira, L.; Agostinho, S.; Barcia, O.; Cordeiro, G.; D'Elia, E. Effect of cysteine on the anodic dissolution of copper in sulfuric acid medium. *J. Electroanal. Chem.* **2004**, *570*, 91–94, doi:10.1016/j.jelechem.2004.03.020.

44. Silva, F.; Do Lago, D.; D'Elia, E.; Senna, L. Electrodeposition of Cu–Zn alloy coatings from citrate baths containing benzotriazole and cysteine as additives. *J. Appl. Electrochem.* **2010**, *40*, 2013–2022, doi:10.1007/s10800-010-0181-z.
45. El-Deab, M.S.; Arihara, K.; Ohsaka, T. Fabrication of Au(111)-Like Polycrystalline Gold Electrodes and Their Applications to Oxygen Reduction. *J. Electrochem. Soc.* **2004**, *151*, E213–E218, doi:10.1149/1.1723499.
46. Dolati, A.; Imanieh, I.; Salehi, F.; Farahani, M. The effect of cysteine on electrodeposition of gold nanoparticle. *Mater. Sci. Eng. B* **2011**, *176*, 1307–1312, doi:10.1016/j.mseb.2011.07.008.
47. El-Deab, M.S.; Sotomura, T.; Ohsaka, T. Size and Crystallographic Orientation Controls of Gold Nanoparticles Electrodeposited on GC Electrodes. *J. Electrochem. Soc.* **2005**, *152*, C1–C6, doi:10.1149/1.1824041.
48. Lin, T.H.; Lin, C.W.; Liu, H.H.; Sheu, J.T.; Hung, W.H. Potential-controlled electrodeposition of gold dendrites in the presence of cysteine. *Chem. Commun.* **2011**, *47*, 2044–2046, doi:10.1039/C0CC03273E.
49. Ebadi, M.; Basirun, W.J.; Sim, Y.L.; Mahmoudian, M.R. Investigation of Electro-Kinetic Behavior of Cysteine on Electrodeposition of Ni Through the AC and DC Techniques. *Metall. Mater. Trans. A* **2013**, *44*, 5096–5105, doi:10.1007/s11661-013-1881-x.
50. Tóth, B.G.; Péter, L.; Révész, A.; Pádár, J.; Bakonyi, I. Temperature dependence of the electrical resistivity and the anisotropic magnetoresistance (AMR) of electrodeposited Ni-Co alloys. *Eur. Phys. J. B* **2010**, *75*, 167–177.
51. Weihnacht, V.; Péter, L.; Tóth, J.; Pádár, J.; Kerner, Z.; Schneider, C.M.; Bakonyi, I. Giant Magnetoresistance in Co-Cu/Cu Multilayers Prepared by Various Electrodeposition Control Modes. *J. Electrochem. Soc.* **2003**, *150*, C507, doi:10.1149/1.1583716.
52. Ribárik, G.; Gubicza, J.; Ungár, T. Correlation between strength and microstructure of ball-milled Al–Mg alloys determined by X-ray diffraction. *Mater. Sci. Eng. A* **2004**, *387*, 343–347.
53. Balogh, L.; Ribárik, G.; Ungár, T. Stacking faults and twin boundaries in fcc crystals determined by x-ray diffraction profile analysis. *J. Appl. Phys.* **2006**, *100*, 1–10.
54. Barna, A. Topographic kinetics and practice of low angle ion beam thinning. *Mater. Res. Soc. Symp. Proc.* **1992**, *254*, 3–22.
55. Pearson, K. X. On the criterion that a given system of deviations from the probable in the case of a correlated system of variables is such that it can be reasonably supposed to have arisen from random sampling. *Lond. Edinb. Dublin Philos. Mag. J. Sci. Ser. 5* **1900**, *50*, 157–175.
56. Arihara, K.; Ariga, T.; Takashima, N.; Arihara, K.; Okajima, T.; Kitamura, F.; Tokuda, K.; Ohsaka, T. Multiple voltammetric waves for reductive desorption of cysteine and 4-mercaptobenzoic acid monolayers self-assembled on gold substrates. *Phys. Chem. Chem. Phys.* **2003**, *5*, 3758–3761, doi:10.1039/B305867K.
57. Solomon, E.I.; Gorelsky, S.I.; Dey, A. Metal–thiolate bonds in bioinorganic chemistry. *J. Comput. Chem.* **2006**, *27*, 1415–1428, doi:10.1002/jcc.20451.
58. Xue, Y.; Li, X.; Li, H.; Zhang, W. Quantifying thiol–gold interactions towards the efficient strength control. *Nat. Commun.* **2014**, *5*, 4348, doi:10.1038/ncomms5348.
59. Pakiari, A.; Jamshidi, Z. Nature and strength of M- S Bonds (M= Au, Ag, and Cu) in binary alloy gold clusters. *J. Phys. Chem. A* **2010**, *114*, 9212–9221, doi:10.1021/jp100423b.
60. Neurock, M.; van Santen, R.A. Theory of carbon-sulfur bond activation by small metal sulfide particles. *J. Am. Chem. Soc.* **1994**, *116*, 4427–4439, doi:10.1021/ja00089a034.
61. Ulman, A. Formation and Structure of Self-Assembled Monolayers. *Chem. Rev.* **1996**, *96*, 1533–1554, doi:10.1021/cr9502357.
62. Reddy, A. Preferred orientations in nickel electro-deposits: I. The mechanism of development of textures in nickel electro-deposits. *J. Electroanal. Chem. (1959)* **1963**, *6*, 141–152, doi:10.1016/S0022-0728(63)80152-7.
63. Li, D.Y.; Szpunar, J.A. Textural evolution in electrodeposits under the influence of adsorbed foreign species: Part I Textural evolution in iron electrodeposits affected by hydrogen co-deposition. *J. Mater. Sci.* **1997**, *32*, 141–152, doi:10.1023/A:1018612205196.
64. Kristinsdóttir, L.; Skúlason, E. A systematic DFT study of hydrogen diffusion on transition metal surfaces. *Surf. Sci.* **2012**, *606*, 1400–1404, doi:10.1016/j.susc.2012.04.028.
65. Ferrin, P.; Kandoi, S.; Nilekar, A.U.; Mavrikakis, M. Hydrogen adsorption, absorption and diffusion on and in transition metal surfaces: A DFT study. *Surf. Sci.* **2012**, *606*, 679–689, doi:10.1016/j.susc.2011.12.017.
66. Yang, H.; Whitten, J.L. Adsorption of SH and OH and coadsorption of S, O and H on Ni(111). *Surf. Sci.* **1997**, *370*, 136–154, doi:10.1016/S0039-6028(96)00968-5.

67. Gubicza, J. *Defect Structure and Properties of Nanomaterials;* Woodhead Publishing: Sawston, UK, 2017.
68. Clausen, B.; Lorentzen, T.; Leffers, T. Self-consistent modelling of the plastic deformation of f.c.c. polycrystals and its implications for diffraction measurements of internal stresses. *Acta Mater.* **1998**, *46*, 3087–3098, doi:10.1016/S1359-6454(98)00014-7.
69. Feng, L.; Ren, Y.Y.; Zhang, Y.H.; Wang, S.; Li, L. Direct Correlations among the Grain Size, Texture, and Indentation Behavior of Nanocrystalline Nickel Coatings. *Metals* **2019**, *9*, 188, doi:10.3390/met9020188.
70. Hu, J.; Shi, Y.N.; Sauvage, X.; Sha, G.; Lu, K. Grain boundary stability governs hardening and softening in extremely fine nanograined metals. *Science* **2017**, *355*, 1292–1296, doi:10.1126/science.aal5166.
71. Zhang, X.; Fujita, T.; Pan, D.; Yu, J.; Sakurai, T.; Chen, M. Influences of grain size and grain boundary segregation on mechanical behavior of nanocrystalline Ni. *Mater. Sci. Eng. A* **2010**, *527*, 2297–2304, doi:10.1016/j.msea.2009.12.005.
72. Chang, L.; Kao, P.; Chen, C.H. Strengthening mechanisms in electrodeposited Ni–P alloys with nanocrystalline grains. *Scr. Mater.* **2007**, *56*, 713–716, doi:10.1016/j.scriptamat.2006.12.036.
73. Zhang, N.; Jin, S.; Sha, G.; Yu, J.; Cai, X.; Du, C.; Shen, T. Segregation induced hardening in annealed nanocrystalline Ni-Fe alloy. *Mater. Sci. Eng. A* **2018**, *735*, 354–360, doi:10.1016/j.msea.2018.08.061.
74. Ebrahimi, F.; Li, H. The effect of annealing on deformation and fracture of a nanocrystalline fcc metal. *J. Mater. Sci.* **2007**, *42*, 1444–1454, doi:10.1007/s10853-006-0969-8.
75. Li, H.; Jiang, F.; Ni, S.; Li, L.; Sha, G.; Liao, X.; Ringer, S.P.; Choo, H.; Liaw, P.K.; Misra, A. Mechanical behaviors of as-deposited and annealed nanostructured Ni–Fe alloys. *Scr. Mater.* **2011**, *65*, 1–4, doi:10.1016/j.scriptamat.2011.03.029.
76. Haj-Taieb, M.; Haseeb, A.S.M.A.; Caulfield, J.; Bade, K.; Aktaa, J.; Hemker, K.J. Thermal stability of electrodeposited LIGA Ni-W alloys for high temperature MEMS applications. *Microsyst. Technol.* **2008**, *14*, 1531–1536, doi:10.1007/s00542-007-0536-5.
77. Rupert, T.J.; Trelewicz, J.R.; Schuh, C.A. Grain boundary relaxation strengthening of nanocrystalline Ni–W alloys. *J. Mater. Res.* **2012**, *27*, 1285–1294, doi:10.1557/jmr.2012.55.
78. Gubicza, J. Annealing-Induced Hardening in Ultrafine-Grained and Nanocrystalline Materials. *Adv. Eng. Mater.* **2020**, *22*, 1900507, doi:10.1002/adem.201900507.
79. Vo, N.; Schäfer, J.; Averback, R.; Albe, K.; Ashkenazy, Y.; Bellon, P. Reaching theoretical strengths in nanocrystalline Cu by grain boundary doping. *Scr. Mater.* **2011**, *65*, 660–663, doi:10.1016/j.scriptamat.2011.06.048.
80. Weissmüller, J. Alloy effects in nanostructures. *Nanostruct. Mater.* **1993**, *3*, 261–272.
81. Donald, A.; Brown, L. Grain boundary faceting in Cu-Bi alloys. *Acta Metallurgica* **1979**, *27*, 59–66, doi:10.1016/0001-6160(79)90056-7.
82. Nieh, T.; Wang, J. Hall–Petch relationship in nanocrystalline Ni and Be–B alloys. *Intermetallics* **2005**, *13*, 377–385, doi:10.1016/j.intermet.2004.07.029.

Article

Characterization of Enamel and Dentine about a White Spot Lesion: Mechanical Properties, Mineral Density, Microstructure and Molecular Composition

Evgeniy Sadyrin [1,*], Michael Swain [1,2], Boris Mitrin [1], Igor Rzhepakovsky [3], Andrey Nikolaev [1], Vladimir Irkha [1,4], Diana Yogina [5], Nikolay Lyanguzov [6], Stanislav Maksyukov [5] and Sergei Aizikovich [1]

1 Research and Education Center "Materials", Don State Technical University, Gagarin Square 1, 344000 Rostov-on-Don, Russia; michael.swain@sydney.edu.au (M.S.); boris.mitrin@gmail.com (B.M.); andreynicolaev@eurosites.ru (A.N.); irkha.vladimir@gmail.com (V.I.); saizikovich@gmail.com (S.A.)

2 Biomaterials and Bioengineering department, Faculty of Dentistry, The University of Sydney, Camperdown, Sydney NSW 2006, Australia

3 Institute of Life Sciences, North Caucasus Federal University, Pushkin Street 1, 355009 Stavropol, Russia; 78igorr@mail.ru

4 Federal Research Centre The Southern Scientific Centre of the Russian Academy of The Sciences, Chehova Street 41, 344006 Rostov-on-Don, Russia

5 Department of dentistry, Rostov State Medical University, Nakhichevansky Lane 29, 344022 Rostov-on-Don, Russia; dianaturbina@mail.ru (D.Y.); kafstom2.rostgmu@yandex.ru (S.M.)

6 Faculty of Physics, Southern Federal University, Bolshaya Sadovaya Street 105/42, 344090 Rostov-on-Don, Russia; n.lianguzov@mail.ru

* Correspondence: e.sadyrin@sci.donstu.ru; Tel.: +7-900-133-83-66

Received: 20 August 2020; Accepted: 17 September 2020; Published: 21 September 2020

Abstract: The study focuses on in vitro tracing of some fundamental changes that emerge in teeth at the initial stage of caries development using multiple approaches. The research was conducted on a mostly sound maxillary molar tooth but with a clearly visible natural proximal white spot lesion (WSL). Values of mineral density, reduced Young's modulus, indentation hardness and creep as well as the molecular composition and surface microstructure of the WSL and bordering dentine area were studied. The results obtained were compared to those of sound enamel and dentine on the same tooth. A decrease of mechanical properties and mineral density both for the WSL and bordering dentine was detected in comparison to the sound counterparts, as well as increase of creep for the enamel WSL. Differences in molecular composition and surface microstructure (including the indenter impressions) were found and described. WSL induces a serious change in the state of not only the visually affected enamel but also surrounding visually intact enamel and dentine in its vicinity. The results provide the basis for future studies of efficacy of minimal invasive treatments of caries.

Keywords: caries; enamel; dentine; nanoindentation; X-Ray microtomography; Raman spectrum analysis

---

1. Introduction

The study of caries from a materials and microstructural point of view, in addition to conventional bitewing X-ray techniques [1,2], provides a better understanding of the changes in the carious enamel and affected dentine. Various mechanical and spectroscopic approaches can be applied in this regard, each with its benefits and drawbacks. Hardness measurements have come a long way from the pioneering studies of over 50 years ago [3] to modern day research involving instrumented estimation of both indentation hardness and reduced Young's modulus at the sub-micron level [4,5].

Computed X-Ray microtomography (micro-CT) is another powerful tool to study the demineralization and remineralization of teeth [6,7], it can also be applied for testing efficiency of caries treatments [8]. Raman spectroscopy enables researchers to study accurately the molecular composition of carious lesions [9]. At the same time, fluorescence spectra due to organic materials may tend to dominate the much weaker Raman signals, therefore, Raman spectroscopic studies are often limited to enamel, which contains only a small fraction of organic components [10]. Atomic force microscopy (AFM) enables quantification of the surface roughness parameters for the carious lesions thus representing another method for testing of different clinical treatments such as bleaching agents [11] or detailed observation of the enamel demineralization process [12]. This tool is mostly suitable for localized regions of interest. Scanning electron microscopy (SEM), although unsuitable to measure the parameters of the surface microgeometry, can be used to observe relatively large areas with serious height changes, such as borders of the lesions. However, additional damage to the sample can occur from exposure to vacuum or to the electron beam [13]. At the same time, the combination of different techniques provides the possibility to overcome limitations and helps to understand the processes occurring inside carious tissues [14,15].

The first clinically visible stage of the carious disease is characterized by enamel demineralization without cavitation. Such diseased enamel usually presents a near intact surface layer of 10–100 μm thickness with a subsurface porous area called the body of the lesion. The pores are formed as a consequence of partial dissolution of carbonated hydroxyapatite crystallites due to etching caused by carbohydrate metabolizing cariogenic biofilm bacteria producing organic acids [16]. Due to the significant difference of the refractive indices of the medium inside the acid-created pores of the demineralization area, a whitish opaque appearance of these lesions can be observed. This phenomenon is called a white spot lesion (WSL) [17,18]. Overall, the number of studies characterizing the fundamental changes that emerge in natural WSLs from a materials and microstructural point of view remains rather low. Huang et al. in [19] using nanoindentation and micro-CT showed excellent correlation between mineral density and elastic modulus for the enamel component of WSLs. Ko et al. [20] and Kinoshita et al. [21] successfully provided Raman spectral imaging characterization for early caries. Mapping of the mechanical properties for natural WSLs supplemented by SEM observations was conducted by Huang et al. in [22]. Metwally et al. used in vivo radiographic tracing of structure to show the stages of the remineralization process for young permanent teeth with WSLs in [23].

The present work aims to characterize the complex of properties (mineral density, reduced Young's modulus, indentation hardness, average roughness, maximum height of roughness) and features (surface structure, molecular composition, indentation creep) of the enamel and dentine about a WSL on a single tooth using all the techniques mentioned in this section. The aim of obtaining such information is to provide the basis for future studies of efficacy of minimal invasive treatments of caries [24–28] and deeper analysis of these properties and features.

2. Materials and Methods

An extracted permanent maxillary molar was collected for orthodontic purposes from an individual (male, 21 years old) in the dental department of Rostov State Medical University clinic. Local independent ethics committee of Rostov State Medical University approved the study (statement 15/9 from 3 October 2019), the patient provided informed consent. The WSL was assessed independently by two experienced clinicians. It was located in the proximal contact zone, visually white in color with no obvious surface damage. Following extraction, the sample was kept in 1 wt. % NaClO solution for 10 min. Then the sample was stored in Hanks Balanced Salt Solution (HBSS) at 4 °C with thymol granules (Unifarm, Slavyansk-na-Kubani, Krasnodar Region, Russia), added to prevent fungal growth and disinfection. The ratio of thymol to HBSS was 1:1000.

According to principal carious lesion classification criteria introduced in [29], the carious WSL under consideration represents a primary active incipient enamel caries lesion located on the smooth-surface of an adult patient. According to FDI World Dental Federation caries matrix [30] there are specific clinical reporting guidelines: I—sound, no obvious dentine caries, a—noncavitated enamel (+), 1—first visual change in enamel (+), where sign (+) indicates the activity of caries lesions as defined in the glossary of terms for caries [31].

We investigated four oval areas (approximately 1.4 mm × 1 mm) of the tooth:

1. natural enamel WSL;
2. dentine bordering the WSL (touching the dentine–enamel junction as close to the WSL as possible);
3. area of sound enamel on the opposite medial side of the tooth;
4. dentine bordering the area of sound enamel (touching the dentine–enamel junction and as close to the area of sound enamel of area 3 as possible).

We concentrated on the different areas of the single sample to avoid variations of experimental data attributed to noncarious factors such as age [32] and influence of the environment [33,34].

A longitudinal section through the region containing a WSL was cut using a precision saw (Isomet 4000, Buehler, Lake Bluff, IL, USA) with an abrasive SiC disc (MetAbrase, Buehler, Lake Bluff, IL, USA). The pulp chamber was cleaned of remnants of soft tissues. The cut surface of the sample close to the WSL was carefully ground using SiC-based abrasive papers of the following grit sizes: P800, P1200 (Siawat 1913, Sia Abrasives, Frauenfeld, Switzerland), P2000, P2500 (Smirdex, Lefki–Xanthi, Greece). Running water was used as lubricant. During grinding the sample was thinned in such a manner that the plane of interest crossed the internal portion of the WSL. Following grinding, polishing was conducted with diamond oil-based suspensions (MetaDi, Buehler, Lake Bluff, IL, USA) with particles of 6 and 1 μm diameters. The lubricants GreenLube (Allied, Rancho Dominguez, CA, USA) and BlueLube (Benchmade, Oregon City, OR, USA) were used respectively. The final polishing was made using a sol-gel suspension (MasterPrep, Buehler, Lake Bluff, IL, USA) based on alumina 0.05 μm diameter particles with distilled water as a lubricant. All suspensions were applied with a soft, synthetic, woven, no-nap cloth (Trident, Buehler, Lake Bluff, IL, USA). Each grinding and polishing step was followed by ultrasonic cleaning of the sample in distilled water (Sonorex RK 31, Bandelin, Berlin, Germany) for 3 min.

Using an optical Greenough stereomicroscope (Stemi 305, Zeiss, Shanghai, China) with a colour video camera (Axiocam 105, Zeiss, Oberkochen, Germany) in reflected light images of the tooth crown, including the four areas under study, were made.

The sample was investigated using a wall mounted dental X-ray device (Xelium Ultra, Swidella International Group Limited, Guangdong, China) in order to qualitatively assess the mineral density variation in the WSL area. The voltage on the X-ray tube was 70 kV ± 10%, current 8 mA ± 20%. The nominal focal spot diameter was 0.8 mm. During the experiment, three images were taken with different exposure times: 0.3 s, 0.4 s, 0.5 s.

Determination of mineral density of the sample was conducted using a micro-CT device (SkyScan 1176, Bruker, Kontich, Belgium). A tooth slice in an Eppendorf-type test tube and calibration phantoms in cylindrical tubes (both filled with distilled water) were placed on the semicylindrical bed adapter. Calibration phantoms were represented by a pair of rods composed of epoxy resin with embedded fine calcium hydroxyapatite powder at concentrations of 0.25 and 0.75 g/cm^3. These phantoms were used for constructing a linear relationship between grey level and mineral density, that allowed conversion of the measured grey value of the tooth slice to an estimated mineral density [35,36]. The micro-CT research was divided into two parts: the first one was conducted using Cu+Al filter and the second one with the Al filter. Within each of the parts, all four tooth areas under study were examined. For the first micro-CT study, a Cu (thickness 0.04 mm) +Al (thickness 0.5 mm) filter of the X-ray beam was used. The following scanning parameters were applied: X-ray tube voltage 80 kV, current 300 μA, pixel size 8.87 μm, sample rotation at each step 0.3°, X-ray tube rotation of 180°, exposure time 1.275 s, ring artefacts correction and Gaussian filter were applied. In the process of scanning, 657 projections for both the tooth slice and the phantoms were obtained using specialized software (CTvox, Bruker, Kontich, Belgium). For the second micro-CT study, a Cu filter (thickness 0.1 mm) of the X-ray beam was used. The following scanning parameters were applied: X-ray tube voltage 90 kV, current 270 μA, pixel size 8.87 μm, sample rotation at each step 0.3°, X-ray tube rotation on 180°, exposure time 1.45 s, ring artefacts correction and Gaussian filter were applied. We generated 657 projections for both the tooth slice and the phantoms.

For the same four tooth areas the mechanical properties were evaluated using a nanoindentation device (NanoTest 600 Platform 3, Micro Materials, Wrexham, UK). The experiments were carried out at a constant temperature of 27.0 ± 0.1 °C in a closed chamber, a calibrated diamond Berkovich indenter was used. The following load profile was applied: the load linearly increased for 20 s, held constant for 30 s, then linearly decreased for 20 s. The thermal drift was recorded and corrected using the device software. The typical and maximum values of the calculated thermal drift rate were 0.4 nm/s and 2 nm/s, respectively. The maximum load P_{max} for all the experiments was 50 mN. To prevent shrinkage of the tooth structure the sample was maintained wet with saline droplets using a syringe pump (Terufusion TE–332, Terumo, Leuven, Belgium). The droplets were applied on the sample between indentations in order to prevent the influence of the droplet impact on the indenter. Values of the reduced Young's modulus E_r and indentation hardness H for each of the areas were obtained using the Oliver–Pharr method [37]. During the entire experiment, the indenter displacement h was recorded while the load P was applied. The maximum value of the displacement (attained at P_{max}) was denoted as h_{max}. Thus, the output of each experiment was a load–displacement curve. Holding load at its maximum value for a time before unloading eliminated the effect of any creep response on the unloading "branch". The unloading "branch" was approximated by the function

$$P = a(h - h_r)^m, \tag{1}$$

where a and m are fitting parameters, h_r is the residual depth of the imprint, i.e., the amount of displacement of the indenter at which the material ceased to resist during unloading. Using this approximation, the derivative called indentation stiffness S was determined at h_{max}:

$$S = \frac{dP}{dh}(h_{max}), \tag{2}$$

To evaluate the reduced Young's modulus the following formula was used:

$$E_r = \frac{S\sqrt{\pi}}{2\beta\sqrt{A_c}}, \tag{3}$$

where A_c is the projected contact area (the projection of the contact surface on a plane orthogonal to the axis of indentation), β is a correction factor. For triangular contact area it is considered to be 1.034 [38]. Indentation hardness was then calculated using the following formula:

$$H = \frac{P_{max}}{A_c}, \tag{4}$$

The nanoindentation research was divided into two parts. In the first part for each tooth area under study 12 identical indentations were performed with 100 μm column and row offset (a four-by-three position matrix was used) and the results were averaged. In the second part a 10 × 20 matrix of identical indentations were made with the same offset in the area, entirely covering the WSL, the enamel around it, the adjacent dentine–enamel junction and associated dentine bordering the enamel WSL region. This matrix was used for the construction of the reduced Young's modulus and indentation hardness maps.

For the surface topography research of the four areas under study an AFM (Nano Compact, Phywe, Göttingen, Germany) was used. The surface scanning was conducted in dynamic mode. The device was equipped with monocrystalline Si probe with an Al coating, a resonance frequency of 190 ± 60 kHz and cantilever stiffness of 48 N/m. Scanning velocity was 0.3 ms per line. The resolution along the axis was 1.1 nm. The quadratic mean noise level in the dynamic mode across the z axis (height) was 0.5 nm.

The Raman spectra were measured using a He-Ne laser (wavelength of the laser excitation was 633 nm) in the Raman spectrometer (inVia Reflex, Renishaw, Wotton-under-Edge, UK) with an Edge filter. The backscattering scheme with a Leica optical microscope was used (the resolution of the spectra was less than 0.5 cm^{-1}). The laser beam diameter on the sample was of 1–2 μm. The Raman spectra measured were corrected using the Bose–Einstein temperature factor.

For the final part of the research a Crossbeam 340 (Zeiss, Oberkochen, Germany) SEM was used. Prior to the study the tooth was ultrasonically cleaned for 7 min and dehydrated according to the protocol proposed by Bertassoni and Swain [39]. This protocol included immersing the tooth slice in reagent grade acetone solutions of 25, 50, and 70% (*v/v*) for 5 min each, followed by 80%, 90%, 95%, and 100% (*v/v*) for 15 min each, and finally by two consecutive immersions in 100% (*v/v*) acetone for 30 min each. After that, the tooth slice was immersed in hexamethyldisilazane (HMDS) overnight which was permitted to evaporate under a gas ventilation hood. As a volatile organosilicon compound HMDS enables further dehydratation of the organic matrix of the tooth. Additionally, after the HMDS treatment the sample was held in the vacuum chamber for ~ 1.5 h at the pressure 2×10^{-4} mbar. The research was conducted using the Everhart–Thornley secondary electron detector with an acceleration voltage of 1 and 2 kV.

3. Results

3.1. Optical Observations

Figure 1 shows the cross-section of the tooth crown through the WSL along with the surrounding enamel, dentine–enamel junction and dentine.

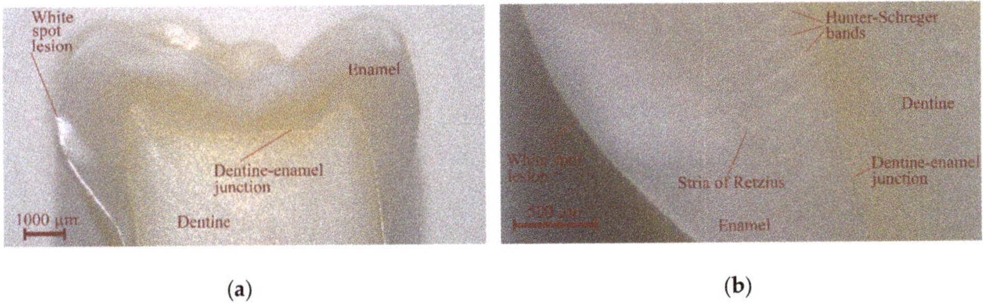

(a) (b)

Figure 1. Optical microscopy of the sample after final polishing: (**a**) the crown overview; (**b**) the white spot lesion (WSL) area.

The Hunter–Schreger bands as well as the stria of Retzius are clearly visible, which testifies to the quality of the sample polishing. The images clearly show the absence of cracks on the surface of the sample, which indicates the carious origin of the lesion (not the consequence of pressing of a dental instrument onto the surface of the tooth during extraction). The borders of the WSL inside the enamel are reasonably sharply defined. The dentine bordering the WSL and the sound dentine area on the opposite medial side of the tooth are visually indistinguishable from each other.

3.2. Bitewing X-ray

None of the obtained images demonstrated an observable decrease in the mineral density of the WSL nor in the bordering dentine area. The typical image obtained with the exposure time 0.5 s is shown in Figure 2a (the WSL area is marked with the burgundy dotted line).

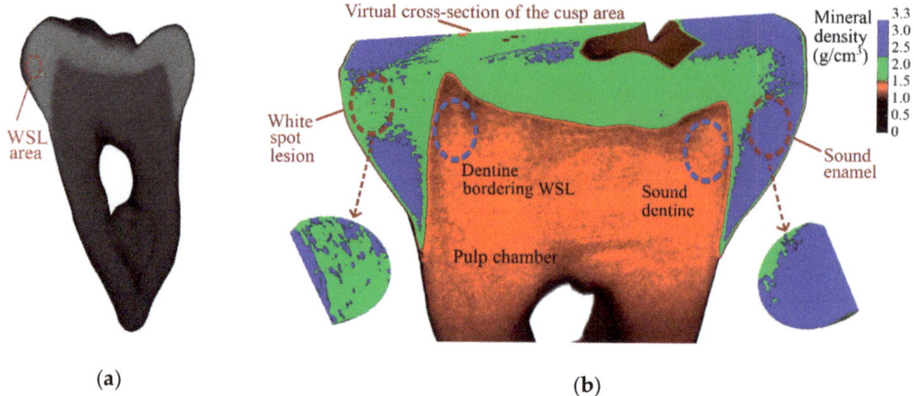

Figure 2. Mineral density research of the sample: (**a**) bitewing X-Ray image (exposure 0.5 s); (**b**) micro-CT (Cu + Al filter) with virtual tablets for the demineralized area of WSL and the sound enamel. Each tablet is cross-sectioned for better visualization of the mineral density by depth. Dentine bordering the WSL and sound dentine is marked with blue dashed ovals.

3.3. Micro-CT

Using software (Nrecon, Bruker, Kontich, Belgium), the projections taken with both Al + Cu and Cu filters were reconstructed into sets of virtual tooth sections. Usage of the calibration phantoms and determining the X-ray attenuation coefficient according to the Bruker recommendations for a mineralized material (such as bone, dentine, and enamel) made it possible to construct a map of the mineral density across the sets of the virtual tooth sections (Figure 2b).

For the WSL area a cylindrical shape region of surface was chosen in the software (CTan, Bruker, Kontich, Belgium) in such a way to cover the main demineralization region in width and length without affecting the bordering enamel. After that a number of projections, where signs of demineralization were visualised, were chosen in such way to cover the main demineralization region in depth without affecting the underlying sound enamel. Thus, a virtual tablet of the WSL area was formed (Figure 2b, left bottom). The same tablets were then constructed for the other three areas of interest for both filters. Results of the measurements of mineral density are summarized in Table 1.

Table 1. Mineral density of the tooth areas.

Filter	Group	Area	Mineral Density, g/cm^3	Standard Deviation, g/cm^3
Al + Cu	WSL	Enamel	2.21	0.06
		Dentine	1.21	0.06
	Sound	Enamel	2.33	0.08
		Dentine	1.21	0.03
Cu	WSL	Enamel	2.35	0.07
		Dentine	1.29	0.05
	Sound	Enamel	2.47	0.08
		Dentine	1.31	0.06

3.4. Nanoindentation

Figure 3 shows the load–displacement curves for each of the tooth areas under study obtained in the first part of the nanoindentation experiments. Each curve contains loading and unloading "branches", and a horizontal segment produced during the holding period at maximum load. For both carious areas the maximum value of the displacement h_{max} is greater than for the sound counterparts.

The character of the load–displacement curves for enamel WSL and sound enamel areas are similar. Comparing the curves for the dentine areas, we observed an obvious shape change of the unloading "branch" for the dentine bordering the WSL. In addition, indentation creep while holding P_{max} constant was recorded for each curve. Determined values of mechanical properties and indentation creep (average and standard deviation) are summarized in Table 2.

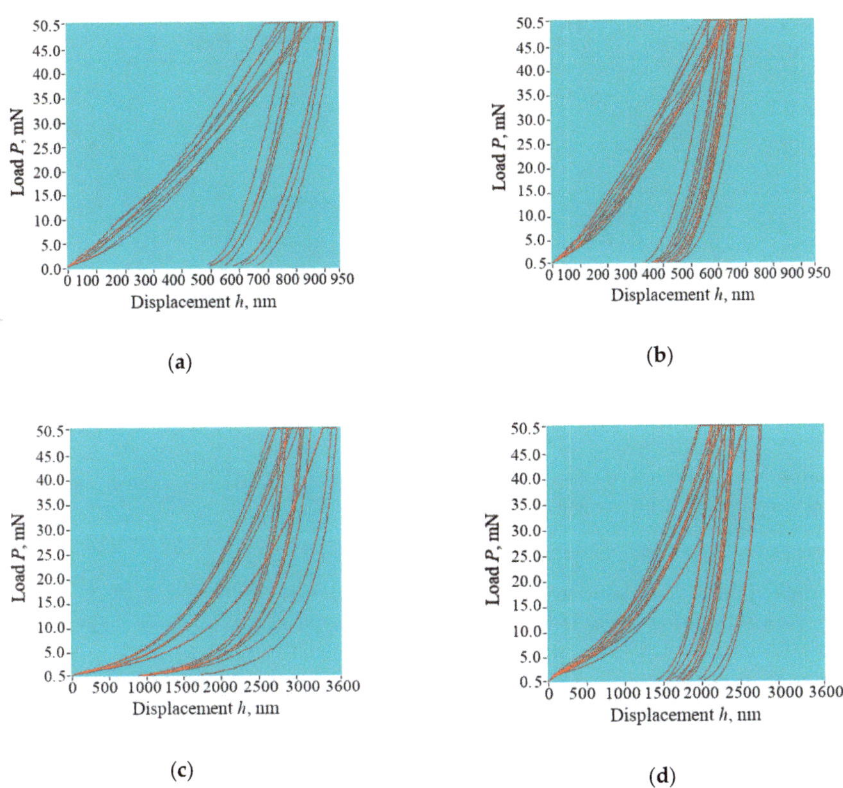

Figure 3. Load-displacement curves for the tooth areas: (**a**) WSL; (**b**) area of sound enamel on the opposite medial side of the tooth; (**c**) dentine bordering the WSL; (**d**) dentine bordering the area of sound enamel. Note the very dissimilar loading and unloading curves for the dentine bordering the WSL in (**c**).

Table 2. Mechanical properties of the tooth areas.

Group	Area	Reduced Young's Modulus E_r, GPa	Indentation Hardness H, GPa	Indentation Creep, nm
WSL	Enamel	69.12 ± 4.97	2.79 ± 0.46	64.60 ± 18.00
	Dentine	6.04 ± 0.78	0.22 ± 0.04	155.22 ± 23.24
Sound	Enamel	111.57 ± 8.95	4.85 ± 0.62	38.16 ± 9.92
	Dentine	13.41 ± 1.55	0.34 ± 0.06	175.96 ± 41.20

Figure 4 demonstrates the results of the second part of the nanoindentation experiments: the maps showing distribution of the reduced Young's modulus and indentation hardness in the area partly covering the WSL and extending to the dentine bordering the WSL. It is seen that the properties of the enamel bordering the WSL are reduced despite its appearance as being sound based on optical images.

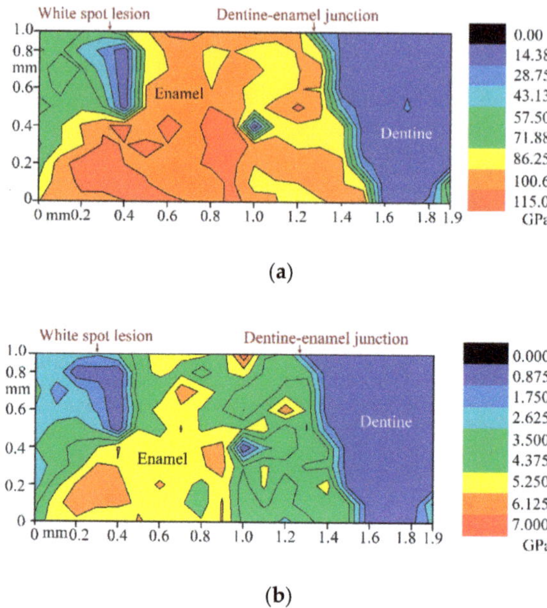

Figure 4. Maps of the mechanical characteristics of WSL and surrounding tooth areas: (**a**) reduced Young's modulus; (**b**) indentation hardness.

3.5. Atomic Force Microscopy

For each of the four areas under study a series of AFM images were taken. Figure 5 shows one image for each of the areas. These images were chosen as they represent the most common features at the nanoscale that we observed among each of the areas. For each image, a 0.8 × 0.8 μm scanning field was chosen for the convenience of visual comparison of the topographies of the areas under study. For comparing the areas affected by caries with their sound counterparts the average surface roughness R_a for each of the images was measured using software (Gwyddion, Czech Metrology Institute, Brno, Czech Republic). Due to the irregularities observed on the visualized surface we considered measurement of the roughness based on a single direction as not being completely indicative.

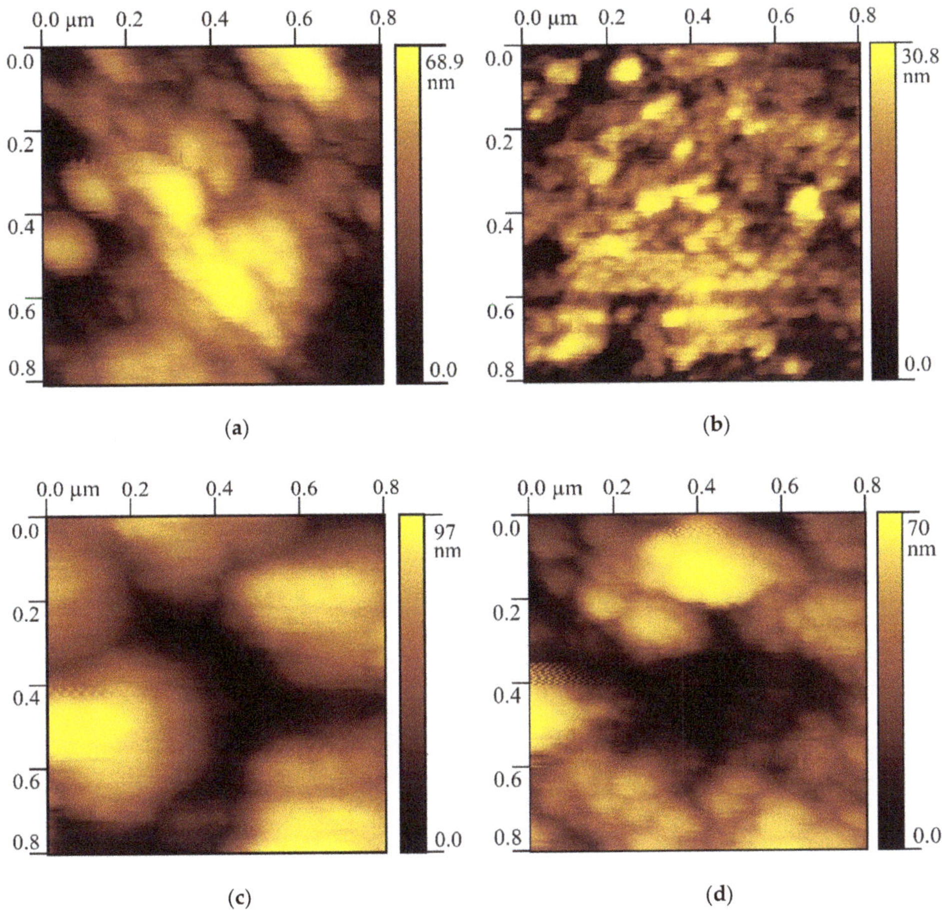

Figure 5. Surface topography of the tooth areas: (**a**) natural enamel WSL; (**b**) area of sound enamel on the opposite medial side of the tooth; (**c**) dentine bordering the WSL; (**d**) dentine bordering the area of sound enamel.

Thus, average roughness R_a was measured in three directions: horizontal, vertical and diagonal. In each of the directions five profiles were constructed. After that the average value (of the 15 profiles) with the standard deviation was calculated. Maximum roughness height was denoted as R_t. The measurement results for the images in Figure 5 are presented in Table 3.

Table 3. Surface roughness of the tooth areas.

Tooth Area	Average R_a by 15 Profiles, nm	Standard Deviation, nm	R_t, nm
Natural enamel WSL	8.9	3.6	68.9
Sound enamel	2.8	1.0	30.8
Dentine bordering WSL	14.2	4.2	97.0
Sound dentine	7.3	3.9	70.0

According to the Student's *t*-test for dependent (paired) samples the difference between the average R_a values measured for carious and sound pairs of enamel and dentine was found to be significant.

The value of *t*-test was significantly higher than the threshold chosen for statistical significance $\alpha = 0.05$ and 14 degrees of freedom (2.145). In the pair of enamel WSL and sound enamel the *t*-test value was 7.889. In the pair of dentine bordering WSL and its counterpart, the *t*-test value was 3.884.

3.6. Raman Spectroscopy

Typical Raman spectra of human sound and carious enamel obtained upon excitation with a 633 nm laser beam are shown in Figure 6a. The rising background visible in the spectra appeared due to fluorescence emission. The Raman bands of enamel observed were related to the phosphate (PO_4^{3-}) and carbonate (CO_3^{2-}) groups, which play the leading role in the structure of hydroxyapatite crystallites. The most intense band was found at 959 cm^{-1} (ν_1 PO_4^{3-}), it is marked with the blue dotted rectangular on Figure 6a. Other bands associated with the ν_2 and ν_4 PO_4^{3-} vibrations were detected in regions from 390 cm^{-1} to 490 cm^{-1} and from 560 cm^{-1} to 625 cm^{-1}, respectively. The bands corresponding to ν_3 PO_4^{3-} were visible in the region from 1010 cm^{-1} to 1060 cm^{-1}, whereas the band assigned to CO_3^{2-} groups is detected near 1071 cm^{-1}. The shift of the most intense band as well as its full width at half maximum (FWHM) were calculated for the most intense ν_1 PO_4^{3-} band. For showing any band shift, the Raman bands are presented in Figure 6b, which was normalized and the background was subtracted manually from each raw spectrum using a polynomial curve. The values of the ν_1 PO_4^{3-} band in case of sound enamel was 959.5 cm^{-1} similar to observations of Ko et al. [20]. This value shifted to 960.5 cm^{-1} for the WSL area. FWHM of sound enamel was 10.85 cm^{-1}. This value reduced to 10.59 cm^{-1} for the WSL area. Besides, on the Raman spectra of WSL we observed a small band at 1295 cm^{-1}, not visible for the sound enamel (marked with the green dotted rectangular on Figure 6a).

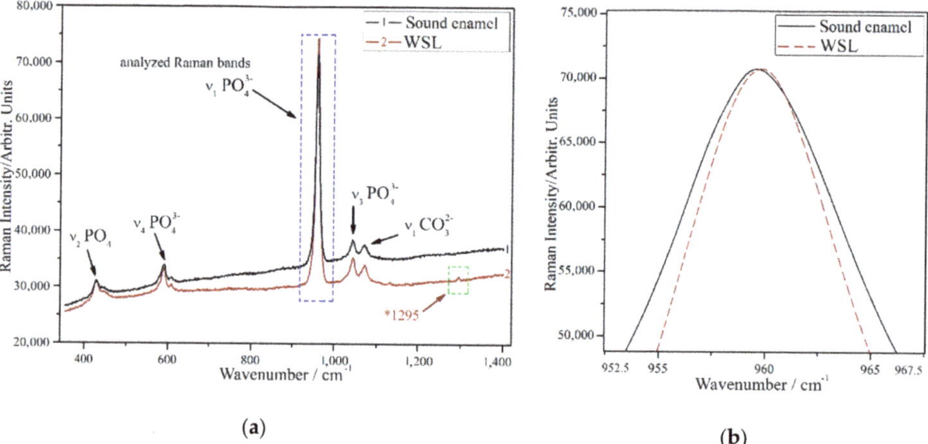

Figure 6. Results of the Raman spectroscopy research: (**a**) spectra sound enamel and WSL (average of four spectra each); (**b**) the shift of the most intense band position due to the carious process.

3.7. Scanning Electron Microscopy

Figure 7 demonstrates the indentation impressions as well as the polished surface nearby for the tooth areas under study. For each area one impression was chosen for the SEM research. The WSL area (Figure 7a) revealed a loss of structure about the interrod areas suggesting that demineralization had weakened this region much more than in the core of the rod structure (the borders of the enamel rods can be clearly distinguished on the figure). The core of the rod structure was weakened as well, although not so severely. The sound enamel shows minimal relief at the rod interface, the surface is smooth and covered with a thin smear layer [40] (Figure 7b). Visual observation of the impressions in dentine (Figure 7c,d) suggests that severe dehydration has occurred in the SEM because of the low vacuum.

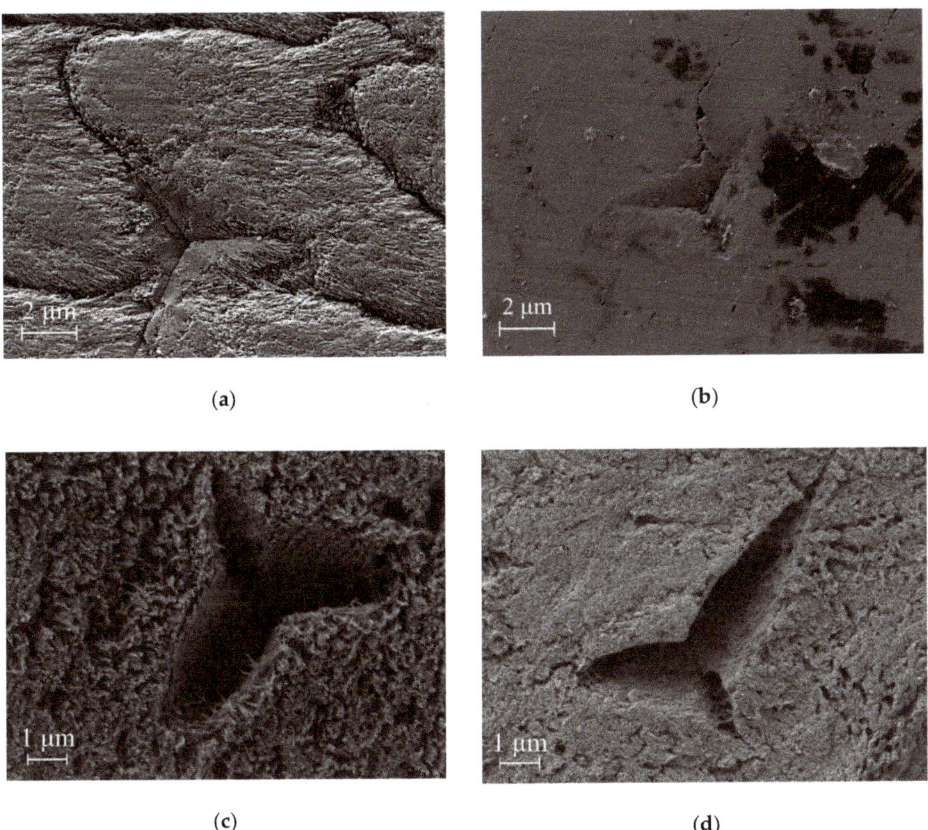

Figure 7. Surface and indenter impressions on the various tooth areas: (**a**) WSL, (**b**) area of sound enamel on the opposite medial side of the tooth, (**c**) dentine bordering the WSL, (**d**) dentine bordering the area of sound enamel. Note that for the sound areas study we rotated the sample 180°.

4. Discussion

Each of the experimental methods in the present work reveals differences between the carious and sound areas of the tooth. Comparing the mineral density derived from micro-CT for these areas, we observed the following: using Al+Cu filter, the WSL showed a small reduction in mineral density compared to sound enamel counterpart (up to 5.2% lower), while dentine areas demonstrated the same values of mineral density. In the case of a Cu filter, the enamel WSL showed practically the same level of reduction in mineral density (up to 4.9% lower), dentine bordering WSL showed also a small reduction in mineral density compared to sound dentine counterpart (up to 1.5% lower). The difference in enamel density is relatively small compared to those found by Huang et al. in [19], where the median value of the relative difference in mineral density reached 28% (2.9 g/cm^3 in sound area and 2.1 g/cm^3 in WSL). It may indicate that the WSL in the studied sample was at an earlier stage of its development.

Reduction in mechanical properties of WSL enamel was found and compared to its sound counterpart: E_r was up to 38.1% lower and H up to 42.5% lower. Values of elastic modulus for sound enamel fitted in the range 47–120 GPa found in literature [22,41–45]. Compared to the findings of

Huang et al. [19,22], the values of elastic modulus for the WSL were greater, which also may be due to the earlier stage of the WSL development. During caries development, the crystallites become partially demineralized generally at the cores and some interfaces as observed during transmission electron microscopic study [46], where the central dark line inside a crystallite represented a stacking fault resembling a dislocation and residual stresses were present about the core. This line possesses a higher concentration of Mg and Na [16] and crystallographic point defects (impurities, vacancies etc.) that would contribute to it being more soluble resulting in greater porosity. It is thus not surprising in this regard that we observed an increase of creep for the enamel WSL area during nanoindentation experiments (up to 69.3% higher, Table 2). Mapping of the mechanical properties revealed a reduction for the enamel adjacent to the WSL (although not as severe as within a WSL). This fact can be important for the selection of area for early caries treatment by a clinician. It should be noted that both indentation hardness and reduced Young's modulus of the inner region of WSL located in the proximity of the dentine–enamel junction are close to those for the dentine bordering the WSL.

Dentine bordering a WSL also demonstrated a decrease in the mechanical properties: E_r was up to 55.0% lower and H up to 35.3% lower (despite the fact that the reduction in mineral density of dentine bordering the WSL was barely visible as well as on the optical and bitewing X-ray images it also did not differ from the sound counterpart). The shape of the unloading "branch" of load–displacement curves for this partially demineralized dentine (Figure 3c) differs from the same "branch" for the sound dentine (Figure 3d), which indicates a change of the mechanism of resistance to loads caused by the mineral loss. A similar effect was reported by Angker et al. [47] and considered to be associated with the greater porosity and permeability of this structure. This leads to an increase in the water content and the breakdown of organic component [48]. It is interesting, however, that these processes did not influence the indentation creep, for which, for dentine bordering a WSL, the data lay within the standard deviation of the sound dentine.

AFM results revealed statistical significance of the R_a changes caused by caries development (at a significance level of $\alpha = 0.05$). This change appears to be higher for enamel than for dentine. The increase in the surface microgeometrical parameters for the carious areas is consistent with the SEM observations, where the same areas appear to be less structurally sound. We observe that both caries-affected areas demonstrate some gaps between the crystallites and even deep grooves, especially at the interrod region. For the WSL enamel, this most likely occurs because of the ability of the interrod enamel to allow diffusion of acid produced by bacteria along this region more readily than for the bulk regions of the rods. This is not surprising as the interrod structure possesses a higher organic content compared to the enamel rods [43] and the crystallites in the interrod regions appear to deviate up to 90° from those of the rod core [49], plus they are less densely packed [50,51]. The microstructure of dentine bordering a WSL is far more open compared to the sound counterpart because of the loss of mineral especially within the collagen-rich intertubular region. In the absence of cavitation and biofilm formation on dentine, the collagen fibers are well preserved by the dehydration protocol. We note the rounding of the indentation induced crack tips in the demineralized dentine versus the sharper cracks in the sound dentine. This is associated with the lower stiffness of the demineralized dentine as the crack shape is related to the stress intensity factor and the modulus of the material. The dehydration of the sample has caused embrittlement of the collagen and the indentation impressions to act as stress concentrations sites where crack initiation occurs (Figure 7d) rather than being simply compressed as for moist dentine.

We should also bear in mind the susceptibility of more fragile partially dissolved crystallites in enamel to breaking up while polishing and forming the smear layer. The bigger aggregates of hydroxyapatite crystallites on the dentine areas (Figure 5c,d) were most likely formed during sample polishing as well and are not related to the caries disease.

The Raman bands of sound enamel observed are in good agreement with those in the literature [41,52,53]. The shift of the most intense band position to a higher wavenumber (Figure 6b) is in a good agreement with the results of Buchwald et al. [54]. This shift along with the change of the

FWHM is a sign of enamel structure relaxing during demineralization as the most strained central line area of the hydroxyapatite during caries dissolves, resulting in less strain in the remaining "skeleton" of the crystallite. The shape of the phosphate-type band ν_4 PO_4^{3-} stretching from 560 cm^{-1} to 625 cm^{-1} appears to be similar for WSL and sound enamel. This observation slightly differs from the one made by Natarjan et al. [55], where flattening was observed for the similar band in carious enamel. We suggest that this difference may be attributed to the more severe stage of caries studied in [55], namely a brown spot lesion (BSL).

The observed band at 1295 cm^{-1} coincides with the amide III δ (=CH) band found in bone [56]. As no such band was found on the sound enamel covered with the remains of the smear layer after grinding and polishing procedures, it may be related to the caries disease. In sound enamel, the protein-rich regions can be found as small clusters between the crystallites [57] in such a way that the protein (presumably amelogenin, that contains amide III group [58] is naturally distributed across the surface of the crystallites. On the other hand, the grooves and gaps between the crystallites separated from each other as a result of crystallites dissolution (in addition to breakdown of some of them during polishing) visualized using AFM may have acted as sites for collecting water and organic material [59] including proteins. Besides, the porous structure of the weakened carious enamel causes more scatter of the Raman laser beam, which generates more signal overall. This phenomena is in a good agreement with [55], where opening of the enamel structure after acid treatment of BSL enabled the infiltrated resin to be detected on Raman spectra (gaps formed by acid on the place of damaged crystallites contributed to the increasing of the signal). This increased volume of the proteins becomes sufficient for the Raman spectroscope to detect an amide group in the carious part of the sample. Previously the amide III band present in Raman spectra was reported to be a sign for a carious disease by Timchenko et al. [60], however that research was conducted at a later and more advanced stage of caries (BSL).

The observed features using Raman spectroscopy are barely noticeable compared to the resolution of the device: the shift for the most intense band was 1.0 cm^{-1}, the change in its FWHM was 0.26 cm^{-1}, the amide band is the least intense among the others present in the spectra. On the other hand, all these features were stable and repeatable on the current sample for all four measurements of carious and sound enamel on the tooth slice. In order to understand the significance of these results for the non-invasive in vivo detection of early caries, various methods already proposed [61–64] and further experiments on the external surfaces of teeth should be conducted.

5. Conclusions

The work reveals some fundamental changes emerging inside human enamel and dentine at the first clinically visible stage of the carious disease from several points of view. The complex of characteristics—mineral density, reduced Young's modulus and hardness, and average roughness—was found for the natural enamel WSL and dentine bordering it. The properties obtained were compared to those of sound counterparts of the aforementioned areas on the same tooth. The results were supplemented by bitewing X-ray, optical, scanning electron microscopy and Raman spectroscopy observations. The significant reduction of the mechanical properties was recorded for both carious areas accompanied by the abnormality of the unloading response and change of the character of indentation marks caused by mineral loss. Increase of indentation creep for WSL area was recorded. The maps of mechanical properties show that the enamel outside the WSL is weakened despite its appearance as being sound on optical and bitewing X-ray images. At the same time, the decrease in mineral density of the WSL area and bordering dentine was rather small. The average roughness of polished surfaces was considerably increased in the caries-affected areas caused by the demineralization process, and the surface relief changed as well. Raman spectroscopy research resulted in detection of interesting repeatable features connected to changes in the molecular composition in the carious areas (the most intense PO_4^{3-} band shift, the change in FWHM and a new band appearance, detected and described for the WSL), which need additional specific studies for possible further implementation in dental practice as caries detection tools.

Author Contributions: Conceptualization, E.S. and M.S.; methodology, M.S.; software, I.R. and B.M.; validation, A.N., D.Y. and S.M.; formal analysis, E.S.; investigation, E.S., B.M., A.N., V.I. and N.L.; resources, S.A.; data curation, M.S.; writing—original draft preparation, E.S.; writing—review and editing, M.S., S.A., B.M.; visualization, I.R., V.I., E.S., B.M., N.L. and D.Y.; supervision, M.S. and S.M.; project administration, E.S.; funding acquisition, S.A. All authors have read and agreed to the published version of the manuscript.

Funding: This research was funded by the Government of the Russian Federation, grant number 14.Z50.31.0046.

Acknowledgments: The research of mineral density of the sample using micro-CT was conducted in Institute of Life Sciences, North Caucasus Federal University; experiments on nanoindentation, optical and electron microscopy were made in Research and Education Center "Materials", Don State Technical University (http://nano.donstu.ru); experiments on AFM were conducted in the "Chemistry" department, Don State Technical University; experiments on Raman spectroscopy were held in the "Nanotechnology" department, Southern Federal University.

Conflicts of Interest: The authors declare no conflict of interest.

References

1. Syriopoulos, K.; Sanderink, G.C.; Velders, X.L.; Van Der Stelt, P.F. Radiographic detection of approximal caries: A comparison of dental films and digital imaging systems. *Dentomaxillof. Radiol.* **2000**, *29*, 312–318. [CrossRef]
2. Shokri, A.; Kasraei, S.; Shokri, E.; Farhadian, M.; Hekmat, B. In vitro effect of changing the horizontal angulation of X-ray beam on the detection of proximal enamel caries in bitewing radiographs. *Dent. Med. Probl.* **2018**, *55*, 29–34. [CrossRef] [PubMed]
3. Avery, J.K. Microradiographic and microhardness studies of developing enamel. *Arch. Oral Biol.* **1962**, *7*, 245–256. [CrossRef]
4. Angker, L.; Swain, M.V.; Kilpatrick, N. Characterising the micro-mechanical behaviour of the carious dentine of primary teeth using nano-indentation. *J. Biomech.* **2005**, *38*, 1535–1542. [CrossRef]
5. Biswas, N.; Mukhopadhyay, A.K.; Dey, A. Nanoindentation of Teeth: A Hard but Tough Hybrid Functionally Graded Composite. In *Nanoindentation of Natural Materials*, 1st ed.; Dey, A., Mukhopadhyay, A.K., Eds.; CRC Press: Boca Raton, FL, USA, 2019; pp. 85–108.
6. Cochrane, N.J.; Cai, F.; Huq, N.L.; Burrow, M.F.; Reynolds, E.C. New approaches to enhanced remineralization of tooth enamel. *J. Dent. Res.* **2010**, *89*, 1187–1197. [CrossRef]
7. Pires, P.M.; Dos Santos, T.P.; Fonseca-Gonçalves, A.; Pithon, M.M.; Lopes, R.T.; De Almeida Neves, A. A dual energy micro-CT methodology for visualization and quantification of biofilm formation and dentin demineralization. *Arch. Oral Biol.* **2018**, *85*, 10–15. [CrossRef]
8. Songsiripradubboon, S.; Hamba, H.; Trairatvorakul, C.; Tagami, J. Sodium fluoride mouthrinse used twice daily increased incipient caries lesion remineralization in an in situ model. *J. Dent.* **2014**, *42*, 271–278. [CrossRef]
9. Wang, Y.; Spencer, P.; Walker, M.P. Chemical profile of adhesive/caries-affected dentin interfaces using Raman microspectroscopy. *J. Biomed. Mater. Res. A* **2007**, *81*, 279–286. [CrossRef]
10. Ramakrishnaiah, R.; Rehman, G.U.; Basavarajappa, S.; Al Khuraif, A.A.; Durgesh, B.H.; Khan, A.S.; Rehman, I.U. Applications of Raman spectroscopy in dentistry: Analysis of tooth structure. *Appl. Spectrosc. Rev.* **2015**, *50*, 332–350. [CrossRef]
11. Park, H.J.; Kwon, T.Y.; Nam, S.H.; Kim, H.J.; Kim, K.H.; Kim, Y.J. Changes in bovine enamel after treatment with a 30% hydrogen peroxide bleaching agent. *Dent. Mater. J.* **2004**, *23*, 517–521. [CrossRef]
12. Sadyrin, E.V.; Kislyakov, E.A.; Karotkiyan, R.V.; Yogina, D.V.; Drogan, E.G.; Swain, M.V.; Maksyukov, S.Y.; Nikolaev, A.L.; Aizikovich, S.M. Influence of Citric Acid Concentration and Etching Time on Enamel Surface Roughness of Prepared Human Tooth: In vitro Study. In *Plasticity, Damage and Fracture in Advanced Materials*, 1st ed.; Altenbach, H., Brünig, M., Kowalewski, Z., Eds.; Springer: Cham, Switzerland, 2020; pp. 135–150.
13. Egerton, R.F.; Li, P.; Malac, M. Radiation damage in the TEM and SEM. *Micron* **2004**, *35*, 399–409. [CrossRef] [PubMed]

14. Lippert, F.; Parker, D.M.; Jandt, K.D. In vitro demineralization/remineralization cycles at human tooth enamel surfaces investigated by AFM and nanoindentation. *J. Colloid Interface Sci.* **2004**, *280*, 442–448. [CrossRef] [PubMed]
15. Zheng, L.; Hilton, J.F.; Habelitz, S.; Marshall, S.J.; Marshall, G.W. Dentin caries activity status related to hardness and elasticity. *Eur. J. Oral Sci.* **2003**, *111*, 243–252. [CrossRef] [PubMed]
16. Yun, F.; Swain, M.V.; Chen, H.; Cairney, J.; Qu, J.; Sha, G.; Liu, H.; Ringer, S.P.; Han, Y.; Liu, L.; et al. Nanoscale pathways for human tooth decay–Central planar defect, organic-rich precipitate and high-angle grain boundary. *Biomaterials* **2020**, *235*, 119748. [CrossRef] [PubMed]
17. Guerra, F.; Mazur, M.; Nardi, G.M.; Corridore, D.; Pasqualotto, D.; Rinado, F.; Ottolenghi, L. Dental hypomineralized enamel resin infiltration. Clinical indications and limits. *Senses Sci.* **2015**, *2*, 135–139.
18. Harris, N.O.; Garcia-Godoy, F. *Primary Preventive Dentistry*, 6th ed.; Pearson Education: Upper Saddle River, NJ, USA, 2004; pp. 64–70.
19. Huang, T.T.; He, L.H.; Darendeliler, M.A.; Swain, M.V. Correlation of mineral density and elastic modulus of natural enamel white spot lesions using X-ray microtomography and nanoindentation. *Acta Biomater.* **2010**, *6*, 4553–4559. [CrossRef]
20. Ko, A.C.T.; Hewko, M.; Sowa, M.G.; Dong, C.C.; Cleghorn, B. Detection of early dental caries using polarized Raman spectroscopy. *Opt. Express* **2006**, *14*, 203–215. [CrossRef]
21. Kinoshita, H.; Miyoshi, N.; Fukunaga, Y.; Ogawa, T.; Ogasawara, T.; Sano, K. Functional mapping of carious enamel in human teeth with Raman microspectroscopy. *J. Raman Spectrosc.* **2008**, *39*, 655–660. [CrossRef]
22. Huang, T.T.Y.; He, L.H.; Darendeliler, M.A.; Swain, M.V. Nano-indentation characterisation of natural carious white spot lesions. *Caries Res.* **2010**, *44*, 101–107. [CrossRef]
23. Metwally, N.; Niazy, M.; El-Malt, M. Remineralization of Early Carious Lesions using Biomimetic Selfassembling Peptides Versus Fluoride agent. (In vitro and In vivo study). *Al-Azhar Dent. J. Girls* **2017**, *4*, 179–188. [CrossRef]
24. Başaran, G.; Veli, İ.; Başaran, E.G. Non-Cavitated approach for the treatment of white spot lesions: A case report. *Int. Dent. Res.* **2011**, *1*, 65–69. [CrossRef]
25. Yuan, H.; Li, J.; Chen, L.; Cheng, L.; Cannon, R.D.; Mei, L. Esthetic comparison of white-spot lesion treatment modalities using spectrometry and fluorescence. *Angle Orthod.* **2014**, *84*, 343–349. [CrossRef] [PubMed]
26. Eckstein, A.; Helms, H.J.; Knösel, M. Camouflage effects following resin infiltration of postorthodontic white-spot lesions in vivo: One-year follow-up. *Angle Orthod.* **2015**, *85*, 374–380. [CrossRef] [PubMed]
27. Kim, S.; Kim, E.Y.; Jeong, T.S.; Kim, J.W. The evaluation of resin infiltration for masking labial enamel white spot lesions. *Int. J. Paediatr. Dent.* **2011**, *21*, 241–248. [CrossRef]
28. Borges, A.B.; Caneppele, T.M.F.; Masterson, D.; Maia, L.C. Is resin infiltration an effective esthetic treatment for enamel development defects and white spot lesions? A systematic review. *J. Dent.* **2017**, *56*, 11–18. [CrossRef] [PubMed]
29. Fejerskov, O.; Kidd, E. *Dental Caries: The Disease and Its Clinical Management*, 2nd ed.; Blackwell Munksgaard: Oxford, UK, 2008; pp. 220–222.
30. Fisher, J.; Glick, M. FDI world dental federation science committee. A new model for caries classification and management: The FDI World Dental Federation caries matrix. *J. Am. Dent. Assoc.* **2012**, *143*, 546–551. [CrossRef] [PubMed]
31. Longbottom, C.L.; Huysmans, M.C.; Pitts, N.; Fontana, M. Glossary of key terms. In *Detection, Assessment, Diagnosis and Monitoring of Caries*, 1st ed.; Pitts, N.B., Ed.; Karger: Basel, Switzerland, 2009; pp. 209–216.
32. Park, S.; Wang, D.H.; Zhang, D.; Romberg, E.; Arola, D. Mechanical properties of human enamel as a function of age and location in the tooth. *J. Mater. Sci. Mater. Med.* **2008**, *19*, 2317–2324. [CrossRef]
33. Wychowanski, P.; Malkiewicz, K. Evaluation of metal ion concentration in hard tissues of teeth in residents of central Poland. *BioMed Res. Int.* **2017**, *2017*, 6419709. [CrossRef]
34. Kamberi, B.; Kqiku, L.; Hoxha, V.; Dragusha, E. Lead concentrations in teeth from people living in Kosovo and Austria. *Coll. Antropol.* **2011**, *35*, 79–82.
35. Oliver, W.C.; Pharr, G.M. An improved technique for determining hardness and elastic modulus using load and displacement sensing indentation experiments. *J. Mater. Res.* **1992**, *7*, 1564–1583. [CrossRef]
36. Alyahya, A.; Alqareer, A.; Swain, M. Microcomputed Tomography Calibration Using Polymers and Minerals for Enamel Mineral Content Quantitation. *Med. Princ. Pract.* **2019**, *28*, 247–255. [CrossRef] [PubMed]

37. Zou, W.; Hunter, N.; Swain, M.V. Application of polychromatic μCT for mineral density determination. *J. Dent. Res.* **2011**, *90*, 18–30. [CrossRef] [PubMed]
38. King, R.B. Elastic analysis of some punch problems for a layered medium. *Int. J. Solids Struct.* **1987**, *23*, 1657–1664. [CrossRef]
39. Bertassoni, L.E.; Swain, M.V. Removal of dentin non-collagenous structures results in the unraveling of microfibril bundles in collagen type I. *Connect. Tissue Res.* **2017**, *58*, 414–423. [CrossRef] [PubMed]
40. Inoue, S.; Abe, Y.; Yoshida, Y.; De Munck, J.; Sano, H.; Suzuki, K.; Lambrechts, P.; Van Meerbeek, B. Effect of conditioner on bond strength of glass-ionomer adhesive to dentin/enamel with and without smear layer interposition. *Oper. Dent.* **2004**, *29*, 685–692.
41. Xue, J.; Li, W.; Swain, M.V. In vitro demineralization of human enamel natural and abraded surfaces: A micromechanical and SEM investigation. *J. Dent.* **2009**, *37*, 264–272. [CrossRef]
42. Cuy, J.L.; Mann, A.B.; Livi, K.J.; Teaford, M.F.; Weihs, T.P. Nanoindentation mapping of the mechanical properties of human molar tooth enamel. *Arch. Oral Biol.* **2002**, *47*, 281–291. [CrossRef]
43. Habelitz, S.; Marshall, S.J.; Marshall, G.W., Jr.; Balooch, M. Mechanical properties of human dental enamel on the nanometre scale. *Arch. Oral Biol.* **2001**, *46*, 173–183. [CrossRef]
44. He, L.H.; Fujisawa, N.; Swain, M.V. Elastic modulus and stress–strain response of human enamel by nano-indentation. *Biomaterials* **2006**, *27*, 4388–4398. [CrossRef]
45. Fong, H.; Sarikaya, M.; White, S.N.; Snead, M.L. Nano-mechanical properties profiles across dentin–enamel junction of human incisor teeth. *Mater. Sci. Eng. C Mater. Biol. Appl.* **1999**, *7*, 119–128. [CrossRef]
46. Yanagisawa, T.; Miake, Y. High-resolution electron microscopy of enamel-crystal demineralization and remineralization in carious lesions. *Microscopy* **2003**, *52*, 605–613. [CrossRef] [PubMed]
47. Angker, L.; Nijhof, N.; Swain, M.V.; Kilpatrick, N.M. Influence of hydration and mechanical characterization of carious primary dentine using an ultra-micro indentation system (UMIS). *Eur. J. Oral Sci.* **2004**, *112*, 231–236. [CrossRef] [PubMed]
48. Stankoska, K.; Sarram, L.; Smith, S.; Bedran-Russo, A.K.; Little, C.B.; Swain, M.V.; Bertassoni, L.E. Immunolocalization and distribution of proteoglycans in carious dentine. *Aust. Dent. J.* **2016**, *61*, 288–297. [CrossRef] [PubMed]
49. Bajaj, D.; Arola, D.D. On the R-curve behavior of human tooth enamel. *Biomaterials* **2009**, *30*, 4037–4046. [CrossRef]
50. Ang, S.F.; Bortel, E.L.; Swain, M.V.; Klocke, A.; Schneider, G.A. Size-dependent elastic/inelastic behavior of enamel over millimeter and nanometer length scales. *Biomaterials* **2010**, *31*, 1955–1963. [CrossRef]
51. Maas, M.C.; Dumont, E.R. Built to last: The structure, function, and evolution of primate dental enamel. *Evol. Anthropol.* **1999**, *8*, 133–152. [CrossRef]
52. Pietrzyńska, M.; Zembrzuska, J.; Tomczak, R.; Mikołajczyk, J.; Rusińska-Roszak, D.; Voelkel, A.; Buchwald, T.; Jampílek, J.; Lukáč, M.; Devínsky, F. Experimental and in silico investigations of organic phosphates and phosphonates sorption on polymer-ceramic monolithic materials and hydroxyapatite. *Eur. J. Pharm. Sci.* **2016**, *93*, 295–303. [CrossRef]
53. Chun-Te Ko, A.; Hewko, M.D.; Leonardi, L.; Sowa, M.G.; Dong, C.C.; Williams, P.; Cleghorn, B. Ex vivo detection and characterization of early dental caries by optical coherence tomography and Raman spectroscopy. *J. Biomed. Opt.* **2005**, *10*, 31118.
54. Buchwald, T.; Okulus, Z.; Szybowicz, M. Raman spectroscopy as a tool of early dental caries detection–new insights. *J. Raman Spectrosc.* **2017**, *48*, 1094–1102. [CrossRef]
55. Natarajan, A.K.; Fraser, S.J.; Swain, M.V.; Drummond, B.K.; Gordon, K.C. Raman spectroscopic characterisation of resin-infiltrated hypomineralised enamel. *Anal. Bioanal. Chem.* **2015**, *407*, 5661–5671. [CrossRef]
56. Penel, G.; Delfosse, C.; Descamps, M.; Leroy, G. Composition of bone and apatitic biomaterials as revealed by intravital Raman microspectroscopy. *Bone* **2005**, *36*, 893–901. [CrossRef] [PubMed]
57. La Fontaine, A.; Zavgorodniy, A.; Liu, H.; Zheng, R.; Swain, M.; Cairney, J. Atomic-scale compositional mapping reveals Mg-rich amorphous calcium phosphate in human dental enamel. *Sci. Adv.* **2016**, *2*, e1601145. [CrossRef] [PubMed]
58. Yang, X.; Wang, L.; Qin, Y.; Sun, Z.; Henneman, Z.J.; Moradian-Oldak, J.; Nancollas, G.H. How amelogenin orchestrates the organization of hierarchical elongated microstructures of apatite. *J. Phys. Chem. B* **2010**, *114*, 2293–2300. [CrossRef] [PubMed]

59. Tohda, H.; Takuma, S.; Tanaka, N. Intracrystalline structure of enamel crystals affected by caries. *J. Dent. Res.* **1987**, *66*, 1647–1653. [CrossRef]
60. Timchenko, E.V.; Timchenko, P.E.; Volova, L.T.; Rosenbaum, A.Y.; Kulabukhova, A.Y.; Zherdeva, L.A.; Nefedova, I.F. Application of Raman spectroscopy method to the diagnostics of caries development. *J. Biomed. Photonics Eng.* **2015**, *1*, 201–205. [CrossRef]
61. Akkus, A.; Yang, S.; Akkus, O.; Lang, L. A portable confocal fiber optic raman spectrometer concept for evaluation of mineral content within enamel tissue. *J. Oper. Esthet. Dent.* **2016**, *1*, 1–5.
62. Yakubu, E.; Li, B.; Duan, Y.; Yang, S. Full-scale Raman imaging for dental caries detection. *Biomed. Opt. Express* **2018**, *9*, 6009–6016. [CrossRef]
63. Okagbare, P.I.; Esmonde-White, F.W.; Goldstein, S.A.; Morris, M.D. Development of non-invasive Raman spectroscopy for in vivo evaluation of bone graft osseointegration in a rat model. *Analyst* **2010**, *135*, 3142–3146. [CrossRef]
64. Ando, M.; Liao, C.S.; Eckert, G.J.; Cheng, J.X. Imaging of demineralized enamel in intact tooth by epidetected stimulated Raman scattering microscopy. *J. Biomed. Opt.* **2018**, *23*, 105005. [CrossRef]

Article

On Nonlinear Bending Study of a Piezo-Flexomagnetic Nanobeam Based on an Analytical-Numerical Solution

Mohammad Malikan [1] and Victor A. Eremeyev [1,2,*]

1 Department of Mechanics of Materials and Structures, Faculty of Civil and Environmental Engineering, Gdansk University of Technology, 80-233 Gdansk, Poland; mohammad.malikan@pg.edu.pl
2 Laboratory of Mechanics of Biomaterials, Research and Education Center "Materials", Don State Technical University, Gagarina sq., 1, Rostov on Don 344000, Russia
* Correspondence: victor.eremeev@pg.edu.pl

Received: 19 July 2020; Accepted: 4 September 2020; Published: 6 September 2020

Abstract: Among various magneto-elastic phenomena, flexomagnetic (FM) coupling can be defined as a dependence between strain gradient and magnetic polarization and, contrariwise, elastic strain and magnetic field gradient. This feature is a higher-order one than piezomagnetic, which is the magnetic response to strain. At the nanoscale, where large strain gradients are expected, the FM effect is significant and could be even dominant. In this article, we develop a model of a simultaneously coupled piezomagnetic–flexomagnetic nanosized Euler–Bernoulli beam and solve the corresponding problems. In order to evaluate the FM on the nanoscale, the well-known nonlocal model of strain gradient (NSGT) is implemented, by which the nanosize beam can be transferred into a continuum framework. To access the equations of nonlinear bending, we use the variational formulation. Converting the nonlinear system of differential equations into algebraic ones makes the solution simpler. This is performed by the Galerkin weighted residual method (GWRM) for three conditions of ends, that is to say clamp, free, and pinned (simply supported). Then, the system of nonlinear algebraic equations is solved on the basis of the Newton–Raphson iteration technique (NRT) which brings about numerical values of nonlinear deflections. We discovered that the FM effect causes the reduction in deflections in the piezo-flexomagnetic nanobeam.

Keywords: flexomagnetic; nanobeam; large deflection; NSGT; Galerkin method; Newton–Raphson method

1. Introduction

To study the flexomagnetic (FM) effect and to better identify it, one can use the family close to it, that is, the piezomagnetic effect. In piezomagnetic, simply by compressing or stretching materials, an internal magnetic field is created in them. The piezomagnetic effect and its application can be seen in many materials and structures. However, in addition to these very useful applications, there is an important drawback that this effect can only exist in about 20 crystal structures with a specific symmetrical classification. However, there is no such limit to the FM effect, and materials with wider classes of symmetry can cause such a phenomenon. The flexomagnetic effect can be very strong and effective, so that it may one day be used in nanosensors or nanometer actuators. As a brief explanation of the FM effect, it can be noted that by bending an ionic crystal, the atomic layers are drawn inside it, and it is clear that the outermost layer will have the most tension. This difference in traction in different layers can cause ions to transfer to the crystal so much that they eventually create a magnetic field. In other words, bending some materials creates a magnetic field, a corresponding phenomenon called flexomagnetic effect. The effect of strain gradients shows that the importance of the FM effect in micro and nano systems is comparable to that of piezomagnetic and even beyond. Additionally, flexomagnetic, unlike piezomagnetic, can be found in a wider class of materials. This means that compared to piezomagnetic, which is invalid and inefficient in materials with central symmetry, there is an FM effect in all biological materials and systems. These traits have led to a growing interest in and research into the flexomagnetic effect in recent years [1,2]. Currently, the role of the flexomagnetic effect in the physics of dielectrics has been investigated in some studies and has shown promising practical applications [3–7]. On the other hand, the difference between theoretical and experimental results shows a limited understanding in this field. This study examines current knowledge of FM in engineering.

The flexomagnetic effect exists in many solid dielectrics, soft membranes, and biological filaments. The flexomagnetic effect is introduced as the effect of size-dependent electromagnetic coupling due to the presence of strain gradients and magnetic fields, and promises many applications in nano-electronic devices (with strong strain gradients). Just as the piezomagnetic effect is expected to have important applications in nano-engines and particles [8–12], so the FM effect can play this role as well. Different fields of science are used to study nanodielectrics by considering the FM effect. These significant parts can be examined from a chemistry and physics point of view, or they can be put under a magnifier in the engineering and industrial aspects. In the engineering aspects, the study of external factors on dielectrics and their mechanical and physical behavioral responses will naturally be the criterion for evaluation. The purpose of this study is to evaluate this aspect in static large deflection analysis of a nano actuator beam. A close look at the history of the study of the mechanical behavior of dielectrics by including the FM effect does not show many studies [13–15]. These studies have generally looked at small deformations (linear strains), which, while important, cannot be the criterion for designing dielectric nanobeams. Definitely, the deformations should be considered as large as possible to obtain a reasonable and reliable safety factor for optimizing these significant nano-electro-magneto-mechanical systems' components.

The present work accounts for the large deflections by adding the nonlinear terms of Lagrangian strain using the von Kármán approach. The constitutive equations are expanded in line with the classical beam theory. It is worth mentioning that the small scale is fulfilled conforming to the second stress and strain gradients. These extra terms should result in two conflict responses, that is softening and hardening in the nanoscale structure based on the literature. We perform the solution of acquired equations, which govern the nonlinear bending of the nanobeam, on the basis of two step solution techniques. The first one is the Galerkin weighted residual method (GWRM) which converts the equations into nonlinear algebraic ones, then the Newton–Raphson technique (NRT), which solves the nonlinear system of algebraic equations and gives the numerical values of displacements into x and z directions. At last, pictorial results are evaluated to show the disagreements and dissimilarities betwixt linear deflection and nonlinear one for the piezo-flexomagnetic nanosize beam.

2. Mathematical Model

Let us consider a piezomagnetic-flexomagnetic nanobeam (PF-NB) with squared cross section of length and thickness L and h; see Figure 1. A uniform vertical static loading acts above the beam. A magnetic potential is joint to the beam to simulate and act as a magnetic field. Moreover, the z-axis is related to the transverse direction, whereas the neutral plane of the beam is coincident with the x-axis.

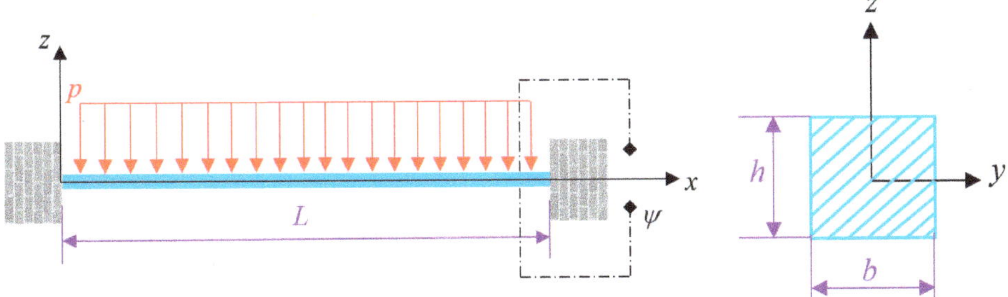

Figure 1. A square ($b = h$) PF-NB clamped at both ends and exposed to a lateral uniform static loading beside an external magnetic potential.

Follow up, the kinematic displacement for each node of the beam is utilized with the aid of the Euler–Bernoulli hypothesis [16,17]. Furthermore, the model is restricted with in-plane deformations. The rectangular displacements correspond with u_1 and u_3, respectively, for axial and transverse directions. However, such displacements for neutral plane are, respectively, regarded with u and w. Thus, one can give accordingly

$$u_1(x\,,z) = u(x) - z\frac{dw(x)}{dx} \tag{1}$$

$$u_3(x,z) = w(x) \tag{2}$$

The Von Kármán assumption tells us that the nonlinear terms related to the u can be excluded from the Lagrangian strain formula because these terms are sufficiently small compared to the other terms [18–24]. The general Lagrangian strain can be mentioned as

$$\varepsilon_{ij} = \frac{1}{2}\left(\frac{\partial u_i}{\partial x_j} + \frac{\partial u_j}{\partial x_i} + \frac{\partial u_k}{\partial x_i}\frac{\partial u_k}{\partial x_j}\right) \tag{3}$$

In regard to this approach, the nonzero nonlinear strain-displacement components can be derived as follows

$$\varepsilon_{xx} = \frac{du}{dx} - z\frac{d^2w}{dx^2} + \frac{1}{2}\left(\frac{dw}{dx}\right)^2 \tag{4}$$

$$\eta_{xxz} = \frac{d\varepsilon_{xx}}{dz} = -\frac{d^2w}{dx^2} \tag{5}$$

where Equations (4) and (5) calculate, respectively, the longitudinal strain and its gradient.

The stress-strain magneto-mechanical coupling relations in the one-dimensional framework can be given owing to [13,14].

$$\sigma_{xx} = C_{11}\varepsilon_{xx} - q_{31}H_z \tag{6}$$

$$\xi_{xxz} = g_{31}\eta_{xxz} - f_{31}H_z \tag{7}$$

$$B_z = a_{33}H_z + q_{31}\varepsilon_{xx} + f_{31}\eta_{xxz} \tag{8}$$

where σ_{xx} is the static stress field component, H_z is the magnetic field component, B_z is the magnetic flux (induction) component, C_{11} is the elastic modulus, f_{31} is the component of the fourth-order flexomagnetic coefficients tensor, a_{33} is the component of the second-order magnetic permeability tensor, q_{31} is the component of the third-order piezomagnetic tensor, g_{31} is the component of the sixth-order gradient elasticity tensor, and ξ_{xxz} is the component of higher-order moment stress tensor.

The variational formulation accurately develops the characteristics relation of PF-NB, thusly

$$\delta U - \delta W = 0 \tag{9}$$

where δ is the symbol of variation, U is the strain energies, and W is created works by outer objects. In such a way, the entire inner energy of the specimen is in the first variation which is equal to zero as well. The strain energy respecting magneto-mechanical composition can be variated just like this (the first variation)

$$\delta U = \int_V (\sigma_{xx}\delta\varepsilon_{xx} + \xi_{xxz}\delta\eta_{xxz} - B_z\delta H_z)dV \tag{10}$$

Equation (10) can be transformed with integration by parts on the basis of the one-dimensional displacement field previously assumed as follows

$$\delta U = \delta\Pi_{U_1}^{Mech} + \delta\Pi_{U_1}^{Mag} + \delta\Pi_{U_2}^{Mech} + \delta\Pi_{U_2}^{Mag} \tag{11}$$

where

$$\delta\Pi_{U_1}^{Mech} = -\int_0^L \left\{ \frac{dN_x}{dx}\delta u + \left[\frac{d^2M_x}{dx^2} + \frac{d}{dx}\left(N_x\frac{dw}{dx}\right) + \frac{d^2T_{xxz}}{dx^2} \right]\delta w \right\} dx \tag{12}$$

$$\delta\Pi_{U_1}^{Mag} = -\int_0^L \int_{-h/2}^{h/2} \frac{dB_z}{dz}\delta\Psi dzdx \tag{13}$$

$$\delta\Pi_{U_2}^{Mech} = \left\{ N_x\delta u - [M_x + T_{xxz}]\frac{d\delta w}{dx} + \left[N_x\frac{dw}{dx} + \frac{dM_x}{dx} + \frac{dT_{xxz}}{dx} \right]\delta w \right\}\Bigg|_0^L \tag{14}$$

$$\delta\Pi_{U_2}^{Mag} = \int_0^L (B_z\delta\Psi)\Bigg|_{-h/2}^{h/2} dx \tag{15}$$

where Ψ is the variable of magnetic potential. The resultants of the stress field can be introduced along the following lines

$$N_x = \int_{-h/2}^{h/2} \sigma_{xx}dz \tag{16}$$

$$M_x = \int_{-h/2}^{h/2} \sigma_{xx}zdz \tag{17}$$

$$T_{xxz} = \int_{-h/2}^{h/2} \xi_{xxz}dz \tag{18}$$

In addition, the magnetic potential was introduced through the relation

$$\frac{d\Psi}{dz} = -H_z \tag{19}$$

External forces (axial force as a result of the longitudinal magnetic field and the lateral loading) create work thermodynamically in the particles so that the mathematical relation in the first variation becomes [25].

$$\delta W = \int_0^L \left[N_x^0\left(\frac{d\delta w}{dx}\frac{dw}{dx}\right) + p(x)\delta w \right] dx \tag{20}$$

in which N_x^0 is the in-plane longitudinal axial force, and p is the lateral load per unit length. Taking into account the closed circuit in conjunction with the inverse piezo case, the electrical boundary conditions can be attributed as below

$$\Psi\left(+\frac{h}{2}\right) = \psi \tag{21}$$

$$\Psi\left(-\frac{h}{2}\right) = 0 \tag{22}$$

in which ψ is the external magnetic potential on the upper surface. Making in hand Equations (8), (13), (15), (21) and (22) practicably expresses the magnetic field component and thereupon the magnetic potential function in line with thickness as follows [13,14]

$$\Psi = -\frac{q_{31}}{2a_{33}}\left(z^2 - \frac{h^2}{4}\right)\frac{d^2w}{dx^2} + \frac{\psi}{h}\left(z + \frac{h}{2}\right) \tag{23}$$

$$H_z = z\frac{q_{31}}{a_{33}}\frac{d^2w}{dx^2} - \frac{\psi}{h} \tag{24}$$

On the basis of Equations (23) and (24), Equations (6)–(8) can be developed as

$$\sigma_{xx} = C_{11}\left[\frac{du}{dx} + \frac{1}{2}\left(\frac{dw}{dx}\right)^2\right] - z\left(C_{11} + \frac{q_{31}^2}{a_{33}}\right)\frac{d^2w}{dx^2} + \frac{q_{31}\psi}{h} \tag{25}$$

$$\xi_{xxz} = -\left(g_{31} + \frac{q_{31}f_{31}z}{a_{33}}\right)\frac{d^2w}{dx^2} + \frac{f_{31}\psi}{h} \tag{26}$$

$$B_z = q_{31}\left[\frac{du}{dx} + \frac{1}{2}\left(\frac{dw}{dx}\right)^2\right] - f_{31}\frac{d^2w}{dx^2} - \frac{a_{33}\psi}{h} \tag{27}$$

Subsequently, Equations (16)–(18) can be rewritten in detail as

$$N_x = C_{11}A\left[\frac{du}{dx} + \frac{1}{2}\left(\frac{dw}{dx}\right)^2\right] + q_{31}\psi \tag{28}$$

$$M_x = -I_z\left(C_{11} + \frac{q_{31}^2}{a_{33}}\right)\frac{d^2w}{dx^2} \tag{29}$$

$$T_{xxz} = -g_{31}h\frac{d^2w}{dx^2} + f_{31}\psi \tag{30}$$

in which N_x, M_x, T_{xxz} show the axial, moment, and hyper stress resultants, and $I_z = \int_A z^2 dA$ is the area moment of inertia.

The resultant magnetic axial stress, which is achieved due to the longitudinal magnetic field, based on Equation (28) can be determined as

$$N^{Mag} = q_{31}\psi \tag{31}$$

This force is supposed to act at both ends of the beam, thus

$$N_x^0 = N^{Mag} \tag{32}$$

Eventually, imposing Equation (9), one can write the governing equations in a combination of mechanical and magnetic conditions as

$$\frac{dN_x}{dx} = 0 \tag{33}$$

$$\frac{d^2M_x}{dx^2} + \frac{d^2T_{xxz}}{dx^2} + \left(N_x^0 + N_x\right)\frac{d^2w}{dx^2} + \frac{dN_x}{dx}\frac{dw}{dx} - p = 0 \tag{34}$$

Due to being the nanobeam a size-dependent particle, the scale-dependent property should be substituted in Equations (33) and (34). In [26], the second strain gradient of Mindlin merged successfully with the nonlocal theory of Eringen. This model (NSGT) was incorporated in a lot of research performed on the nanoparticles in recent years—see e.g., [27–38] and many others—and can be a proper item at the nanoscale.

The model proposed by [26] can be compatible in our case as

$$\left(1 - \mu\frac{d^2}{dx^2}\right)\sigma_{xx}^{NonLocal} = \left(1 - l^2\frac{d^2}{dx^2}\right)\sigma_{xx}^{Local}$$

or as

$$\left(1 - \mu\frac{d^2}{dx^2}\right)\sigma_{xx}^{NonLocal} = \left(1 - l^2\frac{d^2}{dx^2}\right)\left\{C_{11}\left[\frac{du}{dx} + \frac{1}{2}\left(\frac{dw}{dx}\right)^2\right] - z\left(C_{11} + \frac{q_{31}^2}{a_{33}}\right)\frac{d^2w}{dx^2} + \frac{q_{31}\psi}{h}\right\} \tag{35}$$

in which $\mu\left(nm^2\right)$ is the nonlocal parameter, and $l(nm)$ is the strain gradient parameter. Thus, $l > 0$ establishes a nonzero strain gradient into the model, and $\mu = (e_0a)^2$ is the parameter defining nonlocality. It is germane to note that both scale parameters are dependent on the physics of the model and cannot be material constants [39,40]. This means the parameters are not constant values, something like an elasticity modulus for each material.

To implement the influence of size effects into the equations, Equation (35) is plugged to Equations (28)–(30) as

$$N_x - \mu\frac{d^2N_x}{dx^2} = \left(1 - l^2\frac{d^2}{dx^2}\right)\left\{C_{11}A\left[\frac{du}{dx} + \frac{1}{2}\left(\frac{dw}{dx}\right)^2\right]\right\} \tag{36}$$

$$M_x - \mu\frac{d^2M_x}{dx^2} = \left(1 - l^2\frac{d^2}{dx^2}\right)\left\{-I_z\left(C_{11} + \frac{q_{31}^2}{a_{33}}\right)\frac{d^2w}{dx^2}\right\} \tag{37}$$

$$T_{xxz} - \mu\frac{d^2T_{xxz}}{dx^2} = \left(1 - l^2\frac{d^2}{dx^2}\right)\left\{-g_{31}h\frac{d^2w}{dx^2} + f_{31}\psi\right\} \tag{38}$$

Equations (33) and (34) by means of Equations (36)–(38) can be derived in the framework of displacements, respectively, as series of models.

1.1. Piezo-flexomagnetic nanobeam (PF-NB)—Nonlinear case:

$$C_{11}A\left[\frac{d^2u}{dx^2} + \frac{d^2w}{dx^2}\frac{dw}{dx} - l^2\left(\frac{d^4u}{dx^4} + \frac{d^4w}{dx^4}\frac{dw}{dx} + 3\frac{d^3w}{dx^3}\frac{d^2w}{dx^2}\right)\right] = 0 \tag{39}$$

$$\begin{gathered}
-g_{31}h\frac{d^4w}{dx^4} + q_{31}\psi\frac{d^2w}{dx^2} - p - \mu\left(-g_{31}h\frac{d^6w}{dx^6} + q_{31}\psi\frac{d^4w}{dx^4} - \frac{d^2p}{dx^2}\right) \\
-\mu C_{11}A\left[\frac{du}{dx} + \frac{1}{2}\left(\frac{dw}{dx}\right)^2\right]\frac{d^4w}{dx^4} + C_{11}A\mu l^2\left[\frac{d^3u}{dx^3} + \frac{d^3w}{dx^3}\frac{dw}{dx} + \left(\frac{d^2w}{dx^2}\right)^2\right]\frac{d^4w}{dx^4} \\
-\mu C_{11}A\left(\frac{d^2u}{dx^2} + \frac{dw}{dx}\frac{d^2w}{dx^2}\right)\frac{d^3w}{dx^3} + C_{11}A\mu l^2\left(\frac{d^4u}{dx^4} + 3\frac{d^3w}{dx^3}\frac{d^2w}{dx^2} + \frac{dw}{dx}\frac{d^4w}{dx^4}\right)\frac{d^3w}{dx^3} \\
-\mu C_{11}A\left(\frac{d^4u}{dx^4} + \frac{dw}{dx}\frac{d^4w}{dx^4} + 3\frac{d^3w}{dx^3}\frac{d^2w}{dx^2}\right)\frac{dw}{dx} + C_{11}A\left[\frac{du}{dx} + \frac{1}{2}\left(\frac{dw}{dx}\right)^2\right]\frac{d^2w}{dx^2} \\
+C_{11}A\mu l^2\left(\frac{d^6u}{dx^6} + \frac{dw}{dx}\frac{d^6w}{dx^6} + 5\frac{d^5w}{dx^5}\frac{d^2w}{dx^2} + 10\frac{d^4w}{dx^4}\frac{d^3w}{dx^3}\right)\frac{dw}{dx} \\
-C_{11}Al^2\left[\frac{d^3u}{dx^3} + \frac{d^3w}{dx^3}\frac{dw}{dx} + \left(\frac{d^2w}{dx^2}\right)^2\right]\frac{d^2w}{dx^2} + C_{11}A\left(\frac{d^2u}{dx^2} + \frac{dw}{dx}\frac{d^2w}{dx^2}\right)\frac{dw}{dx} \\
-I_z\left(C_{11} + \frac{q_{31}^2}{a_{33}}\right)\left(\frac{d^4w}{dx^4} - l^2\frac{d^6w}{dx^6}\right) - C_{11}Al^2\left(\frac{d^4u}{dx^4} + 3\frac{d^3w}{dx^3}\frac{d^2w}{dx^2} + \frac{dw}{dx}\frac{d^4w}{dx^4}\right)\frac{dw}{dx} = 0
\end{gathered} \tag{40}$$

1.2. Piezo-flexomagnetic nanobeam (PF-NB)—Linear case:

$$\begin{aligned}&-g_{31}h\tfrac{d^4w}{dx^4}+q_{31}\psi\tfrac{d^2w}{dx^2}-p-\mu\Big(-g_{31}h\tfrac{d^6w}{dx^6}+q_{31}\psi\tfrac{d^4w}{dx^4}-\tfrac{d^2p}{dx^2}\Big)\\&-I_z\Big(C_{11}+\tfrac{q_{31}^2}{a_{33}}\Big)\Big(\tfrac{d^4w}{dx^4}-l^2\tfrac{d^6w}{dx^6}\Big)=0\end{aligned}\tag{41}$$

2.1. Piezomagnetic nanobeam (P-NB)—Nonlinear case:

$$C_{11}A\left[\frac{d^2u}{dx^2}+\frac{d^2w}{dx^2}\frac{dw}{dx}-l^2\left(\frac{d^4u}{dx^4}+\frac{d^4w}{dx^4}\frac{dw}{dx}+3\frac{d^3w}{dx^3}\frac{d^2w}{dx^2}\right)\right]=0\tag{42}$$

$$\begin{aligned}&q_{31}\psi\tfrac{d^2w}{dx^2}-p-\mu\Big(q_{31}\psi\tfrac{d^4w}{dx^4}-\tfrac{d^2p}{dx^2}\Big)-\mu C_{11}A\Big[\tfrac{du}{dx}+\tfrac{1}{2}\Big(\tfrac{dw}{dx}\Big)^2\Big]\tfrac{d^4w}{dx^4}\\&+C_{11}A\mu l^2\Big[\tfrac{d^3u}{dx^3}+\tfrac{d^3w}{dx^3}\tfrac{dw}{dx}+\Big(\tfrac{d^2w}{dx^2}\Big)^2\Big]\tfrac{d^4w}{dx^4}\\&-\mu C_{11}A\Big(\tfrac{d^2u}{dx^2}+\tfrac{dw}{dx}\tfrac{d^2w}{dx^2}\Big)\tfrac{d^3w}{dx^3}+C_{11}A\mu l^2\Big(\tfrac{d^4u}{dx^4}+3\tfrac{d^3w}{dx^3}\tfrac{d^2w}{dx^2}+\tfrac{dw}{dx}\tfrac{d^4w}{dx^4}\Big)\tfrac{d^3w}{dx^3}\\&-\mu C_{11}A\Big(\tfrac{d^4u}{dx^4}+\tfrac{dw}{dx}\tfrac{d^4w}{dx^4}+3\tfrac{d^3w}{dx^3}\tfrac{d^2w}{dx^2}\Big)\tfrac{dw}{dx}+C_{11}A\Big[\tfrac{du}{dx}+\tfrac{1}{2}\Big(\tfrac{dw}{dx}\Big)^2\Big]\tfrac{d^2w}{dx^2}\\&+C_{11}A\mu l^2\Big(\tfrac{d^6u}{dx^6}+\tfrac{dw}{dx}\tfrac{d^6w}{dx^6}+5\tfrac{d^5w}{dx^5}\tfrac{d^2w}{dx^2}+10\tfrac{d^4w}{dx^4}\tfrac{d^3w}{dx^3}\Big)\tfrac{dw}{dx}\\&-C_{11}Al^2\Big[\tfrac{d^3u}{dx^3}+\tfrac{d^3w}{dx^3}\tfrac{dw}{dx}+\Big(\tfrac{d^2w}{dx^2}\Big)^2\Big]\tfrac{d^2w}{dx^2}+C_{11}A\Big(\tfrac{d^2u}{dx^2}+\tfrac{dw}{dx}\tfrac{d^2w}{dx^2}\Big)\tfrac{dw}{dx}\\&-I_z\Big(C_{11}+\tfrac{q_{31}^2}{a_{33}}\Big)\Big(\tfrac{d^4w}{dx^4}-l^2\tfrac{d^6w}{dx^6}\Big)-C_{11}Al^2\Big(\tfrac{d^4u}{dx^4}+3\tfrac{d^3w}{dx^3}\tfrac{d^2w}{dx^2}+\tfrac{dw}{dx}\tfrac{d^4w}{dx^4}\Big)\tfrac{dw}{dx}=0\end{aligned}\tag{43}$$

2.2. Piezomagnetic nanobeam (P-NB)—Linear case:

$$q_{31}\psi\frac{d^2w}{dx^2}-p-\mu\left(q_{31}\psi\frac{d^4w}{dx^4}-\frac{d^2p}{dx^2}\right)-I_z\left(C_{11}+\frac{q_{31}^2}{a_{33}}\right)\left(\frac{d^4w}{dx^4}-l^2\frac{d^6w}{dx^6}\right)=0\tag{44}$$

3.1. Nanobeam (NB)—Nonlinear case:

$$C_{11}A\left[\frac{d^2u}{dx^2}+\frac{d^2w}{dx^2}\frac{dw}{dx}-l^2\left(\frac{d^4u}{dx^4}+\frac{d^4w}{dx^4}\frac{dw}{dx}+3\frac{d^3w}{dx^3}\frac{d^2w}{dx^2}\right)\right]=0\tag{45}$$

$$\begin{aligned}&-p+\mu\tfrac{d^2p}{dx^2}-\mu C_{11}A\Big[\tfrac{du}{dx}+\tfrac{1}{2}\Big(\tfrac{dw}{dx}\Big)^2\Big]\tfrac{d^4w}{dx^4}\\&+C_{11}A\mu l^2\Big[\tfrac{d^3u}{dx^3}+\tfrac{d^3w}{dx^3}\tfrac{dw}{dx}+\Big(\tfrac{d^2w}{dx^2}\Big)^2\Big]\tfrac{d^4w}{dx^4}\\&-\mu C_{11}A\Big(\tfrac{d^2u}{dx^2}+\tfrac{dw}{dx}\tfrac{d^2w}{dx^2}\Big)\tfrac{d^3w}{dx^3}\\&+C_{11}A\mu l^2\Big(\tfrac{d^4u}{dx^4}+3\tfrac{d^3w}{dx^3}\tfrac{d^2w}{dx^2}+\tfrac{dw}{dx}\tfrac{d^4w}{dx^4}\Big)\tfrac{d^3w}{dx^3}\\&-\mu C_{11}A\Big(\tfrac{d^4u}{dx^4}+\tfrac{dw}{dx}\tfrac{d^4w}{dx^4}+3\tfrac{d^3w}{dx^3}\tfrac{d^2w}{dx^2}\Big)\tfrac{dw}{dx}+C_{11}A\Big[\tfrac{du}{dx}+\tfrac{1}{2}\Big(\tfrac{dw}{dx}\Big)^2\Big]\tfrac{d^2w}{dx^2}\\&+C_{11}A\mu l^2\Big(\tfrac{d^6u}{dx^6}+\tfrac{dw}{dx}\tfrac{d^6w}{dx^6}+5\tfrac{d^5w}{dx^5}\tfrac{d^2w}{dx^2}+10\tfrac{d^4w}{dx^4}\tfrac{d^3w}{dx^3}\Big)\tfrac{dw}{dx}\\&-C_{11}Al^2\Big[\tfrac{d^3u}{dx^3}+\tfrac{d^3w}{dx^3}\tfrac{dw}{dx}+\Big(\tfrac{d^2w}{dx^2}\Big)^2\Big]\tfrac{d^2w}{dx^2}+C_{11}A\Big(\tfrac{d^2u}{dx^2}+\tfrac{dw}{dx}\tfrac{d^2w}{dx^2}\Big)\tfrac{dw}{dx}\\&-I_zC_{11}\Big(\tfrac{d^4w}{dx^4}-l^2\tfrac{d^6w}{dx^6}\Big)-C_{11}Al^2\Big(\tfrac{d^4u}{dx^4}+3\tfrac{d^3w}{dx^3}\tfrac{d^2w}{dx^2}+\tfrac{dw}{dx}\tfrac{d^4w}{dx^4}\Big)\tfrac{dw}{dx}=0\end{aligned}\tag{46}$$

3.2. Nanobeam (NB)—Linear case:

$$-p+\mu\frac{d^2p}{dx^2}-C_{11}I_z\left(\frac{d^4w}{dx^4}-l^2\frac{d^6w}{dx^6}\right)=0\tag{47}$$

4.1. Classic beam—Nonlinear case:

$$C_{11}A\left(\frac{d^2u}{dx^2}+\frac{d^2w}{dx^2}\frac{dw}{dx}\right)=0\tag{48}$$

$$-p + C_{11}A\left[\frac{du}{dx} + \frac{1}{2}\left(\frac{dw}{dx}\right)^2\right]\frac{d^2w}{dx^2} + C_{11}A\left(\frac{d^2u}{dx^2} + \frac{dw}{dx}\frac{d^2w}{dx^2}\right)\frac{dw}{dx} - C_{11}I_z\frac{d^4w}{dx^4} = 0 \quad (49)$$

4.2. Classic beam—Linear case:

$$-C_{11}I_z\frac{d^4w}{dx^4} = p \quad (50)$$

In what follows, we consider these cases in more details.

3. Solution Approach

The solution process here has two steps. The first step comes with the Galerkin weighted residual method (GWRM) on the basis of the admissible shape functions which satisfy boundary conditions. The second step is imposing the Newton–Raphson technique (NRT) in order to solve the system of nonlinear algebraic equations originated from GWRM. The following displacements were employed [41].

$$u(x) = \sum_{m=1}^{\infty} U_m \frac{dX_m(x)}{dx} \quad (51)$$

$$w(x) = \sum_{m=1}^{\infty} W_m X_m(x) \quad (52)$$

where U_m and W_m are unknown variables that determine displacements through two axes and should be computed, whereas $X_m(x)$ are shape functions, m is the axial half-wave number, and becomes $m = 1, 2, \ldots \infty$. The allowable shape functions given below satisfy end conditions as [41].

$$\mathrm{S-S}: \ X_m(x) = \sin\left(\frac{m\pi}{L}x\right) \quad (53)$$

$$\mathrm{C-C}: \ X_m(x) = \sin^2\left(\frac{m\pi}{L}x\right) \quad (54)$$

$$\mathrm{C-F}: \ X_m(x) = \sin\left(\frac{m\pi}{4L}x\right)\cos\left(\frac{m\pi}{4L}x\right) \quad (55)$$

in which S, C, and F mark one by one the simply-supported, clamped, and free end conditions. Here, e.g., C-F means a side of the beam is inserted in a clamping fixture and the opposite side is free and hanging.

Based on the Fourier sine series, the transverse load can uniformly behave on the nanobeam as the following form [42,43].

$$p(x) = \sum_{m=1}^{\infty} \frac{4p_0}{m\pi}\sin\left(\frac{m\pi}{L}x\right) \quad (56)$$

in which p_0 is density of the lateral load. Inserting Equations (51), (52), and (56) into Equations (39)–(50), and integrating over the axial domain based on the GWRM approach, one can obtain

$$\int_0^L [\eta(x)Y_m]dx = 0 \quad (57)$$

$$\int_0^L [\xi(x)Z_m]dx = 0 \quad (58)$$

in which η and ξ are the first and second equations, respectively, and Y_m and Z_m show the residuals. Then, with ordering and arranging the aforesaid equations, one can receive the nonlinear algebraic system of two equations and two unknown variables (when considering $m = 1$). To solve such a system, there are several methods. As long as the NRT converged the results very quickly and accurately, this technique was employed here. A primary guess (U_0 and W_0) was required for results in this approach. We can express the first iteration as [44].

$$U_1 = U_0 - J^{-1} \times A_0 \tag{59}$$

$$W_1 = W_0 - J^{-1} \times A_0 \tag{60}$$

where J denotes the Jacobian matrix 2×2 and A is a vector 2×1.

$$J = \frac{\partial A_0}{\partial x}, \tag{61}$$

$$A_0 = e\begin{pmatrix} U_0 \\ W_0 \end{pmatrix} \tag{62}$$

where e is the governing equations with placing the first guesses. As a matter of fact, Equations (59) and (60) are iterative equations that are

$$U_{n+1} = U_{n+1} - J^{-1} \times A_{n+1}, \tag{63}$$

$$W_{n+1} = W_{n+1} - J^{-1} \times A_{n+1} \tag{64}$$

where n is the number of iterations to receive the convergence. A few iterations are enough to obtain the desired accuracy. It is worth mentioning that the convergence and the expected accuracy were completely dependent on the value of the primary guesses. Consequently, the solution led to numerical values of displacements along axial and transverse axes. To plot the results for large deflections, we needed to obtain the vertical displacement only, and the other will not be drawn.

4. Numerical Results and Discussion

4.1. Results' Validity

Based on performing some comparative studies, the credit of the present results can be checked. In so doing, in Table 1 a pinned–pinned nanobeam under a distributed uniform force is compared with the linear schema. The maximum deflection which occurred at the center of the beam was in a nondimensional state as proposed by [21,45]. A good harmony among the deflections' values is obviously seen from the Table. It is noteworthy that the classical dimensionless deflection is indicated by $e_0a/L = 0$. From the Table, it is found that the nondimensional maximum deflection increased as the value of the nonlocal parameter increased.

Table 1. Dimensionless maximum deflection for a simply-supported nanobeam exposed to transverse uniform loading.

L/h	e_0a/L	EBT, Linear [21]	EBT, Linear [45]	EBT, Linear [Present]
10	0	0.013021	0.013021	0.013021
	0.05	0.013333	0.013333	0.013333
	0.1	0.014271	0.014271	0.014271
	0.15	0.015833	0.015833	0.015833

For an explicit understanding, another comparison is tabulated by Table 2, for which a typical macroscale beam was utilized under both fixed ends. The present results are validated with those of the finite element method (FEM). Both the current and FEM approach are on the basis of linear analysis. As FEM benefits from shear deformations, it gives higher deflections. It is notable in the Table that enlarging the volume of the load resulted in the discrepancy of deflections. The FEM outcomes can be changeable due to many conditions in its process such as the number of elements, the kind of element, the number of nodes, and the algorithm of meshing, etc.

Table 2. Maximum deflection (mm) for a clamped–clamped macro beam exposed to transverse uniform loading ($E = 210$ GPa, $h = 5$ mm).

L/h	p (kN/mm)	EBT, Linear [Present]	FEM, Linear [ABAQUS]
10	0.01	0.0792	0.0824
	0.02	0.1585	0.1648
	0.03	0.2377	0.2472
	0.04	0.3170	0.3297

4.2. Discussion of the Problem

Here, just employing $n = 4$ gave the convergence in numerical results of the Newton–Raphson solving technique. To the best of the authors' knowledge, no paper exists that has studied large deflections of a piezomagnetic nanosize beam with apparent flexomagneticity, unless otherwise stated. Estimations hereon take the necessary properties for a piezomagnetic nanoparticle accorded by Table 3 as [13,14].

Table 3. Engineering necessary features of a piezomagnetic nanobeam with apparent flexomagneticity.

$CoFe_2O_4$
$C_{11} = 286$ GPa
$q_{31} = 580.3$ N/Ampere.m
$a_{33} = 1.57 \times 10^{-4}$ N/Ampere^2
$L = 10\,h$

In light of the lack of sufficient study on FM, we took $f_{31} = 10^{-9}$ N/Ampere, $f_{31} = 10^{-10}$ N/Ampere as [13,14]. These two values were also theoretically obtained based on some simple assumptions and cannot be the exact numeric values of the flexomagnetic parameter of the aforesaid material presented in Table 3.

An NSGT case was chosen to consider nanoscale impacts. In this model, as can be observed by Equation (31), there were two small scale factors. In point of fact, to determine the results of the bending of the nanoparticle, the amounts of these two parameters are vital. Thus, by exploring within the literature, one can find the 0.5 nm $< e_0a <$ 0.8 nm [46], and $0 < e_0a \leq 2$ nm [47,48], unless otherwise stated. The amount of strain gradient parameter was obtained in a similar size to the lattice parameter of the crystalline structure [49]. This factor for the aforementioned material in Table 3 was obtained in an experiment to change between 0.8 and 0.9 nanometers at a set temperature [50]. Hence, the averaged value of the strain gradient parameter is selected as $l = 1$ nm.

4.2.1. Effect of Nonlinearity

To probe the numerical results, we first show the difference between the results of the linear and nonlinear analyses. Figure 2 is provided for the fixed support, Figure 3 is produced for the hinge support, and lastly, Figure 4 is presented for the cantilever nanobeam. It should be noted that all figures in the results section were plotted in both linear and nonlinear modes for the piezomagnetic nanobeam (P-NB), piezomagnetic-flexomagnetic nanobeam (PF-NB), and common nanobeam (NB). Let us come back to Figures 2–4. First, a comparison of the figures shows a much smaller deflection which resulted from the boundary condition of the fix versus the other ones. For this reason, a larger load amplitude was selected to evaluate the results of the fixed–fixed support to better distinguish between linear and nonlinear analyses. In the first figure, as can be seen, the results of the linear analysis were valid as long as the deflection value did not reach 15% of the thickness, i.e., $w < 0.15\,h$. Of course, it is important to note that according to the second figure and in the boundary condition of the hinge, this value was

$w \leq 0.1\ h$ for NB and $w \leq 0.08\ h$ for PF-NB. This means that if the deflections exceed these values, the linear analysis is no longer valid, and we must use nonlinear analysis to examine the nanobeam's deflections. Considering Figure 4 for a more flexible beam with clamped-free end conditions represents that the allowable value for NB was about $w \leq 0.2\ h$ and for PF-NB, about $w \leq 0.1\ h$. It is relevant to state that due to the C-F case, a very small lateral load was chosen because of the high deflection capacity of the nanobeam in free conditions. Comparing the three figures, it is interesting to note that the difference between the results of the linear and nonlinear analyses was greater in, respectively, C-F > S-S > C-C boundary conditions, and the C-F boundary condition was more sensitive. It may be concluded that nanobeams with end conditions with higher degrees of freedom require a more urgent nonlinear analysis. Another result of these diagrams is that the deflections of magnetic nanobeam in both linear and nonlinear analyses were smaller than that of the conventional nanobeam. In addition, the difference between the results of the linear analysis was greater than that of the nonlinear analysis. These results strongly suggest that nonlinear strains must be used for static deflection analysis in materials, unless the loads are selected so that the deflections are within the range obtained for linear analysis. By carefully examining the results in [14], which is based on linear analysis and a thickness of 10 nm, it can be seen that the deflections in some diagrams of this reference (see Figure 3 of the reference) were within the range, and in some others exceeded the obtained range (see Figure 4 of the reference). Therefore, the linear analysis cannot always be valid, and certainly, nonlinear analysis is a matter of need.

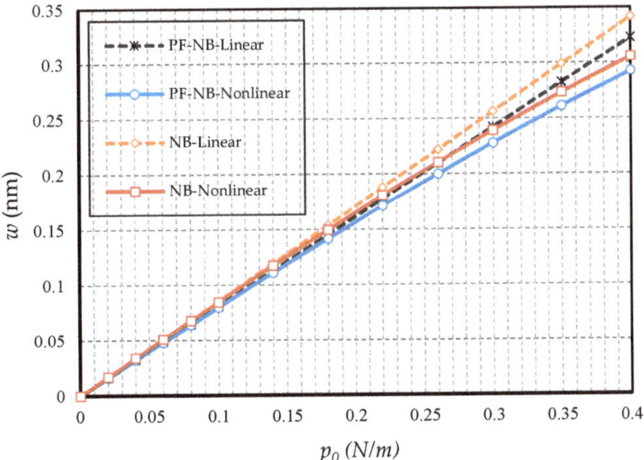

Figure 2. Transverse load vs. different cases of nanobeams ($\Psi = 1$ mA, $l = 1$ nm, $e_0a = 0.5$ nm, C-C).

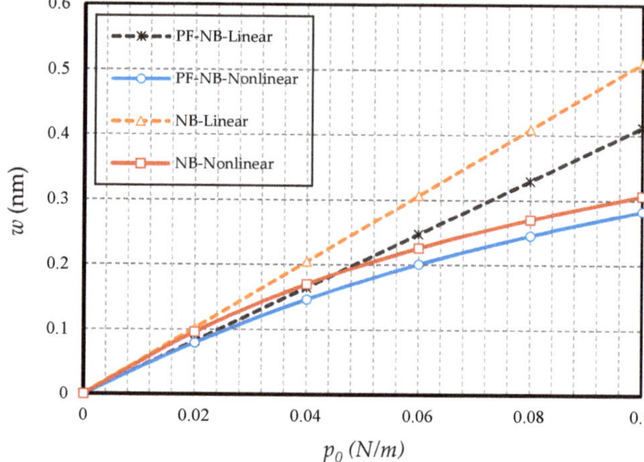

Figure 3. Transverse load vs. different cases of nanobeams (Ψ = 1 mA, l = 1 nm, e_0a = 0.5 nm, S-S).

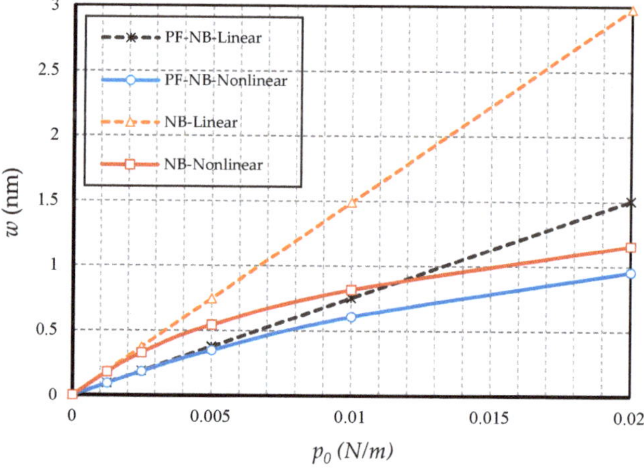

Figure 4. Transverse load vs. different cases of nanobeams (Ψ = 1 mA, l = 1 nm, e_0a = 0.5 nm, C-F).

4.2.2. Effect of Small Scale

In this section, the effect of small-scale parameters is examined, i.e., nonlocal and strain gradient parameters. Figures 5 and 6 show the effect of variations in the value of the nonlocal parameter, respectively, for S-S and C-F, and Figures 7 and 8 exhibit the effect of changes in the value of the strain gradient parameter, respectively, for C-C and S-S. The first and second figure show that as the nonlocal parameters increased, the deflections increased in all four cases examined. As a result, it can be stated that the increase in the nonlocal parameter had a softening effect on the nanobeam material. On the other hand, it is worth noting that as the numerical value of the nonlocal parameter increased, this caused the difference between the linear and nonlinear analyses results. In fact, in the nonlocal analysis of nanobeams, the effect of nonlinear analysis will be greater, and this requires that nonlinear analysis be used to investigate nonlocal deflections. It is important to note that the effect of the nonlocal parameter on the results of magnetic nanobeam was greater than that of the conventional nanobeam. This result is due to the steeper slope of the results of this nanobeam with the increasing nonlocal parameter. It is also interesting to say that the difference between the results of nonlinear and linear analyses in NB was much more than in PF-NB. From the third and fourth figures, which show the effect of changes in the strain gradient parameter in two different boundary conditions, it is clear that increasing this parameter led to a decrease in deflections of all cases and means that the increase in the strain gradient parameter is a tightening effect inside the material. However, it is important to bear in mind that this tightening effect will be greater in the case of a boundary with lower degrees of freedom. As can be observed, in a nanobeam with a double-sided fixed boundary condition, the slope of the reduction in the deflection's results was much faster than in the case of the boundary conditions of the double-sided hinged. It is also interesting to note that increasing the numerical value of the strain gradient parameter will reduce the difference between the results of linear and nonlinear analyses, and in very large values of this parameter, it can be explicitly stated that nonlinear analysis can be ignored provided that small loads are applied.

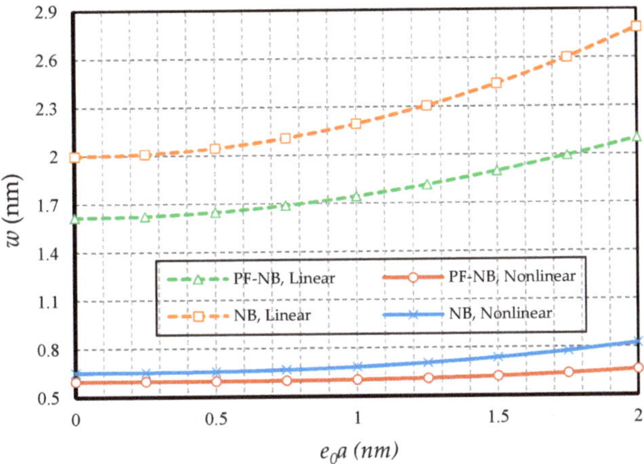

Figure 5. Nonlocal parameter vs. different cases of nanobeams (Ψ = 1 mA, l = 1 nm, p_0 = 0.4 N/m, S-S).

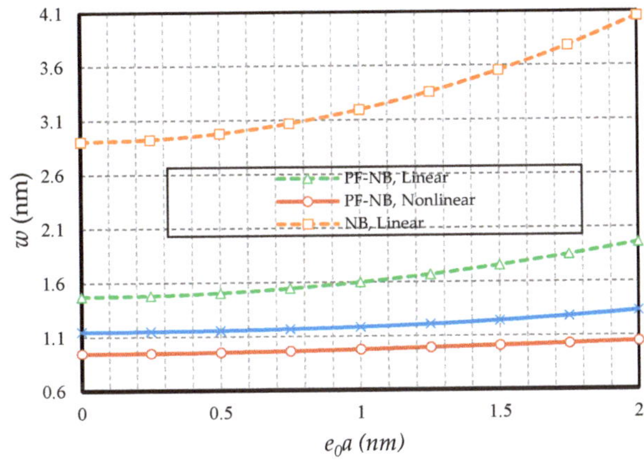

Figure 6. Nonlocal parameter vs. different cases of nanobeams (Ψ = 1 mA, l = 1 nm, p_0 = 0.02 N/m, C-F).

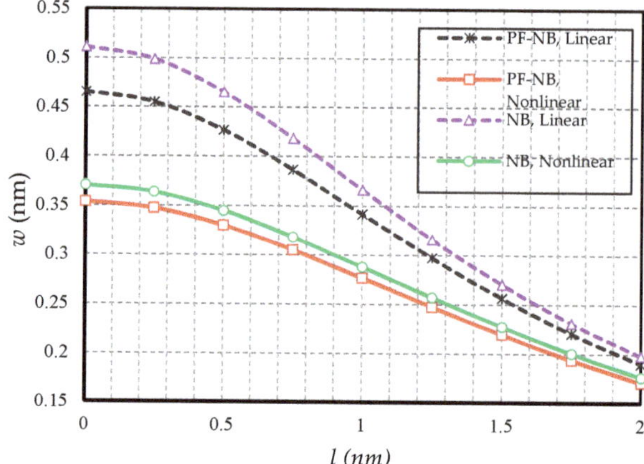

Figure 7. Strain gradient parameter vs. different cases of nanobeams (Ψ = 1 mA, e_0a = 1 nm, p_0 = 0.4 N/m, C-C).

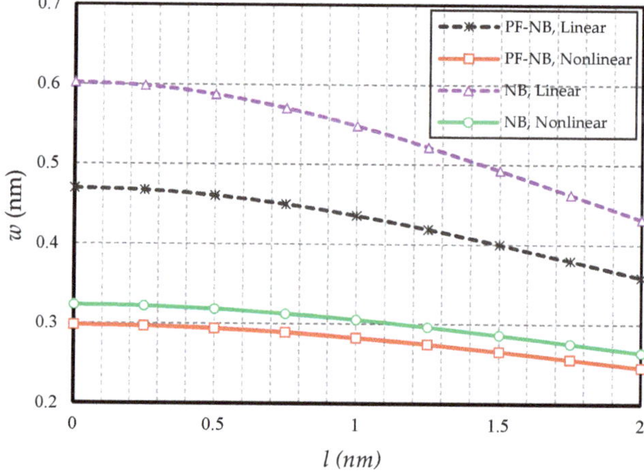

Figure 8. Strain gradient parameter vs. different cases of nanobeams (Ψ = 1 mA, e_0a = 1 nm, p_0 = 0.1 N/m, S-S).

4.2.3. Effect of Magnetic Field

The effect of the external magnetic field was dominant in the mechanical analysis of materials with flexomagnetic capability, while the magnetic effect was inverse. For this purpose, based on Figures 9 and 10, the effect of increasing the magnetic potential in the positive magnetic field is presented in two boundary condition states. Naturally, since the ordinary nanobeam does not have piezomagnetic properties, increasing the magnetic potential will have no effect on this material model. For this reason, the deflections of NB in different values of the external magnetic potential are constant. However, in piezo-flexo nanobeams, with increasing external magnetic potential, the deflections decreased in both linear and nonlinear states in both boundary conditions. Perhaps it can be interpreted that the effect of the magnetic field shrinks the material, and eventually, the material became stiffer and in the case of contraction, most of the deflections became smaller. As can be seen, in the linear analysis case, the difference in results of the conventional and magnetic nanobeams was more visible. In fact, linear analysis showed external effects with a slight exaggeration. Another interesting point is that increasing the potential of external magnetic led to convergence of the results of linear and nonlinear analyses in the piezo-flexomagnetic nanobeam, but this convergence occurred faster in the boundary condition of the hinge, so much so that in small amounts of external magnetic potential, the results of the linear and nonlinear analyses were perfectly matched to each other. Figure 11 is also displayed to show the impact of a negative magnetic field. The general conclusion that can be drawn from these three figures is that in a positive magnetic field the effect of nonlinear analysis decreases and in contrast in a negative magnetic field the influence of nonlinear analysis will be very prominent.

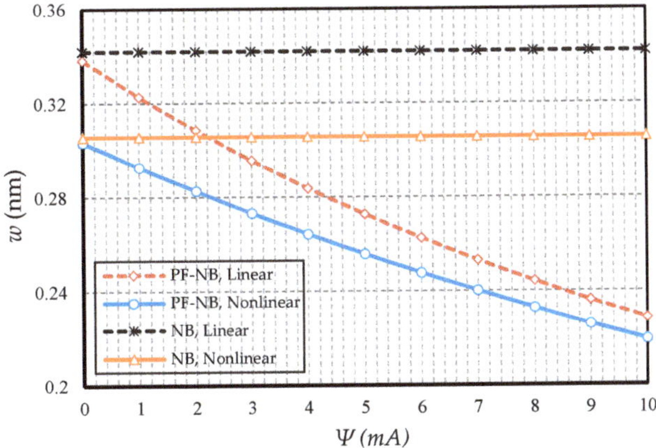

Figure 9. Magnetic potential parameter vs. different cases of nanobeams (l = 1 nm, e_0a = 0.5 nm, p_0 = 0.4 N/m, C-C).

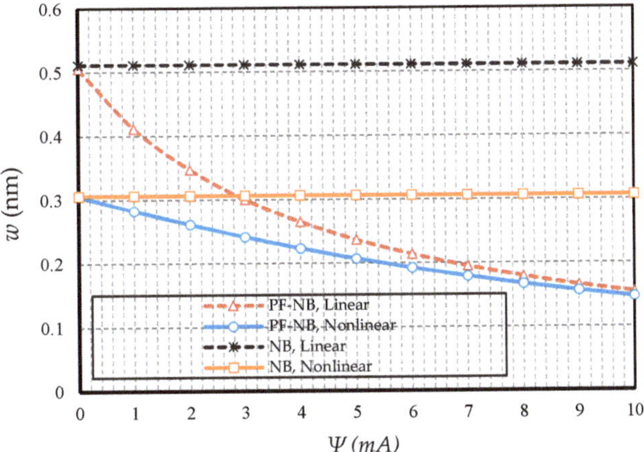

Figure 10. Magnetic potential parameter vs. different cases of nanobeams (l = 1 nm, e_0a = 0.5 nm, p_0 = 0.1 N/m, S-S).

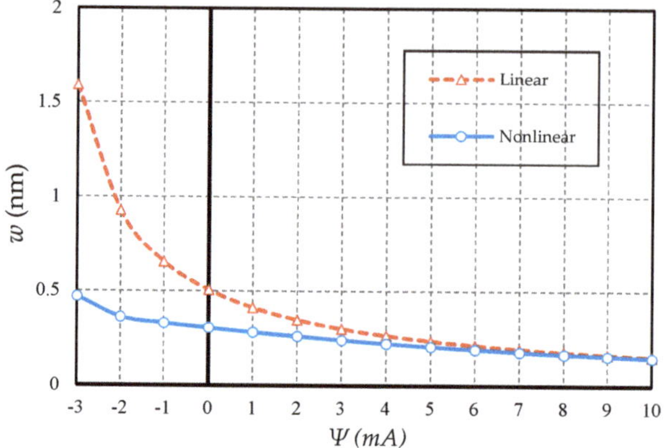

Figure 11. Magnetic potential parameter vs. PF nanobeams (l = 1 nm, e_0a = 0.5 nm, p_0 = 0.1 N/m, S-S).

4.2.4. Effect of Slenderness Ratio

Figures 12 and 13 are drawn by defining the ratio of length to thickness as a slenderness coefficient in the nanobeam. The first figure is reported for the boundary condition of the two heads of fix and the second figure is plotted for the two heads of the hinge. As can be easily seen, increasing the slenderness ratio led to an increase in static deflections in both linear and nonlinear states. Additionally, with increasing this coefficient of the nanobeam, the difference between the results of linear and nonlinear analyses increased significantly. In fact, this suggests that in large quantities of length, the linear analysis presented completely erroneous results. On the other hand, in large quantities of slenderness coefficient, the difference between the results of the magnetic nanobeam and common nanobeam in linear mode were greater than in the nonlinear one, which proves that in large values of length, the linear results showed, with magnification, the mechanical behavior of the magnetic nanobeam versus the conventional nanobeam, and it cannot be true. It should be emphasized that this difference was much greater in the results of the hinge boundary condition even with smaller loads, than in the results of the clamp boundary condition.

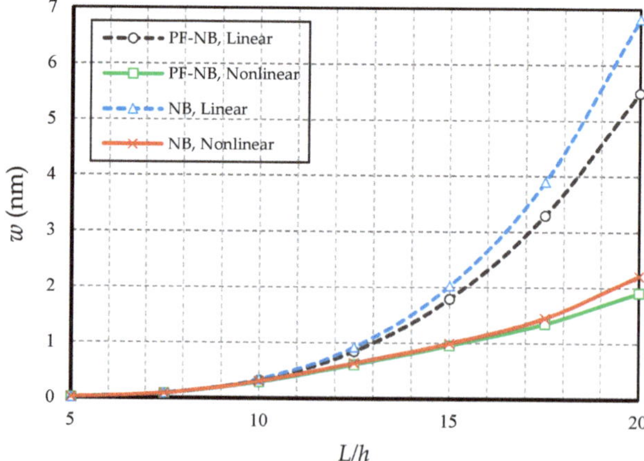

Figure 12. Slenderness ratio vs. different cases of nanobeams (Ψ = 1 mA, l = 1 nm, e_0a = 0.5 nm, p_0 = 0.4 N/m, C-C).

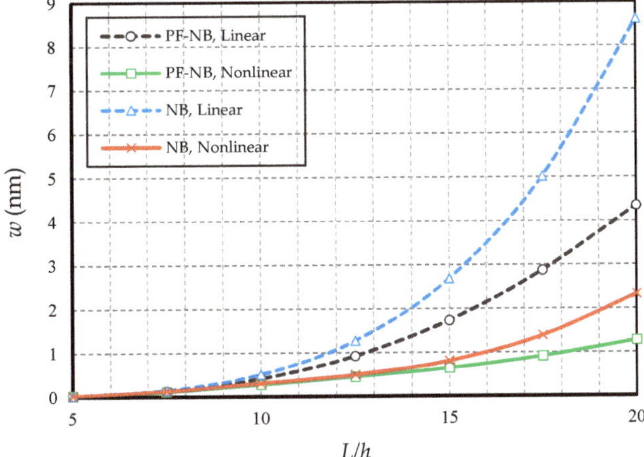

Figure 13. Slenderness ratio vs. different cases of nanobeams (Ψ = 1 mA, l = 1 nm, e_0a = 0.5 nm, p_0 = 0.1 N/m, S-S).

4.2.5. Effect of FM

In this subsection, the aim is to compare the difference in results when the substance has only a piezomagnetic effect when the flexomagnetic effect is added to it. Figure 14 shows the results of the nanobeam with two side clamps; in Figure 15, the nanobeam with two ends of the hinge is presented; finally, Figure 16 shows the cantilever nanobeam. First, as can be seen, the nonlinear analysis reduced the flexomagnetic effect. This result was obtained from the difference between the results of the P-NB and PF-NB in both nonlinear and linear analyses of the figures. On the other hand, as is clear, the results associated with the PF-NB were smaller than those of the P-NB. This finding can be interpreted in such a way that the flexomagnetic effect will lead to more material stiffness, and as a result, the deflections will be smaller while considering this effect. It has to be noted that the slight difference in the results of P-NB versus those of the PF-NB was directly related to the values of the flexomagnetic modulus. According to the references, the value of the parameter was almost based on the assumptions, and due to the novelty, of the discovery of the flexomagnetic effect; the exact values of this parameter have not yet been calculated. For this reason, it is not possible to say why the difference in results between P-NB and PF-NB was high or low. Nevertheless, such a difference was also adequately large on a nanoscale. It should be pointed out that the FM was more remarkable in C-C end conditions. This means that the lower degree of freedom boundary condition increased the impact of FM.

In this study, we end the discussion with Figure 17, in which different values of the flexomagnetic parameter were investigated. To carry out this, the w * was introduced which was the deflections of the PF-NB divided by the deflections of the P-NB. As seen, there was no appreciable change in deflections originated from FM in lower amounts of the FM parameter. The effect of FM on the P-NB became outstanding for large values of FM, and the assumed value $f_{31} = 10^{-10}$ N/Ampere can affect to some extent the behavior of the PF-NB.

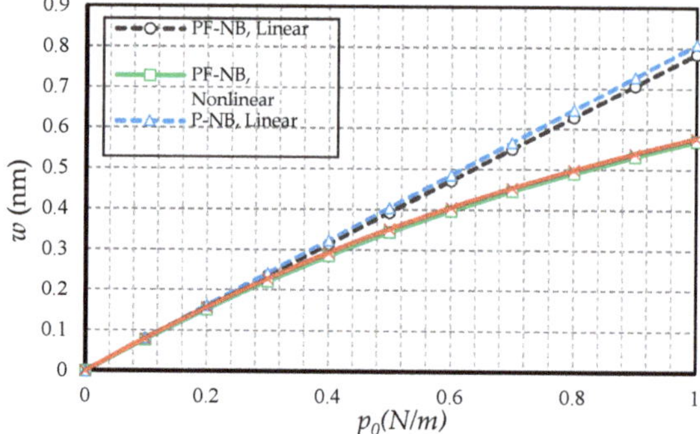

Figure 14. Transverse load vs. deflection for different cases of nanobeams (Ψ = 1 mA, l = 1 nm, e_0a = 0.5 nm, C-C).

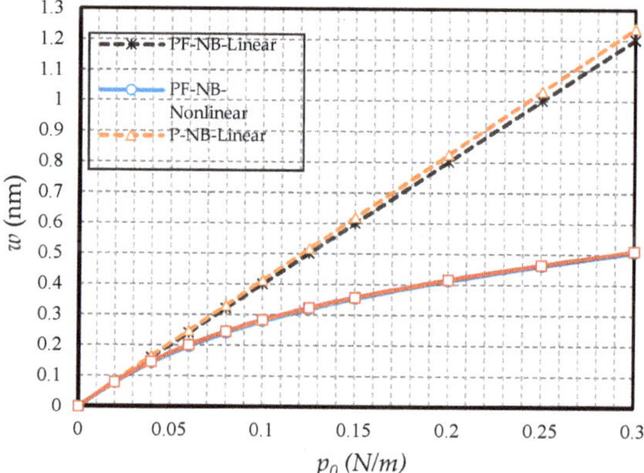

Figure 15. Transverse load vs. deflection for different cases of nanobeams (Ψ = 1 mA, l = 1 nm, e_0a = 0.5 nm, S-S).

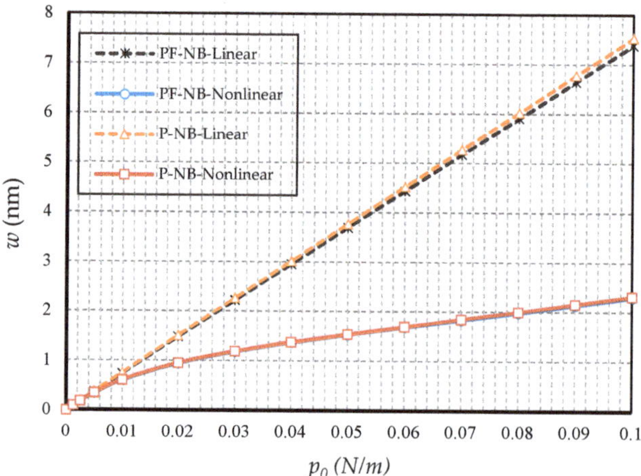

Figure 16. Transverse load vs. deflection for different cases of nanobeams (Ψ = 1 mA, l = 1 nm, e_0a = 0.5 nm, C-F).

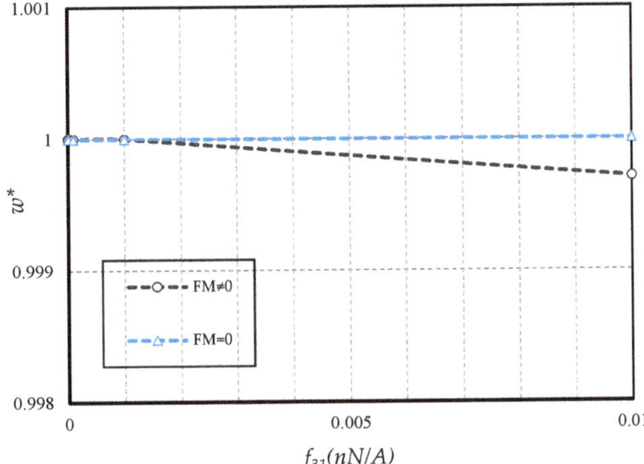

Figure 17. Presence and absence of flexomagnetic modulus for linear bending of a PF-NB ($\Upsilon = 1$ mA, $l = 1$ nm, $e_0a = 0.5$ nm, $p_0 = 0.5$ N/m, S-S).

5. Conclusions

Due to the FM influence being new and interesting, we took into account both piezomagnetic and flexomagnetic effects together for a reduced scale thin beam. The geometrical nonlinearity which induces the large deformations was also assessed. Applying the variational formulation derived the favourable governing equations. To capture the consistent nanoscale effect, the NSGT was inserted into the mathematical model. Transmuting the acquired relations based on the NSGT into the displacement relationship gives an eligible equation, which stands to compute large deflections. The translation and shifting of the nonlinear system of ordinary differential equations into the algebraic ones were performed based on the GRWM. The GRWM concerning an analytical flow estimated clamped, simply-supported, and free end conditions. Afterward, the numerical solution regarding NRT was investigated. From the obtained results, one can briefly write

- In hinged–hinged nanobeams, linear deflections for a NB can be used in the range $w \leq 0.1\,h$, and for a PF-NB, about $w \leq 0.08\,h$. This value in a double-fixed NB and PF-NB is in the range $w < 0.15\,h$. However, for a cantilever case in NB, it is $w \leq 0.2\,h$ and in PF-NB, it is $w \leq 0.1\,h$.
- The difference between the nonlinear analysis and the linear one will be more pronounced in the boundary condition with higher degrees of freedom.
- Increasing the numerical value of the nonlocal parameter leads to a softening effect on the material, and in contrast, increasing the numerical value of the strain gradient parameter leads to the appearance of stiffness in the material.
- The effect of nonlinear analysis is greater in large values of nonlocal parameters and small values of strain gradient parameters.
- The effect of nonlinear analysis on a nonlocal study is greater than a local one.
- The effect of nonlinear analysis in the positive magnetic field decreases. However, the opposite is true in the case of a negative magnetic field.
- For nanobeams with very large lengths, linear analysis gives entirely erroneous results even if the values of lateral loads are not large.
- The flexomagnetic effect leads to more material stiffness, and thus reduces the numerical values of deflections in static analysis.
- The less flexible the boundary condition, the higher the flexomagneticity effect.

Author Contributions: Conceptualization, M.M. and V.A.E.; methodology, M.M. and V.A.E.; software, M.M.; validation, M.M.; formal analysis, M.M.; investigation, M.M. and V.A.E.; resources, M.M. and V.A.E.; data curation, M.M. and V.A.E.; writing—original draft preparation, M.M.; writing—review and editing, V.A.E.; visualization, M.M.; supervision, V.A.E.; project administration, V.A.E.; funding acquisition, V.A.E. All authors have read and agreed to the published version of the manuscript.

Funding: This research was funded by the Government of the Russian Federation (contract No. 14.Z50.31.0046).

Conflicts of Interest: The authors declare no conflict of interest.

References

1. Fahrner, W. *Nanotechnology and Nanoelectronics*, 1st ed.; Springer: Berlin, Germany, 2005; p. 269.
2. Kabychenkov, A.F.; Lisovskii, F.V. Flexomagnetic and flexoantiferromagnetic effects in centrosymmetric antiferromagnetic materials. *Tech. Phys.* **2019**, *64*, 980–983. [CrossRef]
3. Eliseev, E.A.; Morozovska, A.N.; Glinchuk, M.D.; Blinc, R. Spontaneous flexoelectric/flexomagnetic effect in nanoferroics. *Phys. Rev. B* **2009**, *79*, 165433. [CrossRef]
4. Lukashev, P.; Sabirianov, R.F. Flexomagnetic effect in frustrated triangular magnetic structures. *Phys. Rev. B* **2010**, *82*, 094417. [CrossRef]
5. Eliseev, E.A.; Morozovska, A.N.; Khist, V.V.; Polinger, V. Effective flexoelectric and flexomagnetic response of ferroics. In *Recent Advances in Topological Ferroics and their Dynamics, Solid State Physics*; Stamps, R.L., Schultheis, H., Eds.; Elsevier: London, UK, 2019; Volume 70, pp. 237–289.
6. Zhang, J.X.; Zeches, R.J.; He, Q.; Chu, Y.H.; Ramesh, R. Nanoscale phase boundaries: A new twist to novel functionalities. *Nanoscale* **2012**, *4*, 6196–6204. [CrossRef] [PubMed]
7. Zhou, H.; Pei, Y.; Fang, D. Magnetic field tunable small-scale mechanical properties of nickel single crystals measured by nanoindentation technique. *Sci. Rep.* **2014**, *4*, 1–6. [CrossRef]
8. Ebrahimi, F.; Barati, M.R. Porosity-dependent vibration analysis of piezo-magnetically actuated heterogeneous nanobeams. *Mech. Syst. Signal. Pr.* **2017**, *93*, 445–459. [CrossRef]
9. Zenkour, A.M.; Arefi, M.; Alshehri, N.A. Size-dependent analysis of a sandwich curved nanobeam integrated with piezomagnetic face-sheets. *Results Phys.* **2017**, *7*, 2172–2182. [CrossRef]
10. Alibeigi, B.; Beni, Y.T. On the size-dependent magneto/electromechanical buckling of nanobeams. *Eur. Phys. J. Plus* **2018**, *133*, 398. [CrossRef]
11. Malikan, M. Electro-mechanical shear buckling of piezoelectric nanoplate using modified couple stress theory based on simplified first order shear deformation theory. *Appl. Math. Model.* **2017**, *48*, 196–207. [CrossRef]
12. Malikan, M.; Krasheninnikov, M.; Eremeyev, V.A. Torsional stability capacity of a nano-composite shell based on a nonlocal strain gradient shell model under a three-dimensional magnetic field. *Int. J. Eng. Sci.* **2020**, *148*, 103210. [CrossRef]
13. Sidhardh, S.; Ray, M.C. Flexomagnetic response of nanostructures. *J. Appl. Phys.* **2018**, *124*, 244101. [CrossRef]
14. Zhang, N.; Zheng, S.; Chen, D. Size-dependent static bending of flexomagnetic nanobeams. *J. Appl. Phys.* **2019**, *126*, 223901. [CrossRef]
15. Malikan, M.; Eremeyev, V.A. Free Vibration of Flexomagnetic Nanostructured Tubes Based on Stress-driven Nonlocal Elasticity. In *Analysis of Shells, Plates, and Beams*, 1st ed.; Altenbach, H., Chinchaladze, N., Kienzler, R., Müller, W.H., Eds.; Springer Nature: Basel, Switzerland, 2020; Volume 134, pp. 215–226.
16. Song, X.; Li, S.-R. Thermal buckling and post-buckling of pinned–fixed Euler–Bernoulli beams on an elastic foundation. *Mech. Res. Commun.* **2007**, *34*, 164–171. [CrossRef]
17. Reddy, J.N. Nonlocal nonlinear formulations for bending of classical and shear deformation theories of beams and plates. *Int. J. Eng. Sci.* **2010**, *48*, 1507–1518. [CrossRef]
18. Fernández-Sáez, J.; Zaera, R.; Loya, J.A.; Reddy, J.N. Bending of Euler–Bernoulli beams using Eringen's integral formulation: A paradox resolved. *Int. J. Eng. Sci.* **2016**, *99*, 107–116. [CrossRef]
19. Feo, L.; Penna, R. On Bending of Bernoulli-Euler Nanobeams for Nonlocal Composite Materials. *Model. Simul. Eng.* **2016**. [CrossRef]
20. Ghannadpour, S. Ritz Method Application to Bending, Buckling and Vibration Analyses of Timoshenko Beams via Nonlocal Elasticity. *J. Appl. Comput. Mech.* **2018**, *4*, 16–26.
21. Demir, C.; Mercan, K.; Numanoglu, H.; Civalek, O. Bending Response of Nanobeams Resting on Elastic Foundation. *J. Appl. Comput. Mech.* **2018**, *4*, 105–114.

22. Jia, N.; Yao, Y.; Yang, Y.; Chen, S. Size effect in the bending of a Timoshenko nanobeam. *Acta Mech.* **2017**, *228*, 2363–2375. [CrossRef]
23. Marotti de Sciarra, F.; Barretta, R. A new nonlocal bending model for Euler–Bernoulli nanobeams. *Mech Res. Commun.* **2014**, *62*, 25–30. [CrossRef]
24. Zenkour, A.M.; Sobhy, M. A simplified shear and normal deformations nonlocal theory for bending of nanobeams in thermal environment. *Phys. E* **2015**, *70*, 121–128. [CrossRef]
25. Yang, L.; Fan, T.; Yang, L.; Han, X.; Chen, Z. Bending of functionally graded nanobeams incorporating surface effects based on Timoshenko beam model. *Theor. Appl. Mech. Lett.* **2017**, *7*, 152–158. [CrossRef]
26. Lim, C.W.; Zhang, G.; Reddy, J.N. A Higher-order nonlocal elasticity and strain gradient theory and Its Applications in wave propagation. *J. Mech. Phys. Solids* **2015**, *78*, 298–313. [CrossRef]
27. Şimşek, M. Some closed-form solutions for static, buckling, free and forced vibration of functionally graded (FG) nanobeams using nonlocal strain gradient theory. *Compos. Struct.* **2019**, *224*, 111041. [CrossRef]
28. Karami, B.; Janghorban, M. On the dynamics of porous nanotubes with variable material properties and variable thickness. *Int. J. Eng. Sci.* **2019**, *136*, 53–66. [CrossRef]
29. Sahmani, S.; Safaei, B. Nonlinear free vibrations of bi-directional functionally graded micro/nano-beams including nonlocal stress and microstructural strain gradient size effects. *Thin Wall. Struct.* **2019**, *140*, 342–356. [CrossRef]
30. Karami, B.; Janghorban, M.; Rabczuk, T. Dynamics of two-dimensional functionally graded tapered Timoshenko nanobeam in thermal environment using nonlocal strain gradient theory. *Compos. Part B-Eng.* **2020**, *182*, 107622. [CrossRef]
31. Malikan, M.; Dimitri, R.; Tornabene, F. Transient response of oscillated carbon nanotubes with an internal and external damping. *Compos. Part B-Eng.* **2019**, *158*, 198–205. [CrossRef]
32. Malikan, M.; Eremeyev, V.A. On the dynamics of a visco-piezo-flexoelectric nanobeam. *Symmetry* **2020**, *12*, 643. [CrossRef]
33. Malikan, M.; Eremeyev, V.A. Post-critical buckling of truncated conical carbon nanotubes considering surface effects embedding in a nonlinear Winkler substrate using the Rayleigh-Ritz method. *Mater. Res. Express* **2020**, *7*, 025005. [CrossRef]
34. Norouzzadeh, A.; Ansari, R.; Rouhi, H. Nonlinear Bending Analysis of Nanobeams Based on the Nonlocal Strain Gradient Model Using an Isogeometric Finite Element Approach. *IJST-T Civ. Eng.* **2019**, *43*, 533–547. [CrossRef]
35. Hashemian, M.; Foroutan, S.; Toghraie, D. Comprehensive beam models for buckling and bending behavior of simple nanobeam based on nonlocal strain gradient theory and surface effects. *Mech. Mater.* **2019**, *139*, 103209. [CrossRef]
36. Malikan, M.; Nguyen, V.B.; Tornabene, F. Damped forced vibration analysis of single-walled carbon nanotubes resting on viscoelastic foundation in thermal environment using nonlocal strain gradient theory. *Eng. Sci. Technol. Int. J.* **2018**, *21*, 778–786. [CrossRef]
37. Ebrahimi, F.; Dabbagh, A.; Tornabene, F.; Civalek, O. Hygro-thermal effects on wave dispersion responses of magnetostrictive sandwich nanoplates. *Adv. Nano Res.* **2019**, *7*, 157–167.
38. She, G.L.; Liu, H.B.; Karami, B. On resonance behavior of porous FG curved nanobeams. *Steel Compos. Struct.* **2020**, *36*, 179–186.
39. Ansari, R.; Sahmani, S.; Arash, B. Nonlocal plate model for free vibrations of single-layered graphene sheets. *Phys. Lett. A* **2010**, *375*, 53–62. [CrossRef]
40. Akbarzadeh Khorshidi, M. The material length scale parameter used in couple stress theories is not a material constant. *Int. J. Eng. Sci.* **2018**, *133*, 15–25. [CrossRef]
41. Malikan, M.; Eremeyev, V.A. A new hyperbolic-polynomial higher-order elasticity theory for mechanics of thick FGM beams with imperfection in the material composition. *Compos. Struct.* **2020**, *249*, 112486. [CrossRef]
42. Berrabah, H.M.; Tounsi, A.; Semmah, A.; Adda Bedia, E.A. Comparison of various refined nonlocal beam theories for bending, vibration and buckling analysis of nanobeams. *Struct. Eng. Mech.* **2013**, *48*, 351–365. [CrossRef]
43. Ansari, R.; Sahmani, S. Bending behavior and buckling of nanobeams including surface stress effects corresponding to different beam theories. *Int. J. Eng. Sci.* **2011**, *49*, 1244–1255. [CrossRef]

44. Dastjerdi, S.; Jabbarzadeh, M. Nonlinear bending analysis of bilayer orthotropic graphene sheets resting on Winkler–Pasternak elastic foundation based on non-local continuum mechanics. *Compos. Part B-Eng.* **2016**, *87*, 161–175. [CrossRef]
45. Reddy, J.N.; Pang, S.D. Nonlocal continuum theories of beams for the analysis of carbon nanotubes. *J. Appl. Phys.* **2008**, *103*, 023511. [CrossRef]
46. Ansari, R.; Sahmani, S.; Rouhi, H. Rayleigh–Ritz axial buckling analysis of single-walled carbon nanotubes with different boundary conditions. *Phys. Lett. A* **2011**, *375*, 1255–1263. [CrossRef]
47. Duan, W.H.; Wang, C.M. Exact solutions for axisymmetric bending of micro/nanoscale circular plates based on nonlocal plate theory. *Nanotechnology* **2007**, *18*, 385704. [CrossRef]
48. Duan, W.H.; Wang, C.M.; Zhang, Y.Y. Calibration of nonlocal scaling effect parameter for free vibration of carbon nanotubes by molecular dynamics. *J. Appl. Phys.* **2007**, *101*, 24305. [CrossRef]
49. Maranganti, R.; Sharma, P. Length scales at which classical elasticity breaks down for various materials. *Phys. Rev. Lett.* **2007**, *98*, 195504. [CrossRef]
50. Stein, C.R.; Bezerra, M.T.S.; Holanda, G.H.A.; André-Filho, J.; Morais, P.C. Structural and magnetic properties of cobalt ferrite nanoparticles synthesized by co-precipitation at increasing temperatures. *AIP Advan.* **2018**, *8*, 056303. [CrossRef]

MDPI
St. Alban-Anlage 66
4052 Basel
Switzerland
Tel. +41 61 683 77 34
Fax +41 61 302 89 18
www.mdpi.com

Nanomaterials Editorial Office
E-mail: nanomaterials@mdpi.com
www.mdpi.com/journal/nanomaterials

www.ingramcontent.com/pod-product-compliance
Lightning Source LLC
LaVergne TN
LVHW071021170726
843515LV00006B/1179

9783036562919